MW01325481

CHEMICAL ASPECTS OF ENZYME BIOTECHNOLOGY

Fundamentals

INDUSTRY-UNIVERSITY COOPERATIVE CHEMISTRY PROGRAM SYMPOSIA

Published by Texas A&M University Press

ORGANOMETALLIC COMPOUNDS
Edited by Bernard L. Shapiro

HETEROGENEOUS CATALYSIS
Edited by Bernard L. Shapiro

NEW DIRECTIONS IN CHEMICAL ANALYSIS
Edited by Bernard L. Shapiro

CHEMICAL ASPECTS OF ENZYME BIOTECHNOLOGY: Fundamentals
Edited by Thomas O. Baldwin, Frank M. Raushel, and A. Ian Scott

DESIGN OF NEW MATERIALS
Edited by D. L. Cocke and A. Clearfield

FUNCTIONAL POLYMERS
Edited by David E. Bergbreiter and Charles R. Martin

METAL-METAL BONDS AND CLUSTERS IN CHEMISTRY AND CATALYSIS
Edited by John P. Fackler, Jr.

OXYGEN COMPLEXES AND OXYGEN ACTIVATION BY TRANSITION METALS
Edited by Arthur E. Martell and Donald T. Sawyer

CHEMICAL ASPECTS OF ENZYME BIOTECHNOLOGY

Fundamentals

Edited by

Thomas O. Baldwin
Frank M. Raushel
and
A. Ian Scott

Texas A&M University
College Station, Texas

PLENUM PRESS • NEW YORK AND LONDON

Library of Congress Cataloging-in-Publication Data

Texas A & M University, IUCCP Symposium on Chemical Aspects of Enzyme Biotechnology: Fundamentals (8th : 1990)
Chemical aspects of enzyme biotechnology : fundamentals / [edited by] Thomas O. Baldwin, Frank M. Raushel, and A. Ian Scott.
p. cm. -- (Industry-university cooperative chemistry program symposia)
"Proceedings of the Texas A&M University, IUCCP Eighth Annual Symposium on Chemical Aspects of Enzyme Biotechnology: Fundamentals, held March 19-22, 1990, in College Station, Texas"--T.p. verso.
Includes bibliographical references and index.
ISBN 0-306-43815-1
1. Enzymes--Biotechnology--Congresses. 2. Protein folding--Congresses. 3. Protein engineering--Congresses. I. Raushel, Frank M. II. Scott, A. Ian (Alastair Ian) III. Title. IV. Series.
TP248.65.E59T48 1990
660'.63--dc20 91-10371
CIP

Proceedings of the Texas A&M University, IUCCP Eighth Annual Symposium on Chemical Aspects of Enzyme Biotechnology: Fundamentals, held March 19-22, 1990, in College Station, Texas

ISBN 0-306-43815-1

© 1990 Plenum Press, New York
A Division of Plenum Publishing Corporation
233 Spring Street, New York, N.Y. 10013

All rights reserved

No part of this book may be reproduced, stored in a retrieval system, or transmitted in any form or by any means, electronic, mechanical, photocopying, microfilming, recording, or otherwise, without written permission from the Publisher

Printed in the United States of America

PREFACE

The Industry-University Cooperative Chemistry Program has sponsored seven previous international symposia covering a wide variety of topics of interest to industrial and academic chemists. The eighth IUCCP symposium, held March 19-22, 1990, at Texas A&M University, represents a deviation from the former symposia, in that it is the first of a two-symposium series dedicated to the rapidly moving new field of industrial biochemistry that has become known as biotechnology. Biotechnology is really not a new discipline, but rather is a term coined to describe the new and exciting commercial applications of biochemistry. The development of the field of biotechnology is a direct result of recombinant DNA technology, which began in earnest about 15 years ago. Today, we can routinely do experiments that were inconceivable in the early 1970's. Only comparatively simple technology available even in small laboratories is required to synthesize a gene and from it, to produce vast amounts of biological materials of enormous commercial value. These technical developments and others have stimulated increased activities in the field of enzyme biotechnology, using enzymes to catalyze "unnatural" reactions to produce complex molecules with stereochemical precision. It is true today, we can readily produce DNA fragments that will encode any amino acid sequence that we might desire, but at this point, our foundation of basic knowledge falls short. The dream of "designer enzymes" is still a fantasy, but the current wave of research activity and exciting new developments suggest that in the future the dream may become a reality.

Unlike the other subdisciplines of chemistry, biochemistry does not have a long history of industrial activity. Until comparatively recently, biochemistry was considered to be a basic science discipline, and the information produced in biochemical laboratories was of little immediate commercial value. While academic organic chemists have for years had their industrial counterparts, biochemists have remained largely isolated from industrial applications. To a large extent, this was probably due to the time lag that existed btween basic biochemical discoveries and commercial applications. As recently as one decade ago, there were few biochemists in industrial laboratories. Today, however, universities have difficulty producing enough well-educated biochemists to satisfy the needs of industry.

The Eighth International IUCCP Symposium entitled "Chemical Aspects of Enzyme Biotechnology: Fundamentals" is especially significant from the point of view of both Texas A&M University and the IUCCP for several reasons. First, Texas A&M University has made a major commitment to development of excellent research programs in the field of biotechnology. That commitment, formulated in the early 1980's is gaining momentum. Second, the building which will house the new Institute of Biosciences and Technology, a component of Texas A&M University, is now being constructed in the Medical Complex in Houston. The main thrust of activities of the IBT will be in the area of biotechnology and applications of biotechnology to agriculture and human nutrition, with the intent that it will become a major center for biotechnology research. Third, the faculties of several academic departments at Texas A&M University, including the Chemistry Department, the Department of Biochemistry & Biophysics, the Biology Department and the Department of Chemical Engineering, are working to incorporate aspects of biotechnology into the academic curricula. Success in these various ventures at Texas A&M University will require significant input from our industrial colleagues. The primary functions of the university are to generate a new and expanding knowledge base through research and to educate students, some of whom will become the leaders of the next generation of industrial activities. The open exchange of ideas fostered by the IUCCP forum is vital for the educational mission of the University. The Eighth International IUCCP Symposium served as a good first step toward fostering closer interaction between faculty and students of Texas A&M University working in the field of biotechnology and our industrial colleagues. The Ninth International IUCCP Symposium will continue this process and broaden the definition of the discipline to include new and exciting applications of the basic research that were presented in the Eighth International IUCCP Symposium.

The co-chairman of the conference were Professor A.I. Scott, Frank M. Raushel and Thomas O. Baldwin of the Texas A&M University Chemistry Department. The program was developed by an academic steering committee consisting of the co-chairman and members appointed by the sponsoring chemical companies Dr. James Burrington, BP America; Dr. Robert Durrwachter, Hoechst-Celanese; Dr. Barry Haymore, Monsanto Chemical Company; Dr. Mehmet Gencer, BF Goodrich; Dr. Paul Swanson, Dow Chemical Company; and Professor Arthur Martell, Texas A&M IUCCP Coordinator. We also thank Merck for financial assistance to the minisymposium on vitamin B_{12}.

In closing, the organizers of the Eighth International IUCCP Symposium must recognize the contributions that have been made to the symposium by Mrs. Mary Martell, who was responsible for and dealt with the innumerable details that accompany such an activity. Her efficiency and pleasant sense of humor are greatly appreciated.

Thomas O. Baldwin
Frank M. Raushel
A. Ian Scott

CONTENTS

ENZYME MECHANISMS

PROTEIN FOLDING

DESIGN AND REDESIGN OF ENZYMES AND PROTEINS

NEW DRUGS BASED ON ENZYME MECHANISMS

ORGANIC SYNTHESIS WITH ENZYMES

VITAMIN B_{12}

APPENDIX

EFFECT OF METAL IONS AND ADENYLYLATION STATE ON THE ENERGETICS OF THE E. COLI GLUTAMINE SYNTHETASE REACTION

Lynn M. Abell and Joseph J. Villafranca

Department of Chemistry
The Pennsylvania State University
University Park, PA 16802

Abbreviations used: glutamine synthetase, GS; low adenylylation state enzyme, GS n=2; high adenylylation state enzyme, GS n=12; Hepes, N-(2-hydroxyethyl)piperazine-N'-2-ethanesulfonic acid; PEP, phosphoenol pyruvate; EDTA, ethylene diamine tetraacetic acid; Pipes, piperazine-N,N'-bis(2-ethanesulfonic acid).

Glutamine synthetase (GS) from E. coli requires two divalent metal ions per active site in order to catalyze the ATP-dependent formation of glutamine from glutamate and ammonia and plays a central role in ammonia metabolism in both prokaryotic and eukaryotic systems.

$$\text{MgATP} + \text{L-Glu} + NH_4^{\oplus} \xrightleftharpoons{2\,M^{+2}} \text{MgADP} + \text{L-Gln} + \text{Pi} \qquad (1)$$

The enzyme has a molecular weight of 600,000 and is composed of 12 identical subunits arranged in two hexameric rings which are stacked on top of each other. The x-ray crystal structure indicates that the active site is located at the interface of two adjacent subunits in a given hexameric ring (Almassy et al., 1987; Yamashita et al., 1989).

Glutamine synthetase activity in E. coli is highly regulated by a number of feedback inhibitors, by a covalent adenylylation modification and at the level of transcription by the level of ammonia and other nitrogen sources in the growth media. The adenylylation reaction is catalyzed by the enzyme adenyltransferase which results in transfer of the AMP moiety of ATP in tyrosine 397. The adenylylation modification is reversible and is in turn controlled by a closed bicyclic cascade mechanism (Rhee et al., 1985). The effect of adenylylation on a given subunit is to change the divalent metal ion requirement for optimal activity. Unadenylylated subunits show maximal activity with Mg^{+2} at pH 7.5 and this is the physiologically important metal. Adenylylated subunits on the other hand show maximal activity with Mn^{+2} at pH 6.5.

The chemical mechanism which has been proposed for this enzyme involves the initial formation of a γ-glutamyl phosphate intermediate from ATP and

Chemical Aspects of Enzyme Biotechnology
Edited by T.O. Baldwin *et al.*, Plenum Press, New York, 1990

glutamate followed by displacement of the activated phosphate group by ammonia. The catalytic role proposed for the two divalent metal ions in this reaction are quite different and they are distinguished from one another by their dissociation constants. The tightly bound n_1 ion is thought to keep the enzyme in its catalytically active conformation. The less tightly bound n_2 metal ion is thought to be involved in nucleotide binding as well as phosphate transfer.

The presence of the acid labile γ-glutamyl phosphate intermediate was previously detected from observation of a burst of acid labile phosphate in a rapid quench experiment with low adenylylation state enzyme and Mg^{+2} as the activating metal (Meek et al., 1982). Use of this technique provided not only evidence for the intermediacy of γ-glutamyl phosphate but also offered some insight into the energetics of phosphate transfer. Analysis of the burst rate and the slower steady-state rate which followed showed that the internal equilibrium for phosphate transfer had a ratio of 1.9 in the forward direction and that product release and not phosphate transfer was rate-limiting.

In this study, we wished to examine the effect of metal ion and adenylylation on the energetics of phosphate transfer using rapid quench kinetic techniques. Steady-state kinetic measurements were conducted to determine the kinetic properties of both the adenylylated and unadenylylated form of the enzyme with respect to ATP, glutamate and both Mg^{+2} and Mn^{+2}. The results of these experiments were used to design rapid quench experiments with both enzyme forms and both metal ions.

EXPERIMENTAL PROCEDURES

Materials. [$\gamma^{32}P$]ATP (10-50 Ci/mmol) was obtained from New England Nuclear and purified prior to use according to Lewis & Villafranca (1989). All other biochemicals were from Sigma and were of the highest purity available.

Enzyme. Glutamine synthetase was purified from *E. coli* YMC10. GS is overexpressed in this strain by the plasmid pgln6 which contains *glnA*, the structural gene for glutamine synthetase. GS of low adenylylation state (GS n=2) was grown on minimal media with glucose as the carbon source and glutamine as the sole nitrogen source. In the final growth $MnCl_2$ was added to a final concentration of 1 mM to inhibit oxidation of GS (Roseman & Levine, 1987). The enzyme was purified using a variation of the zinc precipitation method (Miller et al., 1974). High adenylylation state enzyme (GS n=12) was prepared *in vitro* using the enzyme adenylyl transferase and the method of Hennig and Ginsburg, (1971) and then repurified by zinc precipitation and an ammonium sulfate precipitation. Protein concentrations were determined by either the BCA assay (Pierce) in the absence of Mn^{+2} or by spectrophotometric methods (Ginsburg et al., 1970). Adenylylation state was determined spectrophotometrically (Shapiro & Stadtman, 1970) and by the transferase assay method (Stadtman et al., 1979).

The Mn^{+2} form of the enzyme was converted to the Mg^{+2} form by dialysis against several changes of buffer containing EDTA for 12 hours and then dialyzed extensively for 36 hours against 50 mM Hepes or Pipes, 100 mM KCl and 25 mM $MgCl_2$ at the desired pH. The specific activity of the enzyme was found to be unchanged after this procedure.

Steady-State Kinetics. The biosynthetic activity at pH 6.5 and 7.5 using either Mg^{+2} or Mn^{+2} as the activating metal ion was monitored using the lactate dehydrogenase and pyrvate kinase coupling system. The assay mix contained 100 mM Hepes, 100 mM Pipes, 100 mM KCl,1 mM PEP, 190 ug NADH,

33 ug/ml puruvate kinase and 33 ug/ml L-lactate dehydrogenase. A 5-fold increase in coupling enzymes was required for studies containing Mg^{+2} at pH 6.5. The amount of metal required for optimal activity was determined for each enzyme at a given pH by holding [ATP] and [glutamate] constant and varying the metal ion concentration. When the optimal concentration was determined a fixed concentration of excess metal ion over ATP was held constant when ATP concentration was varied (Morrison, 1979). Assays were performed in 1.0 cm cuvettes with 1.00 ml total volume in a Cary 2200 UV/VIS spectrophotometer thermostatted at 25°C. The change in absorbance at 340 nm vs time was monitored. Initial rates of reaction were determined using spectra calc software.

Rapid Quench Experiments. The rapid quench experiments were performed at either 10°C or 17°C on an apparatus designed and built by Johnson (1986). The reactions were initiated by the simultaneous mixing of two solutions; one containing enzyme and activating metal ion at the appropriate pH (0.039 ml), and the other containing all substrates (0.042 ml) and trace amounts of [$\gamma^{32}P$]ATP (5000 cpm/nmol). The reactions were quenched with 190 ul of 0.6 N HCl and then immediately neutralized with 45 ul of 1 M Tris containing 4 N KOH. [^{32}P]Pi was isolated as described by Johnson (1986). All reactions carried out at pH 6.5 contained 100 mM Pipes, 100 mM KCl, 8 mM $MnCl_2$ or 25 mM $MgCl_2$ and all substrates present at 5-10x their Km's in the final reaction mixture. PEP and pyruvate kinase were included in the reaction mixture in order to remove any inhibition by ADP.

Data Analysis. Steady-state kinetic data were analyzed using the computer program HYPERO (Cleland, 1979). Rapid quench data were fit to equation 2 by a nonlinear least squares fit.

$$Pi/E = A(1-e^{-\lambda t}) + k_{ss}t \qquad (2)$$

where Pi = concentration of radioactive inorganic phosphate, E = concentration of adenylylated or unadenylylated subunits/reaction, A = burst amplitude, k_{ss} = steady-state rate constant, λ = transient phase rate constant and t = time.

The biosynthetic reaction path way was simulated to fit the rapid quench data using the computer program KINSIM (Barshop et al., 1983). The program was modified by K.A. Johnson to allow the input of data from rapid quench experiments as x,y pairs and to calculate the sum square errors in fitting the data.

RESULTS

Steady-state Kinetics. The steady-state kinetic data obtained for various forms of the enzyme and summarized in Table 1. The Km for either ATP or glutamate was determined by keeping all of the substrates saturating except the variable substrate. With either form of the enzyme, the pH optimum seems to depend on the activating metal ion and is pH 7.5 with Mg^{+2} and pH 6.5 with Mn^{+2} (Parakh et al., 1990; Ginsburg et al., 1970). When Mn^{+2} is used as the activating metal with GS n=2, a sharp decrease in substrate Km's was observed compared to those measured with Mg^{+2}. The activity observed for GS n=2 with Mn^{+2} is not due to activity from a small amount of adenylylated subunits because the Km's observed for GS n=12 are much different and much larger. It has been found that when enzyme with different average adenylylation states are used in steady-state kinetic measurements, the Km's measured with Mn^{+2} as the activating metal are independent of the adenylylation state and the kcat values remain the same when calculated based on the concentration of unadenylylated subunits. Thus, the Km values for enzyme of low and intermediate adenylylation state reflect the values of the unadenylylated subunits since these are smaller (Abell & Villafranca, 1990). GS n=12 has a small but measurable amount of activity with Mg^{+2} as the activating metal. With either activating metal, the effect of adenylylation is to elevate the Km's for both substrates when compared to unadenylylated subunits under identical conditions. Steady-state kinetic determinations were conducted on GS n=2 with Mg^{+2} at pH 6.5 as a control to show that the observed decreases in Km's is indeed due to the metal ion and is not a pH affect.

Rapid-Quench Experiments. Rapid quench-flow experiments were conducted with all substrates present at 5 to 10x's the Km values measured in the steady-state experiments. Experiments were conducted at 10°C unless otherwise specified in order to allow more accurate measurement of the burst rates. The time course of the reaction was followed from 5 ms to 4 s with duplicate measurements made at all time points between 5 and 80 ms. The data were fit to equation 2 and the results are summarized in Table 2. The time course for the quench-flow experiment with GS n=2 with Mn^{+2} at pH 6.5 shows a well defined burst with an amplitude of 0.68. This amplitude is quite similar to that of 0.57 previously observed for GS n=2 with Mg^{+2} at pH 7.5. The concentration of substrates used in this case was chosen so as to maintain saturating conditions for the unadenylylated subunits but also minimize any contribution to the observed activity from the adenylylated subunits. The Km's for ATP and glutamate for GS n=2 and GS n=12 with Mn^{+2} are different enough to allow a fairly complete separation of the two activites.

In contrast to these results, the time course for the quench flow experiment with GS n=12 with Mn^{+2} at pH 6.5 shows a very small burst amplitude of 0.056, an order of magnitude smaller than that observed for other forms of the enzyme. To ensure that this result was not due to the temperature of the experiment, the experiment was repeated at 17°C. The burst rate at 17°C was difficult to measure accurately by fitting the data to equation 2, however, accurate values for the burst amplitude could be obtained and were found to be identical to that measured at 10°C. As a control, a quench flow experiment was carried out with GS n=2 Mg^{+2} at pH 6.5 so that the effect of pH could be determined. The burst rate was too fast to measure accurately, but accurate values for the burst amplitude and the steady-state rate could be obtained by fitting the data to equation 2. Rapid quench experiments could not be conducted on GS n=12 with Mg^{+2} as the activating metal because the Km for ATP for this form of the enzyme was too large for the experiment to be conducted under saturating conditions.

The results of the rapid quench experiments may be related to rate constants in the biosynthetic reaction for glutamine synthetase initially by the analysis utilized by Meek et al. (1982). The rate constants for individual steps in the biosynthetic reaction can be represented as shown in Scheme I.

Scheme I

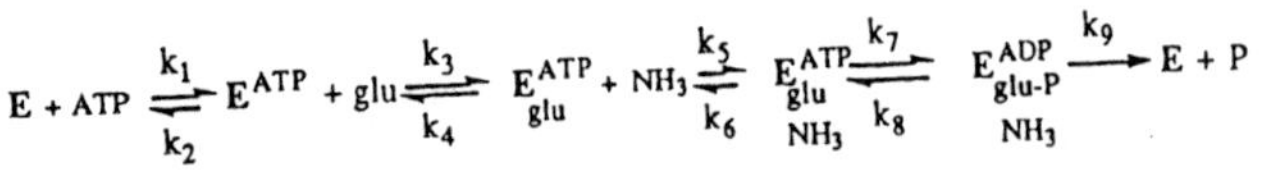

In this scheme k7 represents the phosphate transfer step and k9 includes the ammoniolysis step as well as product release steps. Since the substrate concentrations were saturating in each experiment and therefore the association rates for substrates are presumably fast, equations 3-5 may be used to relate k7, k8 and k9 to the experimental values obtained from rapid quench experiments.

$$k_{ss} = \frac{k_7 k_9}{k_7 + k_8 + k_9} \tag{3}$$

$$A = \frac{k_7(k_7 + k_8)}{(k_7 + k_8 + k_9)^2} \tag{4}$$

$$\lambda = k_7 + k_8 + k_9 \tag{5}$$

The results of these calculations are summarized in Table 3. In cases where large errors in the burst rate calculated using equation 2 made accurate rate constant calculations impractical using equations 3-5, kinetic constants were determined by simulating the mechanism shown in Scheme I using the kinsim program. The simulation results could be overlaid on the experimental data so that the best fit could be determined. For initial data fits, the association rates for all substrates were assumed to be fast, and values for k7, k8 and k9 were then obtained. Where accurate values for the burst rate were obtained from fitting the experimental data to equation 2, kinsim simulations were found to be in complete agreement with the rate constants calculated from equations 3-5.

Table 1. Steady-State Kinetic Parameters[a].

Enzyme	metal	pH	Km(ATP) (uM)	Km(glu) (mM)	kcat (s-1)
GS n=2	Mg	7.5	150 ± 7	3.3 ± 0.6	33 ± 3
GS n=2	Mg	6.5	235 ± 5	6.6 ± 1.3	7.7 ± 0.5
GS n=2	Mn	6.5	1.5 ± 0.3	0.070 ± 0.005	2 ± 0.2
GS n=12[b]	Mn	6.5	55 ± 3	3.4 ± 0.02	5 ± 1
GS n=12	Mg	7.5	657± 53	6.6 ± 1.0	0.9 ± 0.3

[a]For GS n=2, kcat was calculated using the unadenylylated subunit concentration. All measurements were conducted at 25 C in buffer containing 100 mM KCl,50 mM NH4Cl, 100 mM Pipes at pH 6.5, 100 mM Hepes at pH 7.5[b] Sager & Villafranca, unpublished results.

Table 2. Rapid-Quench Kinetic Data

Enzyme	λ (s^{-1})	A	kss (s^{-1})
GS n=2 Mg pH 7.5 10°C[a]	88	0.57	4.0
GS n=2 Mg pH 6.5 10°C	40 ± 20	0.016 ± 0.02	1.20 ± 0.04
GS n=2 Mn pH 6.5 10°C	6.4 ± 0.4	0.68 ± 0.02	0.47 ± 0.01
GS n=12 Mn pH 6.5 10°C	24.0 ± 8.8	0.056 ± 0.01	0.43 ± 0.01
GS n=12 Mn pH 6.5 17°C	40 ± 20	0.056 ± 0.021	1.58 ± 0.04

[a]Previously measured by Meek et al. 1982.

Table 3. Calculated Kinetic Constants from Rapid Quench Data

Enzyme	k_7 (s^{-1})	k_8 (s^{-1})	k_9 (s^{-1})	k_7/k_8
GS n=2 Mg pH 7.5 10°C[a]	54	28	9	1.9
GS n=2 Mg pH 6.5 10°C	8.0	25	5.7	0.32
GS n=2 Mn pH 6.5 10°C	4.8	1.0	0.62	4.8
GS n=12 Mn pH 6.5 10°C	1.8	16.4	5.8	0.11
Gs n=12 Mn pH 6.5 17°C	3.7	19.1	17.3	0.19

[a] Previously measured by Meek et al., 1982.

Table 4. Comparison of kcat/Km Values with Association Rates from Kinsim Fits

Enzyme	$kcat/Km_{ATP}$	k_1	$kcat/Km_{glu}$	k_3	$kcat/Km_{NH3}$
GS n=2 Mn pH 6.5	0.31	1	0.007	0.5	0.005
GS n=12 Mn pH 6.5	0.008	0.5	0.00013	0.01	0.005

All rate constants are in $uM^{-1}s^{-1}$. k_1 and k_3 were the association rate constants obtained from the kinsim fits. Km_{NH3} for both enzyme forms is 100 uM. Kcat's observed at 10°C were used in calculations. Simulations were not very sensitive to dissociation rates. These rates were constrained so that kr/kf was equal to dissociation constants where known or to Km.

DISCUSSION

The results in Table 3 show that in the case of GS n=2 with Mg^{+2} at pH 7.5 that the rate-limiting step is product release. Changing the pH from 7.5 to 6.5 and keeping the metal ion constant, results in the phosphate transfer step becoming rate limiting but the product release step is still rate limiting at lower pH. The effect of changing the metal ion from Mg^{+2} to Mn^{+2} for GS n=2, appears to result in a rate decrease for all of the kinetic steps especially the dephosphorylation step in the back reaction, however product release is still rate-limiting. This result is consistent with the dramatically decreased Km's observed for this form of the enzyme in steady-state kinetic measurements. The internal equilibrium for phosphate transfer is shifted in the forward direction by a factor of two when the metal is changed from Mg^{+2} to Mn^{+2}.

In contrast to these results adenylylation of this enzyme was found to make phosphate transfer the rate limiting step instead of product release. The internal equilibrium for this enzyme is shifted so that the dephosphorylation reaction is favored over the forward reaction.

The energetic differences between the adenylylated and unadenylylated form of the enzyme with Mn^{+2} can best be illustrated by construction of a difference-free-energy diagram. Construction of this diagram is facilitated by further Kinsim fits of the experimental data. Using the rate constants determined in Table 3, the substrate association rates could now be varied and lower estimates of these rates obtained. Kcat/Km for ammonia provided a good lower limit estimate for the association rate for NH_3 however kcat/Km for ATP and glutamate severely underestimated the association rates for these substrates as judged by poor kinsim fits. The rate constants which resulted in the best fits are summarized in Table 4. Stopped-flow fluorescence measurements are currently underway to measure the association rate of ATP. The difference-free-energy diagram constructed from these data and the data in Table 3 are shown in Figure 1. Figure 1 shows that the largest free energy differences between the two enzyme forms occurs in the binding of glutamate and the stabilization of the γ-glutamyl phosphate intermediate. Use of Mn^{+2} as the activating metal ion with GS n=2 results in much tighter binding of substrates and products as reflected in k_9, the rate limiting product release step and a slower kcat compared with the adenylylated enzyme. The effect of adenylylation is to loosen the binding of substrates and products which is reflected in an increase in kcat and Km. The substrate Km's for GS n=2 with Mg^{+2} and GS n=12 with Mn^{+2} are quite similar, and so other considerations being equal one might expect kcat for these two enzymes to be similar. However, kcat for GS n=12 is severely reduced due to the inability of this form of the enzyme to satisfactorily stabilize the γ-glutamylphosphate intermediate. These results strongly suggest a significant influence on the geometry of the active site due to the covalent adenylylation modification.

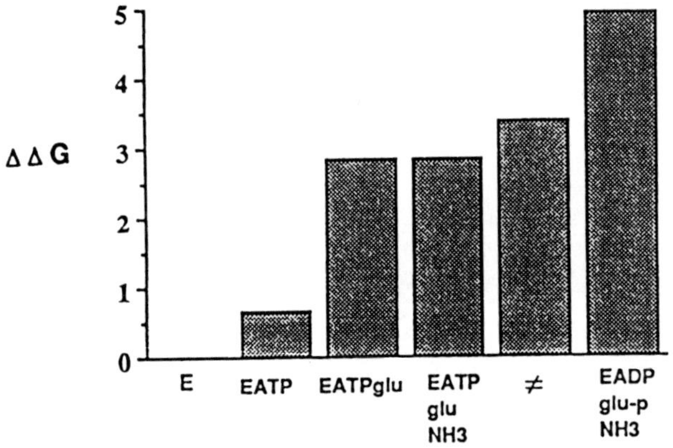

Figure 1. Difference-Free-Energy Diagram for GS n=12 - GS n=2 with Mn^{+2}.

REFERENCES

Abell, L.M. and Villafranca, J.J., 1990, Manuscript in preparation.

Almassy, R.J., Jason, C.A., Hamlin, R., Xuong, N-H., and Eisenberg, D., 1986, Nature, 323:304.

Barshop, B.A., Wrenn, R.F., and Frieden, C., 1983, Anal. Biochem., 130:134.

Cleland, W.W., 1979, Methods Enzymol., 63:103.

Ginsburg, A., Yeh, J., Hennig, S.B., and Denton, M.D., 1970, Biochem., 9:633.

Hennig, S.B., and Ginsburg, A., 1971, Arch. Biochem. Biophys., 144:611.

Johnson, K.A., 1986, Methods Enzymol., 134:677.

Lewis, D.A. and Villafranca, J.J., 1989, Biochem., 28:8454.

Meek, T.D., Johnson, K.A., and Villafranca, J.J., 1982, Biochem., 21:2158.

Miller, R.E., Shelton, E., and Stadtman, E.R., 1974, Arch. Biochem. Biophys., 163:155.

Morrison, J.F., 1979, Methods Enzymol., 63:257.

Parakh, C.R., Abell, L.M., and Villafranca, J.J., 1990, unpublished results.

Rhee, S.G., Chock, P.B., and Stadtman, E.R., 1985, Methods Enzymol., 113:213.

Roseman and Levine, 1987, J. Biol. Chem., 262:2101.

Shapiro, B.M., Kingdon, H.S., and Stadtman, E.R., 1967, Proceedings Natl. Acad. Sci., U.S.A., 58:642.

Stadtman, E.R., Smyfniotis, P.Z., Davis, J.N., and Wittenberger, M.E., 1979, Anal. Biochem., 85:275.

Yamashita, M.M., Almassy, R.J., Janson, C.A., Cascio, D., and Eisenberb, D., 1989, J. Biol. Chem., 264:17681.

STRUCTURE-FUNCTION RELATIONSHIPS IN MANDELATE RACEMASE AND MUCONATE LACTONIZING ENZYME

John A. Gerlt,[1] George L. Kenyon,[2] John W. Kozarich,[1] David T. Lin,[1] David C. Neidhart,[3] Gregory A. Petsko[3,4] Vincent M. Powers,[2] Stephen C. Ransom,[1] and Amy Y. Tsou[1]

Department of Chemistry and Biochemistry,[1] University of Maryland, College Park, MD 20742
Department of Pharmaceutical Chemistry,[2] University of California, San Francisco, CA 94143
Department of Chemistry,[3] Massachusetts Institute of Technology, Cambridge, MA 02139
Rosensteil Basic Medical Sciences Research Center[4] Brandeis University, Waltham, MA 02254

INTRODUCTION

Strains of Pseudomonads show considerable metabolic diversity by utilizing a large variety of organic compounds as sole carbon and energy sources. This metabolic diversity has both provided enzymologists with many new and unusual enzymes for mechanistic scrutiny and raised questions regarding the evolutionary pathways by which the metabolic diversity could have arisen. Our laboratories are interested in mechanistic, structural, and molecular biological aspects of the metabolism of mandelic acid by *Pseudomonas putida*. Both enantiomers of mandelic acid are oxidatively degraded to benzoic acid by the enzymes of the mandelate pathway (Figure 1), and benzoic acid, in turn, is oxidatively degraded to intermediates of the citric acid cycle by the enzymes of the β-ketoadipate pathway (Figure 2). In this chapter we summarize our recent observations on the mechanism of the reaction

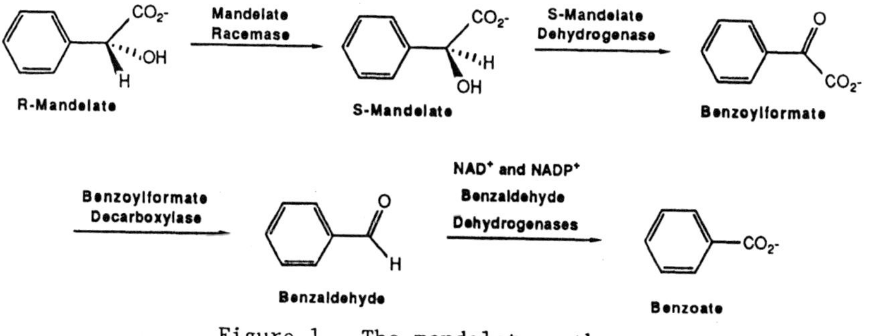

Figure 1. The mandelate pathway.

Chemical Aspects of Enzyme Biotechnology
Edited by T.O. Baldwin *et al.*, Plenum Press, New York, 1990

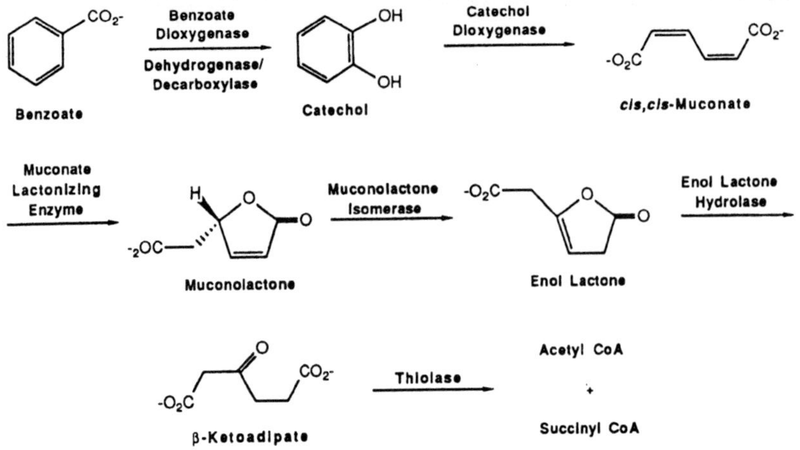

Figure 2. The β-ketoadipate pathway.

catalyzed by mandelate racemase and the structural and mechanistic relationship of this enzyme to muconate lactonizing enzyme, an enzyme central to the catabolism of benzoic acid and, therefore, mandelic acid. We have made the unexpected discovery that these enzymes are certainly divergently related and anticipate that molecular biological methods can be used to change muconate lactonizing enzyme into mandelate racemase and *vice versa*.

MECHANISM OF THE REACTION CATALYZED BY MANDELATE RACEMASE

Evidence for a Carbanionic Intermediate

We have obtained unambiguous chemical evidence that the inversion of configuration at the α-carbon of mandelate catalyzed by the racemase occurs *via* the generation of a carbanionic intermediate (or a mechanism involving concerted abstraction of a proton from one face and reprotonation on the opposite face). While early evidence for the carbanionic character of the reaction pathway was obtained by indirect experiments (the effect of electron donating and withdrawing groups on the rate of the reaction; Kenyon and Hegeman, 1970), recent experiments are much more direct.

The first unequivocal evidence that the reaction involves carbanionic intermediates was obtained by investigating the processing of p-halomethyl substituted mandelates by the racemase (Lin and Kozarich, 1988; Lin and Kozarich, unpublished observations). The strategy for these studies was that abstraction of the α-proton would generate a carbanion which could undergo elimination of the halide ion and generation of a p-xylylene intermediate; rearomatization of this species would produce p-methylbenzoylformate as product (Figure 3). Thus, elimination of halide would lead to oxidation of the substrate and would constitute persuasive evidence for the intermediacy of a carbanion intermediate. The bromo-, chloro-, and fluoromethyl substituted mandelates were prepared. Bromide ion was eliminated from p-bromomethyl mandelate in competition with racemization, with the apparent partitioning of the carbanionic intermediate being 500 in favor of racemization. Chloride ion was eliminated from p-chloromethyl mandelate, with the apparent partitioning of the carbanionic intermediate being

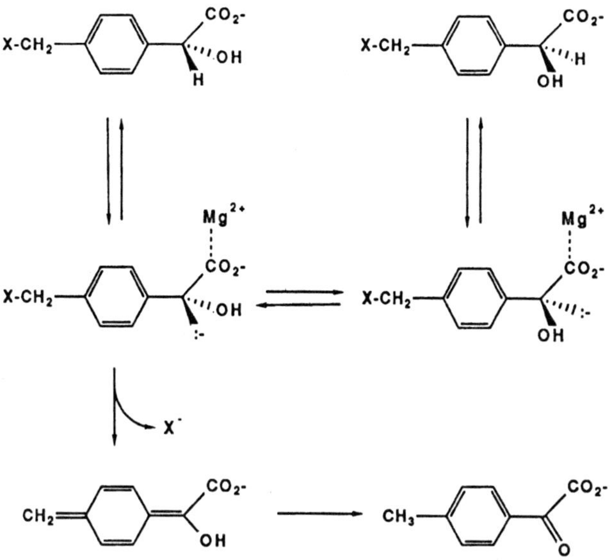

Figure 3. The processing of p-halomethyl substituted mandelates by mandelate racemase.

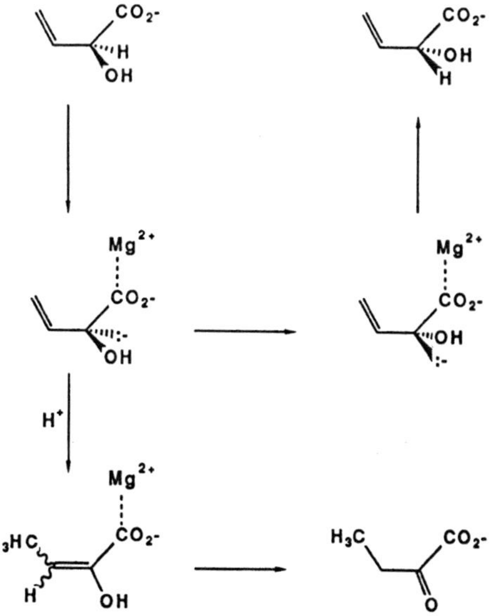

Figure 4. The processing of vinylglycolate by mandelate racemase.

5000 in favor of racemization. No detectable fluoride ion was eliminated from p-fluoromethyl mandelate. These results can be explained only by the abstraction of a proton to generate a carbanion.

The smallest known substrate for mandelate racemase is vinyl glycolate (2-hydroxy-3-butenoate) (Powers and Kenyon, unpublished observations). Both enantiomers are substrates for racemization with essentially equivalent values for k_{cat} and K_m, 250 sec^{-1} and 3.5 mM, respectively. The value for k_{cat} is approximately 35% that measured for racemization of mandelate, 700 sec^{-1}, and the value for K_m is approximately 1000% larger than that observed with mandelate, 0.37 mM. The saturated analog 2-hydroxybutyrate is not a substrate for the racemase. While these results demonstrate the importance of vinyl unsaturation (consistent with resonance stabilization of a carbanionic intermediate), the more important observation is that vinyl glycolate is slowly tautomerized to α-ketobutyrate, albeit at a rate approximately 0.01% of that observed for racemization (Figure 4). This isomerization to the most stable tautomer can be explained only by initial proton abstraction to generate a carbanion that can occasionally undergo protonation at the γ-carbon, generating an enol. This enol then tautomerizes to α-ketobutyrate. The stereochemistry of this allylic rearrangement has not been determined, so it is unknown whether the γ-protonation and enol tautomerization are catalyzed by the racemase.

Evidence for a Two Base Mechanism

Given the intermediacy of a carbanion, the question then arises as to the precise mechanism of the 1,1-proton shift in which a proton is removed from the substrate and delivered to the opposite face to give product with the inverted configuration. Two general types of mechanisms have been proposed: the one-base mechanism in which a single general base abstracts the proton and the conjugate acid so formed protonates the carbanion on the opposite face (Figure 5); and the two-base mechanism in which one general base abstracts the proton and the conjugate acid of a

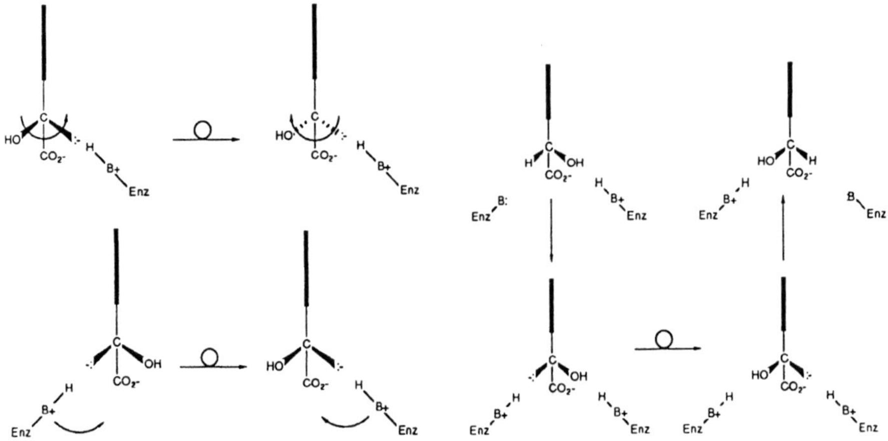

Figure 5. Left panel, the one-base mechanism; right panel, the two-base mechanism

second base protonates the carbanion (Figure 5). The one-base mechanism predicts that the proton which is abstracted may be delivered to the opposite face in the absence of exchange with solvent; the two-base mechanism predicts that the proton which is abstracted is not the same proton used for quenching the carbanionic intermediate. Whereas the one-base mechanism is economical in terms of the number of required functional groups, it necessarily must be accompanied by significant motion of either the conjugate acid or the carbanion; either of these is potentially difficult to envision. The two-base mechanism does not require significant motion and would allow the mechanism of proton abstraction-reprotonation to have at least some concerted character, thereby minimizing the lifetime of the highly reactive intermediate.

Two different experimental methods have now been used to provide evidence in support of the two-base mechanism (Powers and Kenyon, unpublished observations). Both of these involve reaction of one enantiomer of protio mandelate in D_2O: the first is the observation of a kinetic overshoot, and the second is the incorporation of deuterium into both unreacted substrate and product.

In the first experiment the circular dichroism of an enantiomer of mandelate is monitored as a function of extent of reaction. If a one-base mechanism were rigorously operative, no solvent hydrogen would be incorporated into product, and the molar ellipticity would monotonically decrease to zero. If a two-base mechanism were rigorously operative, deuterium would necessarily be incorporated into product and no deuterium could be incorporated into substrate. If a primary hydrogen kinetic isotope effect were observed in the overall reaction and when equal concentrations of the two enantiomeric mandelate were achieved, the racemase would discriminate against the deuterium containing product and continue to produce product. Eventually, the system would come to equilibrium as both enantiomers become completely deuterated.

Although initial attempts at these experiments using the limited amounts of the racemase that were available at the time did not reveal a kinetic overshoot (Whitman and Kenyon, 1986), more recent experiments using larger amounts of enzyme that are now available *via* an expression system utilizing the cloned and sequenced gene have demonstrated the existence of a kinetic overshoot. This observation eliminates a one-base mechanism in which no exchange of the abstracted proton with solvent deuterium can occur.

In the second experiment, the exchange of deuterium into protiated substrate and the incorporation of deuterium into product have been quanitated at low extents of reaction. These experiments not only provide strong evidence for a two-base mechanism, i.e., essentially no protium is incorporated into product, but also provide clues as the identity of the general basic functional groups in the active site of the racemase.

These experiments were conducted in both the **R** to **S** and the **S** to **R** directions, and the results obtained are summarized in Table 1. In both directions, essentially no protium is transferred from substrate to product; this implies that either the two-base mechanism is correct or the single base in the one-base mechanism can exchange rapidly with solvent hydrogen. In the **R** to **S** direction, essentially no deuterium is incorporated into substrate; however, in the **S** to **R** direction, significant deuterium is incorporated into substrate (given the low extents of reaction). If the one-base mechanism were operative, equivalent extents of deuterium incorporation into substrate might have been anticipated; thus, the one-base mechanism is unlikely.

Table 1. Exchange of substrate hydrogen with solvent hydrogen

Direction	% turnover	% H in product	% D in substrate
R to S	5.6	1.3	0.1
	8.0	1.5	0.3
S to R	6.1	2.7	3.4
	8.2	2.5	5.0

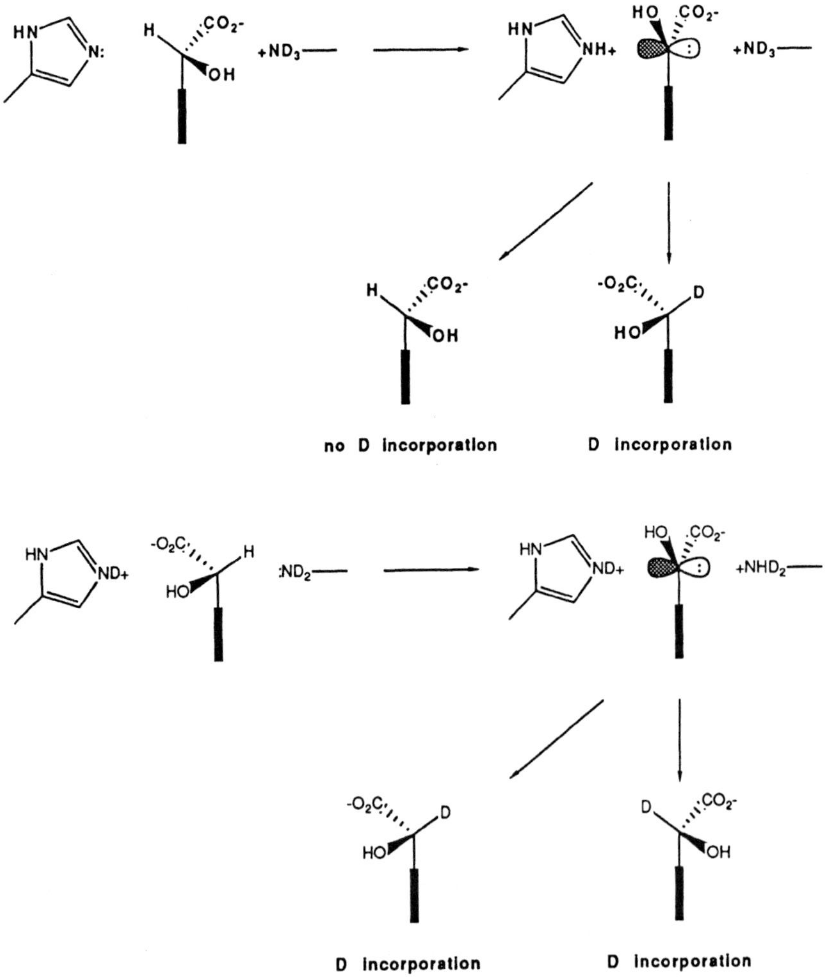

Figure 6. Top panel, an explanation for the observed exchange in the R to S direction; bottom panel, an explanation for the observed exchange in the S to R direction.

The most attractive explanation for the lack of incorporation of deuterium into R-mandelate substrate and the significant incorporation of deuterium into S-mandelate substrate is that the two acid/base catalysts present in the active site differ in valency. For example, if the base in the R to S direction were histidine, the conjugate acid would have no choice but to return the protium that was abstracted to the intermediate carbanion (Figure 6). If the base in the S to R direction were lysine, the conjugate acid could return either protium or deuterium, depending upon the magnitude of the primary hydrogen effect, to the intermediate carbanion (Figure 6). In either case, the product would contain significant deuterium since the general acid in the active site would be deuteriated.

In summary, the available chemical evidence indicates that the mechanism of the reaction catalyzed by mandelate racemase involves formation of an intermediate having carbanionic character, and the proton abstraction and reprotonation occur by a two-base mechanism.

CLONING AND DNA SEQUENCE ANALYSIS OF THE GENES OF THE MANDELATE PATHWAY

We are interested in further characterizing the mechanism of the reaction catalyzed by mandelate racemase by site-directed mutagenesis as well as investigating the mechanisms of other enzymes in the mandelate pathway, including benzoylformate decarboxylase. We, therefore, sought to clone the genes for the entire mandelate pathway so that they could be sequenced and expressed at high levels. We have succeeded in cloning a 10.5 kb *Eco* RI restriction fragment that encodes all five genes as evidenced by enzymatic assays as well as by its ability to confer R-mandelate utilization to a strain of *P. putida* (ATCC 17453) that does not synthesize any enzymes of the mandelate pathway (Tsou et al., 1990). A limited restriction map of this restriction fragment is shown in Figure 7 along with the locations of various genes that were established by deletion analysis and/or DNA sequence analysis.

The genes for mandelate racemase (*mdl*A) (Ransom et al., 1988), S-mandelate dehydrogenase (*mdl*B), and benzoylformate decarboxylase (*mdl*C) have been sequenced and are arranged in an operon whose transcription is positively regulated by mandelate (Tsou et al., 1990). The genes for the NAD^+- (*mdl*D) and $NADP^+$-dependent (*mdl*E) benzaldehyde dehydrogenases are separated from this operon by approximately 2 kb, thereby demonstrating that the genes encoding the enzymes in the mandelate pathway do not comprise a single operon as was hypothesized based upon their coordinate expression and cotransduction. The DNA located between the *mdl*CBA operon and the genes for the benzaldehyde dehydrogenases contains at least two open reading frames (*mdl*X and *mdl*Y) that appear to be involved in

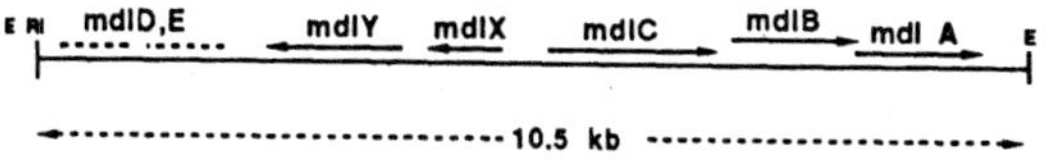

mdl A = mandelate racemase (359 aa)
mdl B = S-mandelate dehydrogenase (393 aa)
mdl C = benzoylformate decarboxylase (499 aa)
mdl D = NAD^+ benzaldehyde dehydrogenase (? aa)
mdl E = $NADP^+$ benzaldehyde dehydrogenase (? aa)

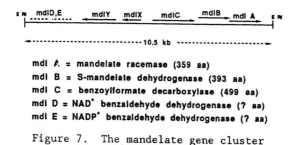

Figure 7. The mandelate gene cluster

mandelate metabolism, although their precise catalytic activities have not yet been described. Thus, it is likely that the mandelate pathway as it has been known for nearly 30 years is a subset of enzymes within a larger pathway.

The genes for both mandelate racemase (Tsou et al., 1989) and benzoylformate decarboxylase (Tsou et al., 1990) have been expressed in both *Escherichia coli* and strains of *P. putida*. In the case of mandelate racemase the most efficient expression was obtained by isolating an overproducing mutant that is partially resistant to α-phenylglycidate, an active site directed affinity label; in the case of benzoylformate decarboxylase, the most efficient expression was obtained with the *trc* promoter (IPTG inducible) in *E. coli*.

The primary sequence of mandelate racemase deduced from the DNA sequence reveals that the racemase polypeptide is 359 amino acids in length; *in vivo*, the N-terminal methionine is cleaved. The polypeptide has two cysteine residues (Ransom et al., 1988).

STRUCTURE OF MANDELATE RACEMASE

Mandelate racemase isolated to homogeneity from the overproducing mutant of *P. putida* has been crystallized in three different crystal forms (Neidhart et al., 1988; Neidhart and Petsko, unpublished observations). In two of these grown from polyethylene glycol, two polypeptides occupy the asymmetric unit. In the third crystal form, grown from $(NH_4)SO_4$, one polypeptide occupies the asymmetric unit, and diffraction is observed to beyond 1.8 Å. Four useful heavy atoms derivatives were obtained, and the structure has now been solved and refined at 2.5 Å resolution with R = 0.18 (Neidhart and Petsko, unpublished observations).

The structure reveals that mandelate racemase crystallizes as an octamer of eight identical polypeptide chains. The quaternary structure can be described as two square planar tetramers stacked on top of one another.

The secondary and tertiary structure of a polypeptide within the octamer is shown in Figure 8. The first 130 amino acids form an

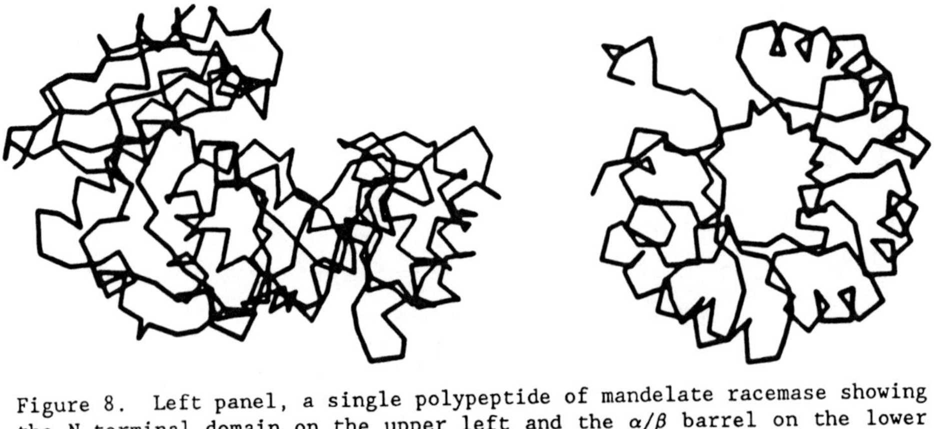

Figure 8. Left panel, a single polypeptide of mandelate racemase showing the N-terminal domain on the upper left and the α/β barrel on the lower right; right panel, a side view of the α/β barrel domain.

N-terminal domain composed of three β-sheets and four α-helices; the last 30 amino acids constitute a C-terminal meander. The middle 200 amino acids form an α/β barrel with eight internal β-sheets and seven external α-helices. This α/β barrel is similar to but differs from the protypic α/β barrel first observed in triose phosphate isomerase (Banner et al., 1975); the barrel in the isomerase has eight β-sheets and eight α-helices. In mandelate racemase the volume occupied by the eighth α-helix in the isomerase is occupied by the polypeptide connections to the N- and C-terminal segments. The binding site for the catalytically essential Mg^{2+} and the residues that comprise the active site are all located within the α/β barrel domain.

ACTIVE SITE OF MANDELATE RACEMASE

The structure of the active site of mandelate racemase is shown in Figure 9. The essential Mg^{2+} is coordinated to the carboxylates of Asp 195, Glu 221, and Glu 247. Both the carboxylate and α-hydroxyl groups of mandelate are coordinated to the metal ion; presumably this bidentate coordination facilitates abstraction of the α-proton by charge neutralization of the carboxylate and stabilizes an intermediate with enolate (i.e., carbanionic) character. The ε-amino functional groups of Lys 164 and 166 are present on one side of the active site cleft, and the imidazolium functional group of His 297 is on the the other side. The latter residue is closely approximated to the carboxylate groups of Asp 270 and Glu 327.

Based on this active site structure, we hypothesize that the reaction catalyzed by the racemase involves two base-conjugate acid pairs, Lys 166 and His 297. The functional groups of these residues are in accord with the valency deduced in the deuterium exchange experiments; in addition, the structure of the active site is also in accord with the previous proposal that a monoprotic functional group

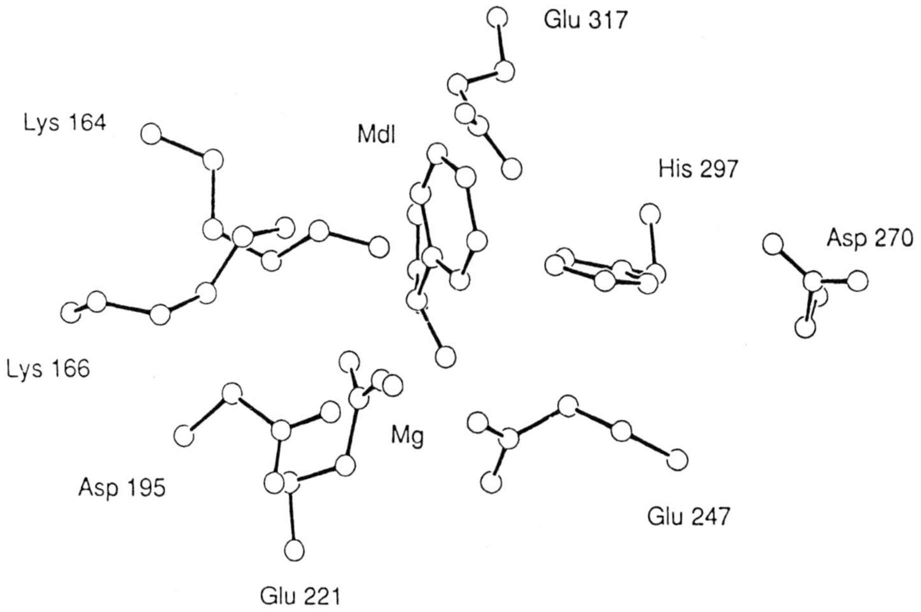

Figure 9. The active site of mandelate racemase.

(imidazole) abstracts the proton from the R enantiomer and a polyprotic functional group (amine) abstracts the proton from the S enantiomer. Thus, the structure of the active site revealed by the x-ray analysis confirms the expectations derived from the solution experiments.

Irrespective of the direction of the reaction, the k_{cat} dependence on pH suggests the presence of an essential acid having a pK_a of approximately 7.5; the pK_as of Lys 166 and His 297 are apparently equal. The abnormally low pK_a for Lys 166 may be the result of its close proximity to the positively charged functional group of Lys 164. [The close proximity of His 297 to the essential Mg^{2+} may require the presence of two carboxylate anions to produce the apparent pK_a associated with its functional group.]

STRUCTURE AND ACTIVE SITE OF MUCONATE LACTONIZING ENZYME

Comparison of the α-carbon backbone of mandelate racemase with those of other enzymes containing an α/β barrel structure or domain revealed a striking similarily to muconate lactonizing enzyme, an enzyme of the β-ketoadipate pathway that is required for the catabolism of mandelate *via* benzoate; the backbones of the racemase and lactonizing enzyme (Goldman et al., 1987) are compared in Figure 10. The lactonizing enzyme is a polypeptide 371 amino acids in length (Houghton and Ornston, personal communication; Taylor, Kozarich and Gerlt, unpublished observations) compared to 358 residues for the racemase (Ransom et al., 1988). The lactonizing enzyme has the same structural arrangement found in the racemase, i.e., an N-terminal domain composed of three α-helices and four β-sheets, a C-terminal meander, and a central α/β barrel. Approximately 60% of the residues of the racemase can be superimposed to within 1.3 Å of a structurally homologous residue in the lactonizing enzyme; these structurally conserved residues are 30% conserved in identity. Both enzymes crystallize in the I422 space group as octamers.

The structure of the active site of the lactonizing enzyme is shown in Figure 11. The essential Mn^{2+} is coordinated to the carboxylate groups of Asp 198, Glu 224, and Glu 250. The carboxylate of α-ketoglutarate, a competitive inhibitor that was cocrystallized with the lactonizing enzyme, is also coordinated to the metal ion; this

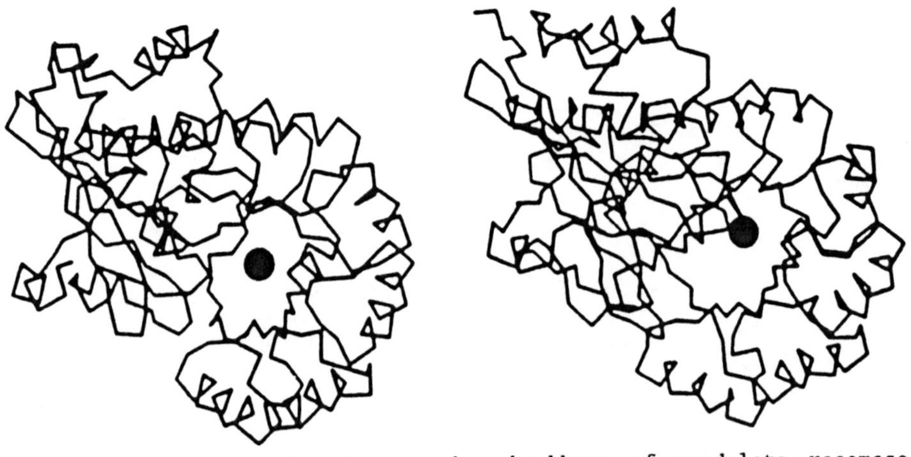

Figure 10. Left panel, the α-carbon backbone of mandelate racemase; right panel, the α-carbon backbone of muconate lactonizing enzyme.

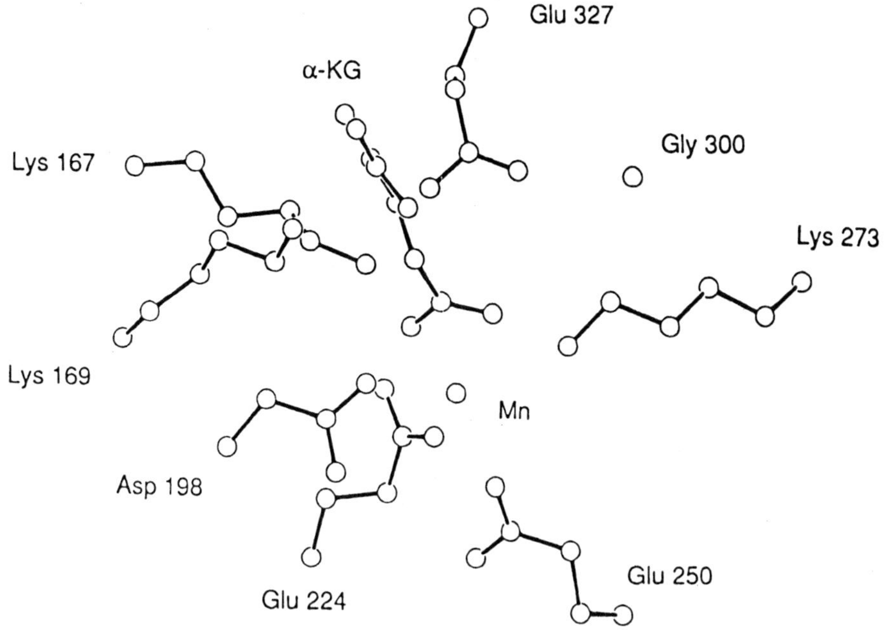

Figure 11. The active site of muconate lactonizing enzyme.

coordination should facilitate the generation of carbanion character at the adjacent carbon that would accompany addition of the remote carboxylate of the substrate to the adjacent double bond. The ε-amino functional groups of Lys 167 and 169 are present in the active site, and we hypothesize that these together constitute the general acid that would protonate the carbanion. The argument made earlier about the effect of the close proximity of the two lysine residues in the active site of the racemase on the pK_a of the lysine presumably also applies to the active site of the lactonizing enzyme. The other side of the active site cleft has no functional groups that would participate as general acid/base catalysts: the structurally conserved residue corresponding to His 297 in the racemase is Gly 300, and the volume occupied by the imidazole functional group of His 297 in the racemase is occupied by Met 302 in the lactonizing enzyme.

RELATIONSHIP BETWEEN MANDELATE RACEMASE AND MUCONATE LACTONIZING ENZYME

Although the racemase and lactonizing enzyme differ in the number and exact identity of amino acid residues, the backbones of their α/β barrel domains are strikingly conserved. Accordingly, comparison of the residue numbers of the metal ion ligands and the active site lysines reveals that these have been exactly conserved in sequence space (Figures 9 and 11). This homology as well as previously discussed superpositioning constitute persuasive evidence that the racemase and lactonizing enzyme are related by divergent evolution.

Presumably, Nature realized that within the catabolism of aromatic acids the same structural solution to the problem of stabilizing and protonating a carbanionic intermediate could be used for both the racemase and lactonizing enzyme despite the fact that the mechanisms by which the carbanions are generated (abstraction of a proton by the racemase and addition of a carboxylate group across a double bond by the lactonizing enzyme) differ. Whereas numerous examples are known of enzyme pairs having homologous structures that catalyze mechanistically identical reactions (e.g., trypsin/chymotrypsin and lactate/malate dehydrogenases), to the best of our knowledge mandelate racemase and muconate lactonizing are the first examples of enzymes having homologous structures that catalyze different chemical reactions.

Since the lactonizing enzyme is essential to the catabolism of benzoate, we assume that the structural gene for mandelate racemase evolved by duplication of the gene for the lactonizing enzyme followed by divergent evolution directed toward mandelate utilization. Given the similarity in active site structures, we are currently attempting to mimic evolution with site-directed mutagenesis by converting the lactonizing enzyme into the racemase. The α-carbon of Gly 300 in the lactonizing enzyme is superimposable on the α-carbon of His 297 in the racemase; thus, we assume that the mechanistically most important substitution will be G297H in the lactonizing enzyme. In addition, we note that the α-carbon of Lys 273 in the lactonizing enzyme is superimposable on the α-carbon of Asp 270 in the racemase and that the space occupied by the imidazole functional group of His 297 in the racemase is occupied by the side chain of Met 302 in the lactonizing enzyme; substitutions at these three positions may be required for the successful generation of racemase activity in the overall architecture of the lactonizing enzyme. Even if we are successful in generating only a small level of racemase activity in the lactonizing enzyme (we do not anticipate that three amino acid changes will produce a "perfect" racemase from the lactonizing enzyme), our results will be important in understanding the principles of active site design.

ACKNOWLEDGEMENTS

We wish to thank Professor Adrian Goldman for supplying us with coordinates of muconate lactonizing enzyme in advance of publication. This research was supported by AR-17323 (to G.L.K.), GM-26788 (to G.A.P.), GM-34572 (J.A.G.), GM-37210 (to J.W.K.), and GM-40570 (to J.A.G., G.L.K., J.W.K., and G.A.P).

REFERENCES

Banner, D. W., Bloomer, A. C., Petsko, G. A., Phillips, D. C., Pogson, C. I., and Wilson, I. A. (1975) *Nature 255*, 609.

Goldman, A., Ollis, D. L., and Steitz, T. A. (1987) *J. Mol. Biol. 194*, 143.

Kenyon, G. L., & Hegeman, G. D. (1970) *Biochemistry 9*, 4036.

Lin, D. T., Powers, V. M., Reynolds, L. J., Whitman, C. P., Kozarich, J. W., and Kenyon, G. L. (1988) *J. Amer. Chem. Soc. 110*, 323.

Neidhart, D. J., Powers, V. M., Kenyon, G. L., Tsou, A. Y., Ransom, S. C., Gerlt, J. A., and Petsko, G. A. (1988) *J. Biol. Chem. 263*, 9268.

Ransom, S. C., Gerlt, J. A., Powers, V. M., and Kenyon, G. L. (1988) *Biochemistry 27*, 540.

Tsou, A. Y., Ransom, S. C., Gerlt, J. A., Buechter, D. D., Babbitt, P. C., and Kenyon, G. L. (1990) *Biochemistry*, in press.

Tsou, A. Y., Ransom, S. C., Gerlt, J. A., Powers, V. M., and Kenyon, G. L. (1989) *Biochemistry 28*, 969.

Whitman, C. P., and Kenyon, G. L. (1986) in *Mechanisms of Enzymatic Reactions: Stereochemistry*, P. A. Frey, Ed., Elsevier, New York, p.191.

AN ENZYME-TARGETED HERBICIDE DESIGN PROGRAM BASED ON EPSP SYNTHASE: CHEMICAL MECHANISM AND GLYPHOSATE INHIBITION STUDIES

James A. Sikorski*, Karen S. Anderson#, Darryl G. Cleary, Michael J. Miller, Paul D. Pansegrau, Joel E. Ream, R. Douglas Sammons and Kenneth A. Johnson#

Technology Division
Monsanto Agricultural Company
A Unit of Monsanto Company
800 N. Lindbergh Blvd.
St. Louis, MO 63167

#Departments of Molecular & Cell Biology and Chemistry
301 Althouse Laboratory
The Pennsylvania State Univ.
University Park, PA 16802

INTRODUCTION

The herbicide markets of the late 1990's and beyond will demand high performance products with stringent environmental acceptability requirements. We have initiated a multi-disciplinary herbicide discovery program directed toward the inhibition of key plant enzymes as one approach to meet these challenges. Plants contain a variety of biosynthetic pathways which are essential for their growth. Effective, plant-specific enzyme inhibitors offer the opportunity to satisfy the herbicide performance needs of the marketplace while exhibiting favorably low mammalian toxicity properties.[1] Our enzyme-targeted research effort systematically integrates mechanistic biochemistry, molecular biology, modeling, inhibitor recognition, and structural biology information with organic synthesis through a series of collaborations within and outside Monsanto. A thorough understanding of the chemical mechanism for a particular enzyme target is an essential first step for the design of potent inhibitors.

An attractive enzyme system on which to test this approach is the biological target for N-phosphonomethylglycine (glyphosate), 1, the active ingredient in ROUNDUP ® herbicide.[2] Glyphosate is a broad-spectrum, non-selective herbicide that controls most of the world's worst weeds while exhibiting very low mammalian toxicity. Glyphosate has achieved worldwide

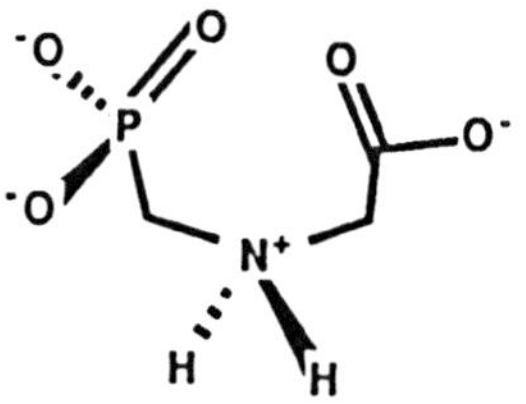

1

acceptance as an environmentally friendly herbicide. It is relatively non-toxic to mammals, birds, fish, insects and most bacteria. Glyphosate is one of a very few compounds currently registered in the U.S. to control undesired vegetation on surface waterways. For this application it is known as RODEO ® herbicide.

EPSP SYNTHASE (EPSPS)

It is widely accepted that glyphosate kills plants by inhibiting the enzyme EPSP (5-enolpyruvoylshikimate-3-phosphate) synthase (EC 2.5.1.19). EPSPS is a critical enzyme in the biosynthesis of the essential aromatic amino acids and other important aromatic compounds. As the sixth step in the shikimate pathway, EPSPS provides the immediate precursor to chorismate, an important branch point intermediate.[3] This pathway is present in all plants and micro-organisms, but is completely absent in mammals, birds, fish and insects. The highly specific interaction of glyphosate with this enzyme provides effective weed control while retaining very favorable environmental acceptability. This mode of action is only known for glyphosate-based herbicides. No other commercial herbicides exhibit this mode of action.

Experimental evidence supporting inhibition of EPSPS as the primary mode of action for glyphosate includes the following: (a) The herbicidal effects of glyphosate-treated plant cells can be reversed by aromatic amino acids.[4] (b) Plants treated with glyphosate accumulate shikimate-3-phosphate.[5] (c) All plant and bacterial EPSPS's isolated and characterized to date are inhibited by glyphosate.[6] (d) Plant cells exposed to non-lethal doses of glyphosate can increase their tolerance to glyphosate by over-producing EPSPS.[7] (e) Plants which are genetically engineered with modified EPSPS exhibiting reduced glyphosate affinity are more tolerant to glyphosate applications.[6,8]

EPSPS catalyzes a very unusual transfer reaction of the enolpyruvoyl portion of PEP regiospecifically to the 5-OH of S3P, producing EPSP and P_i (Scheme 1).[9] EPSPS has been isolated and purified to homogeneity from fungal[10], bacterial[11] and plant sources.[12]

Scheme 1. EPSP Synthase

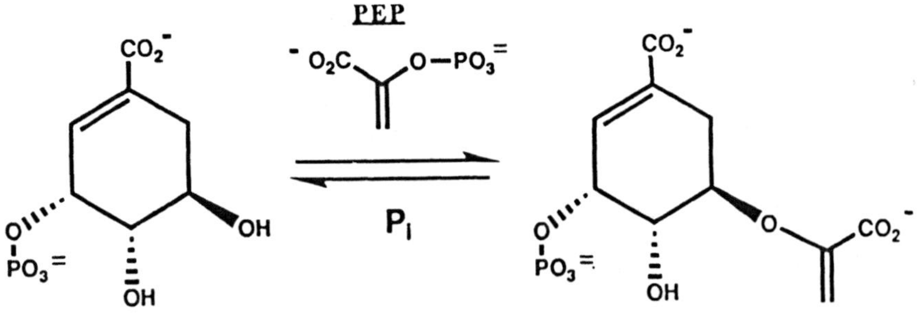

S3P, 2 EPSP, 3

While the fungal enzyme is part of a multi-enzyme complex, plant and bacterial enzymes are monomeric with molecular weights ranging between 44,000 and 48,000. No metal ions or co-factors are associated with these homogenous proteins.

Each of the EPSPS substrates is a multiply-charged anion, suggesting that the enzyme provides a correspondingly rich electropositive active site for substrate recognition. Glyphosate acts as a competitive inhibitor to PEP and an uncompetitive inhibitor with S3P.[10,13] Neither PEP nor glyphosate has any significant affinity for EPSPS in the absence of S3P.[14] Anion recognition at the active site must therefore be quite specific. EPSP synthases have been isolated and characterized from a sufficient variety of plant and bacterial sources to indicate that the over-expressed *E. coli* enzyme[15] can function as a suitable model for plant enzymes.[12a] Consequently, the biochemistry of the *E. coli* enzyme has been characterized in greater detail.

This highly specific enzymatic reaction between PEP and secondary alcohols currently has no solution chemistry analogy. As an enzyme inhibitor, glyphosate displays a unique specificity for this enzyme, and minor structural changes in the glyphosate backbone lead to a significant loss of inhibitor potency. This will be discussed in more detail below. A detailed description of the chemical mechanism for this system is required to understand the fundamental chemistry of this unique enzymatic transformation, provide direction for potent inhibitor design and characterize the molecular level details of glyphosate inhibition. A clearly defined glyphosate binding site also offers new opportunities for the design of EPSPS's with lower glyphosate sensitivity for genetically engineered herbicide tolerant plants.[6]

EPSPS MECHANISM

Background

UDP-N-acetylglucosamine enolpyruvoyl transferase (EC 2.5.1.7), an important initial step in cell wall peptidoglycan biosynthesis, is the only other enzyme known to catalyze a similar enolpyruvoyl transfer reaction of PEP.[17] The mechanistic evidence to date suggests that this transferase proceeds through a covalently bound enolpyruvoyl intermediate. While fosfomycin is a potent irreversible inhibitor of this transferase, it does not inhibit EPSPS. On the other hand, glyphosate does not inhibit this transferase or any other PEP dependent enzyme.[18] Unlike EPSPS, this enzyme does display some affinity for PEP in the absence of co-substrate. A number of investigators have therefore attempted to correlate these results with the EPSPS mechanism.

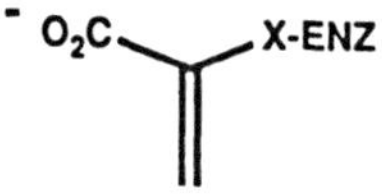

Covalent Intermediate

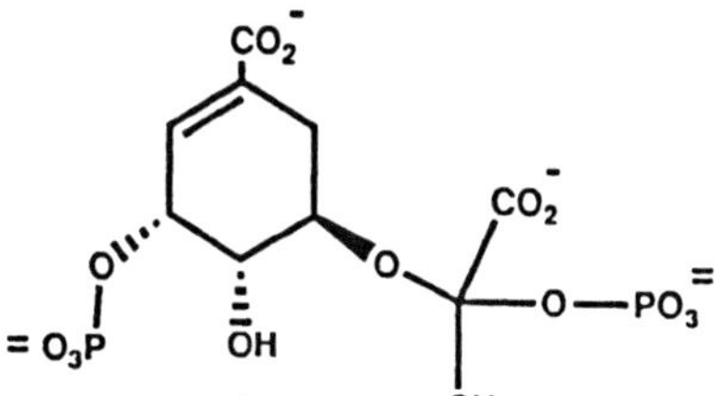

4

Tetrahedral Intermediate

Mechanistic studies of EPSPS have attracted considerable attention based on the unique chemistry of this enzyme. Rather than present an exhaustive review of the literature, only key results will be highlighted. Early studies by Sprinson demonstrated that the enolpyruvoyl transfer reaction proceeds exclusively with C-O bond cleavage.[19] Most enzymatic reactions of PEP involve P-O bond cleavage processes. In addition, a facile exchange of deuterium or tritium into the methylene protons of PEP is observed when the enzymatic reaction is performed in either D_2O or ^3H_2O. These results suggest that a transient methyl group is produced at the terminal carbon of PEP sometime during catalysis. Based on these results, Sprinson and co-workers first suggested that EPSPS proceeds through a direct addition-elimination sequence via the tetrahedral intermediate, **4**, shown above.

Subsequent studies by Knowles[20] and Floss[21] confirm that a significant primary deuterium isotope effect operates at the C-3 position of PEP. Using PEP stereospecifically labeled at the methylene positions with one deuterium and one tritium, the transient methyl group produced during the EPSPS reaction is chiral and the addition-elimination steps proceed with overall retention of chirality at C-3.

EPSPS slowly catalyzes tritium exchange into PEP from ^3H_2O in the presence of 4,5-dideoxy-S3P, a substrate analog lacking the 5-OH required for product formation.[18] Abeles and co-workers proposed that EPSPS utilizes a covalently bound enolpyruvoyl enzyme intermediate similar to that proposed for UDP-N-acetyl-glucosamine enolpyruvoyl transferase. At the time we began this investigation, no single mechanistic model unambigously explained all of the experimental results. Some resolution of the controversy between covalently bound enzyme intermediates and 4 was required to provide direction for inhibitor design.

E. coli EPSPS Kinetics

The tetrahedral intermediate mechanism should be kinetically distinguishable from that involving a covalently bound enzyme intermediate, based on the number of distinct chemical species involved. The former requires a single kinetically competent intermediate, while the latter invokes at least three kinetically distinct entities. While a detailed steady state kinetic study of *E. coli* EPSPS is available, it provides no information to distinguish between these pathways.[22] A rapid-quench kinetic analysis of *E. coli* EPSPS was completed in collaboration with Professor Kenneth A. Johnson to determine the important enzymatic rate constants during EPSPS catalysis. A complete kinetic profile for EPSPS is now available based on these studies.[23]

Substrate trapping experiments with radiolabeled PEP and EPSP confirm that EPSPS utilizes a compulsory ordered binding process in the forward direction with S3P binding first and PEP second. Similarly, product release is also fully ordered with P_i first and EPSP second. The kinetic analysis is consistent with the formation of a single intermediate during catalysis. The buildup and decay of this intermediate can be observed on a millisecond timescale in the form of pyruvate, based on chemical breakdown during the acidic rapid quench procedure. Under these conditions, EPSP and PEP are perfectly stable. Based on this comprehensive kinetic model, the presence of this intermediate can be maximized under "internal equilibrium" conditions using high concentrations of PEP and P_i to lock the shikimate species in a bound state with EPSPS (Scheme 2).

Scheme 2. EPSPS Internal Equilibrium

$$E^{S3P} \overset{\pm PEP}{\rightleftharpoons} E^{S3P}_{PEP} \rightleftharpoons E^{I} \rightleftharpoons E^{EPSP}_{PI} \overset{\pm PI}{\rightleftharpoons} E^{EPSP}$$

While these results are consistent with the formation of Sprinson's tetrahedral intermediate, a direct detection method was sought which would avoid the breakdown to pyruvate.

EPSPS NMR Studies

If one compares the chemistry which occurs at the C-2 carbon of PEP during EPSPS catalysis, this center changes from an sp^2 center in PEP to an sp^3 quaternary carbon in the tetrahedral intermediate and back to a different sp^2 carbon in EPSP. Consequently, this system seems ideally suited to a ^{13}C NMR study to directly observe the key enzymatically generated intermediates utilizing [2-^{13}C]PEP as a spectroscopic probe.[24]

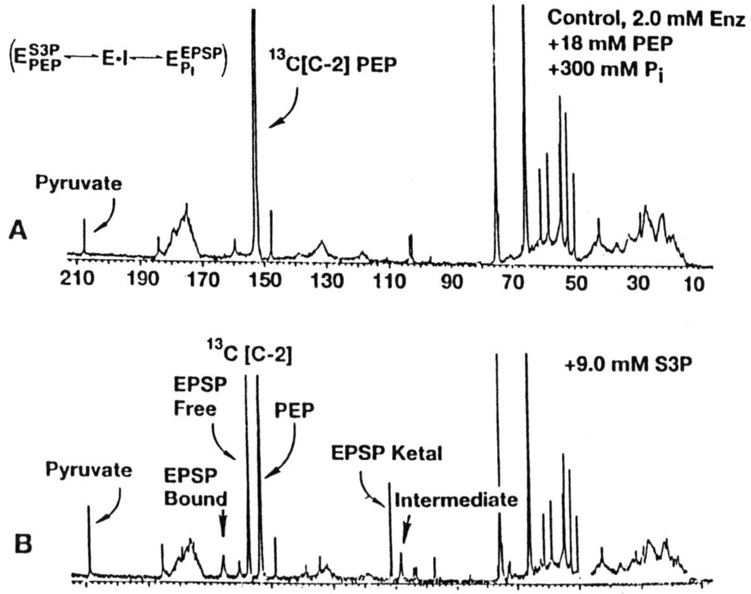

FIGURE 1. CARBON NMR SPECTRA OF THE EPSPS INTERNAL EQUILIBRIUM

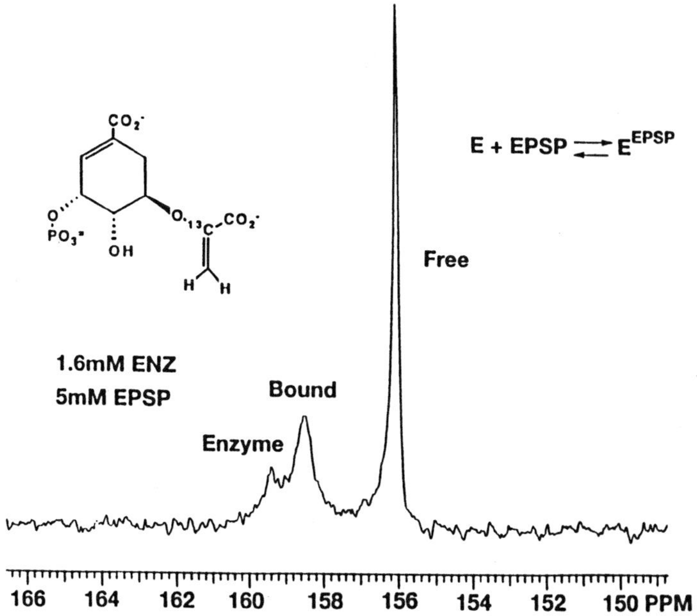

FIGURE 2. CARBON NMR SPECTRUM OF THE EPSP BINARY COMPLEX

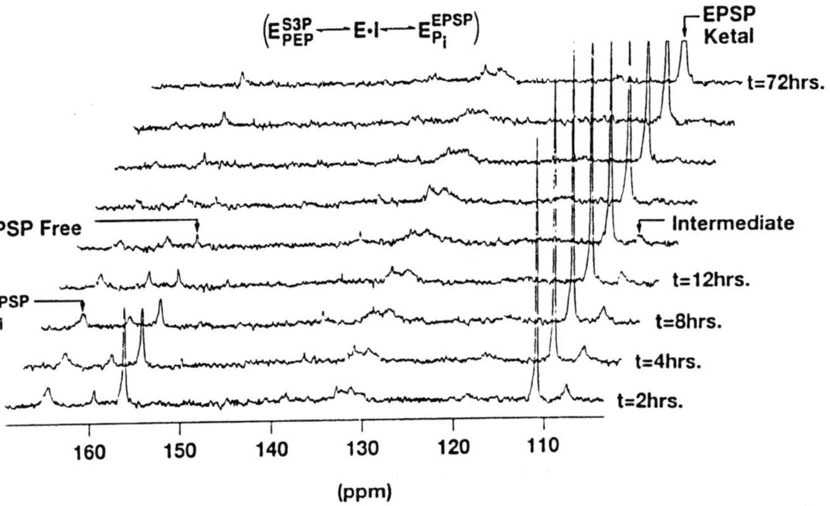

FIGURE 3

TIME DEPENDENT CARBON NMR SPECTRA OF EPSP KETAL FORMATION

The ^{13}C NMR of EPSPS with [2-^{13}C]PEP alone shows no significant change in chemical shift relative to the PEP solution spectrum in the absence of enzyme. A signal at 147.2 ppm is present as a minor impurity in the original labeled PEP sample. However, under internal equilibrium conditions with added S3P, a surprisingly large number of non-protein signals are observed (Figure 1A). Two broadened signals at 107.4 and 164.5 ppm are readily apparent as potential enzyme-bound species while sharp signals at 207.1, 110.6 and 96.4 ppm are more consistent with enzyme free entities. Indeed, the latter species can be readily separated from enzyme by ultrafiltration without denaturation.

EPSP Hydrolysis These results suggest that significant side reactions may be occurring, particularly under conditions employing high concentrations of EPSPS. Prolonged incubations of either S3P or PEP alone with enzyme show no detectable changes by NMR or HPLC. However, EPSPS slowly hydrolyzes EPSP in the absence of added phosphate, producing S3P and pyruvate. This hydrolysis is quite slow relative to catalysis, but becomes a significant side reaction during the time course required for NMR investigations. Presumably, hydrolysis occurs as water binds in the phosphate site under these conditions. This production of pyruvate in the keto (207.1 ppm) and hydrated ketal (96.4 ppm) forms is consistent with the ^{13}C NMR signals observed. EPSP hydrolysis is effectively blocked by glyphosate, presumably by sequestering enzyme in the very stable S3P*glyphosate*EPSPS ternary complex.[14]

EPSP hydrolysis is surprising in that no comparable hydrolysis of PEP occurs after prolonged incubation with EPSPS in the presence of either 4,5-dideoxy-S3P[18] or 5-deoxy-S3P.[25] Under these conditions, PEP hydrolysis might be expected since deuterium exchange into PEP is observed in D_2O when EPSPS is saturated with these substrate analogs. While vinyl ether hydrolysis is mechanistically well characterized, no reversible protonation of vinyl ethers in solution has been observed.[16] A comparable solution study of EPSP hydrolysis has been initiated in collaboration with Professor A. J. Kresge for comparison with this EPSPS-catalyzed event.

EPSP Ketal Formation Anion exchange chromatography of the ultrafiltered enzyme-free mixture produces a new compound easily separated from EPSP, PEP and S3P. When the internal equilibrium reaction stands with added excess PEP, this new entity becomes the only detectable shikimate species separable from the mixture without denaturing EPSPS. Radiolabel incorporation experiments with [U-^{14}C]S3P and [1-^{14}C]PEP indicate that all ten carbon atoms from both molecules are present. This material can be isolated by preparative ion-exchange chromatography and is isomeric with EPSP by mass spectral analysis of the corresponding permethylated derivatives. This EPSP ketal, 5, is isolated as a single diastereomer whose structure is fully consistent with the observed ^{1}H, ^{31}P and ^{13}C NMR spectral results.[26]

The stereochemistry at the C-2' position of **5** is exclusively **R** by 1-D NOE measurements in which an NOE is observed between the C-3' methyl signal and the H-5 proton. The ^{13}C NMR of the ketal quaternary carbon is essentially identical to the signal observed in the internal equilibrium at 110 ppm. While no significant chemical shift change occurs with **5** in the presence of EPSPS, a slight line-broadening is observed suggesting that only a relatively weak, fast exchange process takes place. The slow rate of ketal formation coupled with its reversible enzyme affinity strongly suggests that **5** is an unexpected side reaction product and clearly not an EPSPS intermediate.

The overall rate of ketal formation is many orders of magnitude slower than catalysis. Yet ketal production is quite efficient and specific, under conditions where enough PEP is added to make up for its loss as pyruvate.

EPSPS is clearly required for ketal production since prolonged reactions of EPSP alone or with pyruvate fail to produce the ketal. Yet, prolonged incubation of ketal and phosphate with EPSPS does not produce S3P or EPSP. Glyphosate also effectively blocks ketal formation by sequestering enzyme in the stable ternary complex with S3P.[14]

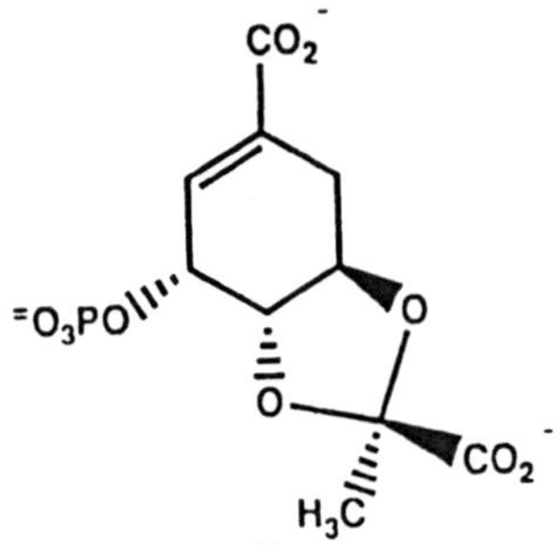

5

EPSP Ketal

EPSP hydrolysis products and EPSP ketal formation clearly account for all of the enzyme-free species observed by ^{13}C NMR at internal equilibrium (Figure 1B). The corroboration of ^{13}C NMR chemical shift assignments with HPLC analyses and isolation of chemical species contributed significantly to the successful identification of these side reaction processes. Another group has independently investigated the ^{13}C NMR of EPSPS using [2-^{13}C]PEP and confirmed these spectral data.[24b]

<u>EPSPS Bound Species</u> Two broadened signals remain to be identified in the ^{13}C NMR spectrum of the internal equilibrium. The EPSP ketal provides a convenient reference point at 110.6 ppm for a quaternary carbon bearing two oxygens and a carboxyl group. Sprinson's tetrahedral intermediate would be expected to occur in nearly the same place. The broadened signal observed at 107.4 is consistent with this structure, considering that it is tightly bound to enzyme.

The broadened signal at 164.5 occurs in a region expected for an EPSP vinylic carbon. However, a significant downfield chemical shift is observed relative to that of EPSP alone in solution or in the enzyme binary complex (Figure 2). This is consistent with the EPSP*P_i*EPSPS ternary complex as predicted from the rapid quench kinetic results. However, the magnitude of the downfield shift is surprising and suggests that there may be substantial buildup of positive charge at the C-2' position of EPSP in this ternary complex.

These two distinct bound species (107 ppm, 164 ppm) are closely linked in equilibrium during the time course required for ketal production (Figure 3). As one monitors the internal equilibrium over time, the signals at 107 and 164 remain fairly constant as the overall concentration of free EPSP in solution drops. The consumption of free EPSP is then followed by a simultaneous loss of both bound species with a concomitant increase in ketal production. Eventually, the ketal becomes the only shikimate species present. The characterization of the internal equilibrium species[26] and the time dependence of these processes have also been confirmed using ^{31}P NMR (data not shown).

While all of these spectral results are consistent with the internal equilibrium complexes summarized in Scheme 2, they remain ambiguous by themselves. A proper reference point for the postulated tetrahedral intermediate can only be established with a carefully isolated and fully characterized species. Consequently, procedures were examined to pursue the isolation of this elusive intermediate.[27]

EPSPS Tetrahedral Intermediate

Isolation and Characterization Several attempts were made to identify species with ionic character greater than EPSP by anion exchange chromatography after rapid quenching with acid (aq. HCl). Under these conditions, only pyruvate[23] was detected, and no new ionic species were observed. A search of the literature then uncovered N-acetyl-β-D-neuraminic acid 2-phosphate, **6**, as the one known chemical entity that is structurally similar to **4**.[28] This neuraminic acid derivative possesses a quaternary center with all of the key structural features of **4**. Compound **6** is stable to mild base in either aqueous triethylamine or ammonium hydroxide, but easily eliminates phosphate following treatment with strong acid or base.

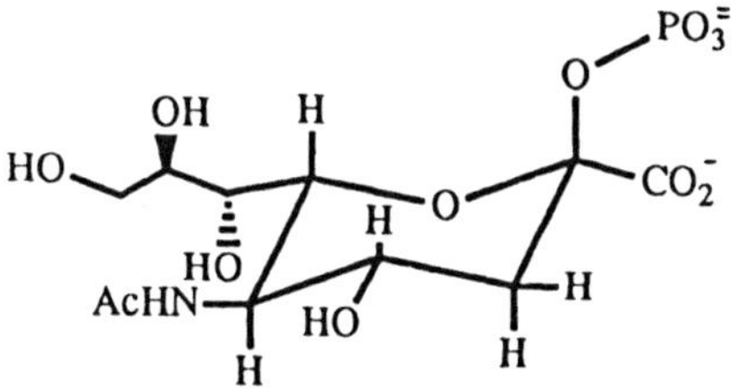

6

N-Acetyl-β-D-neuraminic acid 2-phosphate

Based on this chemical model, aqueous triethylamine seemed an ideal choice as a quenching agent for the internal equilibrium. Indeed, following treatment with neat triethylamine, the internal equilibrium provides a mixture of **3** (25%) and **4** (75%) which can be separated from S3P, PEP and P_i using anion-exchange chromatography with an ammonium bicarbonate gradient. Final stage purification can be achieved in a second chromatographic step with a phosphate gradient.[27] Radiolabel incorporation experiments with [U-^{14}C]S3P, [1-^{14}C]PEP and [^{32}P]PEP indicate that all of the key structural features of **4** must be present in a single molecule. The intermediate can be isolated within five seconds after mixing substrates with enzyme. It presumably forms as a single diastereomer, although the stereochemistry at the quaternary 5'-carbon is currently not known. A complete spectral characterization by ^{13}C, ^{31}P and ^{1}H NMR is fully consistent with the Sprinson tetrahedral intermediate, **4**.[27]

Improved procedures cleanly separate **4** from EPSP and the other components of the internal equibrium in a single step.[24a] This provides a quantitative technique to analyze the relative concentration of individual species and compare the results with those predicted from the kinetic model. When S3P*EPSPS binary complex is admixed with PEP in a single turnover experiment, **4** forms within fifteen milliseconds. While S3P and EPSP are also present, no other shikimate species are detected. Quantitation of the individual components of the internal equilibrium are in excellent agreement with those predicted from the kinetic model (Scheme 3).

Scheme 3

$$E^{S3P}_{PEP} \rightleftharpoons E^{I} \rightleftharpoons E^{EPSP}_{Pi}$$

Calcd:	4% (S3P)	40%	56% (EPSP)
Found:	6%	33%	61%

^{13}C NMR. This quantitation of the internal equilibrium at 20°C agrees qualitatively with the relative concentrations of enzyme-bound species observed by ^{13}C NMR at 6 °C. The low concentration of PEP in a ternary complex with S3P makes this complex difficult to detect by NMR. The solution ^{13}C NMR signal for the quaternary center in **4** appears at 102 ppm. This agrees very closely to the assigned tetrahedral intermediate signal (107.4 ppm) observed in an enzyme-bound form under internal equilibrium conditions. Some downfield chemical shift for the tetrahedral intermediate is expected based on the dynamic equilibria with the EPSP*P_i*EPSPS complex observed in Figure 3. The bound signal remaining at 164 ppm must therefore correspond to EPSP in a ternary complex with P_i since EPSP is the only other shikimate product observed after rapid quenching. The magnitude of the 9 ppm downfield chemical shift is surprising for this complex and is consistent with a buildup of positive charge at the C-2' position of EPSP. Sprinson's tetrahedral intermediate is therefore the single key mechanistic intermediate required for EPSPS catalysis. No other species are detected by NMR or rapid quench procedures.

Solution Biochemistry. Sufficient quantities of the tetrahedral intermediate are now available to define its fundamental properties in solution.[29] Compound **4** is reasonably stable at pH ≥7. Under acidic conditions this intermediate readily decomposes forming S3P and pyruvate. This is consistent with the transient production of pyruvate observed during the rapid quench procedures as the acidic breakdown product of **4**.[23] After standing at neutral pH, **4** forms S3P & pyruvate (76%), EPSP & P_i (8%) along with the EPSP ketal & P_i (16%). Breakdown to S3P and pyruvate is clearly preferred in solution. EPSPS has identified a unique chemical means to overcome this inclination for pyruvate production that favors the P_i elimination in the forward direction to EPSP.

The low concentration of the EPSP ketal following solution breakdown of **4** raises another interesting question regarding the origin of **5**. Does the ketal form directly from **4** at the enzyme active site or only in solution following an infrequent release of intermediate from the enzyme? This issue is still unresolved and the subject of ongoing studies. Ketal production is maximized under internal equilibrium conditions where the concentration of **4** is optimum. Since ketal forms as a single diastereomer under internal equilibrium conditions, it is tempting to suggest that **5** forms from chiral **4** by an S_N2 displacement of phosphate by the shikimate 4-OH. This can conceivably occur either at the active site or in solution.

While one might expect to obtain a mixture of ketal isomers via an S_N1 process in solution, the solvated nature of the anionic 3-phosphate and pyruvoyl carboxylate groups may favor formation of the single ketal R isomer by electrostatic and steric repulsions. The stereochemistry of the ketal produced from solution breakdown of **4** has not been determined. EPSPS is clearly required in some role for ketal production, yet reversal of **5** back to S3P or EPSP is not observed following prolonged incubation of **5** with EPSPS and P_i. This lack of reversibility suggests that EPSPS is not catalyzing the formation of **5** directly. Rather, the high concentration of enzyme-bound **4** in the internal equilibrium allows its occasional release into solution and breakdown to **5**. The rate of ketal production observed during solution breakdown of 4 is not significantly increased by the addition of EPSPS. However, the unexpectedly fast rebinding of 4 with EPSPS described below suggests that no free intermediate is present in solution in the presence of enzyme. Therefore, observation of the same rate of formation of 5 with or without EPSPS implies conversion of 4 to 5 at the active site.

Rebinding with EPSPS. Upon exposure to EPSPS, 4 immediately forms all of the expected enzyme products and substrates in relative concentrations which agree with the kinetic model. In a single turnover experiment upon addition to EPSPS, 4 partitions in the forward and reverse directions within fifty milliseconds. The rate of rebinding of 4 with EPSPS is surprisingly fast. The K_d of the intermediate is estimated to be in the picomolar range.[29] Presumably, the tetrahedral intermediate binds tightly because it effectively incorporates all of the functionality present in S3P and PEP into a single compound. The estimated dissociation constant for 4 (5×10^{-11} M) agrees reasonably well with the multiplicative products of the experimentally determined binding constants for S3P (7×10^{-6} M) and PEP (15×10^{-6} M).[14,23] The tetrahedral intermediate, therefore, offers the first opportunity to successfully validate the concept of multi-substrate inhibition with an enzymatically generated species rather than a synthetic inhibitor.[32]

Summary. All of the available experimental evidence indicates that EPSPS utilizes a simple addition-elimination reaction through a single, kinetically competent tetrahedral intermediate as originally proposed by Sprinson.[19]

Mechanism-Based EPSPS Inhibitors

The tight interaction displayed by 4 in the presence of EPSPS strongly suggests that compounds which effectively mimic this structure should be potent EPSPS inhibitors. Indeed, two classes of shikimate-based tetrahedral intermediate mimics have recently been synthesized by Professor Paul Bartlett and co-workers.[30,31] Nanomolar inhibition constants are observed in each series. These compounds represent the first non-glyphosate structures to display significant EPSPS inhibition. The best derivative in each class exhibits a factor of $10\text{-}10^2$ improvement in K_i relative to glyphosate (K_i = 200 nM).

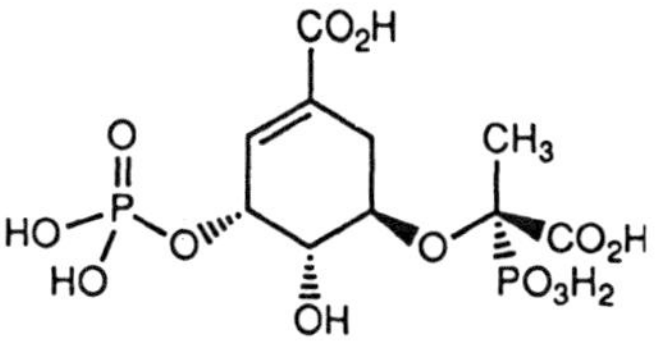

P. Bartlett (R)[30]

K_i= 3 nM

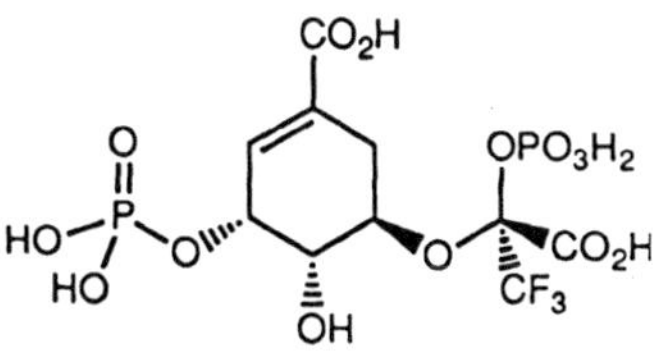

P. Bartlett (R)[31]

K_i= 30 nM

GLYPHOSATE AS A TRANSITION STATE INHIBITOR

The isolation of the tetrahedral intermediate coupled with Sprinson's original experiments demonstrating EPSPS-catalyzed deuterium and tritium exchange into PEP are two results which strongly support protonation of the PEP double bond during catalysis. Amrhein and Streinrucken first proposed that glyphosate binds as a transition state inhibitor which effectively occupies the transient PEP oxonium ion (PEP^+) site[10] (Figure 4). Since glyphosate does not inhibit any other PEP dependent enzyme, it can not function as a ground-state mimic of PEP.

Evidence supporting this model for glyphosate binding includes the following:
(a) Like PEP, glyphosate has no detectable interaction with the enzyme alone and only binds to the pre-formed S3P*EPSPS binary complex.[14] (b) EPSP partially protects several active site amino acids (Lys-22[33], Cys-408[34], Arg-27[35]) from inactivating reagents, while S3P and glyphosate afford complete protection. This suggests at least partial overlap between the 5'-enolpyruvoyl portion of EPSP and glyphosate. (c) EPSPS undergoes a fluorescence change upon binding to EPSP, but not when S3P binds alone. An even larger fluorescence change is observed with S3P and glyphosate.[14a] (d) A molecular modeling comparison of PEP versus glyphosate using SYBYL[36] demonstrates that the conformational flexibility of glyphosate permits a low energy conformation that overlaps quite well with PEP. When the phosphonic acid of glyphosate is fitted to the PEP phosphate and the two carboxyl groups are superimposed, glyphosate adopts a conformation with one methylene group above and one below the plane of the PEP double bond. Although somewhat larger in total volume, glyphosate can adopt a conformation with excellent

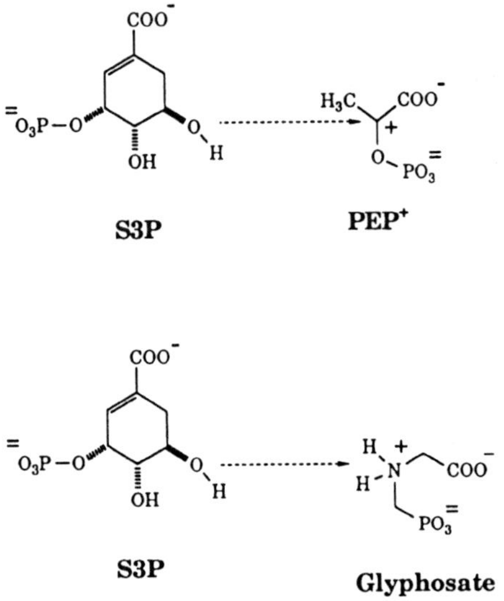

FIGURE 4. GLYPHOSATE AS A TRANSITION STATE INHIBITOR.

overlap versus PEP.[37] (e) Electrostatic potential calculations of glyphosate versus PEP+ show an excellent agreement in charge distribution for each of these species. These studies were performed in collaboration with Dr. James P. Snyder of the G. D. Searle Drug Design Group.[38] (f) A careful comparison of rate constants for PEP versus glyphosate shows that glyphosate binds some fifty-fold more tightly and dissociates from EPSPS much more slowly than PEP.[14,23] However, the utility of these types of measurements in characterizing transition state inhibitors has recently been called into question.[29]
(g) As an enzyme inhibitor, glyphosate and its enzyme-active derivatives form a very narrow chemical class. Minor structural changes in the glyphosate backbone dramatically reduce enzyme affinity.[39] Only two closely related analogs, N-hydroxy-N-phosphonomethylglycine[40] and N-amino-N-phosphonomethylglycine,[41] inhibit EPSPS to about the same degree as glyphosate (Figure 5).

The structural specificity for glyphosate analog binding is apparent from comparison of these N-heterosubstituted glyphosates with the corresponding N-alkyl derivatives Similarly, simple heteroatom substitutes for the NH in the glyphosate backbone produce analogs with no significant interaction with EPSPS.[39]

While most of the available data supports the transition state model for glyphosate binding, some discrepancies remain unanswered. Glyphosate does not bind many orders of magnitude more tightly than PEP. A growing body of evidence indicates that various mutant EPSPS's can be identifed with altered

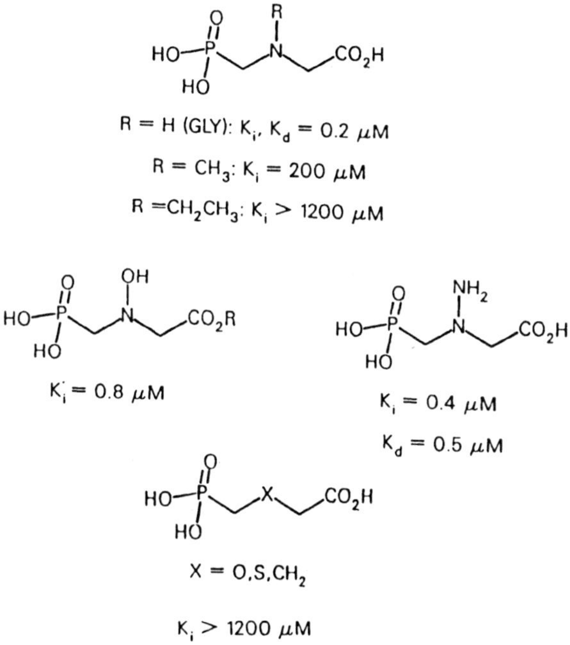

FIGURE 5. N-SUBSTITUTED GLYPHOSATES AS EPSPS INHIBITORS.

PEP and glyphosate affinites which are more selective in lowering binding at the glyphosate site. This ability to distinguish the PEP site from that of glyphosate argues strongly against the transition state model. A three-dimensional x-ray structure of the herbicidal ternary complex offers the best opportunity to define the molecular detail at the glyphosate binding site and unambiguously address this issue.

EPSPS X-RAY STRUCTURE

A concerted effort in collaboration with the Monsanto groups of Dr. Sherin Abdel-Meguid, Dr. Ganesh M. Kishore, Dr. Henry E. Dayringer and Dr. Wm. C. Stallings recently provided the first three-dimensional x-ray crystal structure for native *E. coli* EPSPS.[42] Crystals of native *E. coli* EPSPS, grown from aqueous ammonium sulfate, are suitable for a high resolution structure determination. An α-carbon backbone ribbon tracing of the structure at three Angstroms resolution is shown in Figure 6.

The structure is made up of two distinct hemi-spherical domains, each of approximately equal size with a radius of ~25 Å. Interestingly, both the N-amino terminus and the carboxyl terminus of the polypeptide chain are located in the top domain. The two hemispheres are connected by a double stranded hinge. Each domain contains six parallel α-helices with their corresponding macrodipoles oriented to create an electropositive attraction for anionic substrates. Presumably, the active site is near the hinge region and the two domains collapse to form a solvent-excluded cage during catalysis. The magnitude of the distance separating the two domains suggests that crystal packing forces may have caused crystallization in a more open form than normally exists in solution. Efforts are continuing to further refine this structure and extend this approach to the herbicidal ternary complex.

FIGURE 6. AN α-CARBON RIBBON TRACING OF *E. Coli* EPSPS AT 3Å.

CONCLUSIONS

A variety of kinetic, biophysical and spectroscopic techniques coupled with HPLC-based isolation procedures and structural methods have recently defined many new aspects of EPSP synthase biochemistry. The results of these studies have raised our understanding of EPSPS to a level of sophistication rarely attained for a target of herbicide action. The isolated tetrahedral intermediate unambiguously defines the key mechanistic questions for EPSPS and provides clear direction for new inhibitor design and synthesis. The commonality in binding sites between PEP and glyphosate remains as an important issue to resolve. The three-dimensional x-ray crystallographic studies currently underway offer the first opportunity to define the molecular level mode of action for glyphosate as an active site targeted herbicide. A clear distinction in three dimensional binding modes between PEP and glyphosate has significant potential for the rational design of new small molecule inhibitors and novel EPSPS mutants for glyphosate tolerant crops.[6]

REFERENCES

1. Schloss, J.V. (1989) *Target Sites for Herbicide Action*, (Boger, P. & Sandmann, G., eds.) pp 165-245, CRC Press Inc., Boca Raton, FL.

2. Franz, J. E. (1985) *The Herbicide Glyphosate* (Grossbard, E. & Atkinson, D., eds.) pp 1-17, Butterworth, Boston, MA.

3. (a) Amrhein, N.; Schab, J. and Steinrucken, H. (1980) *Naturwissenschaften*, *67*, 356. (b) Amrhein, N. (1985) *Recent Adv. Phytochem.* *20*, 83-117. (c) Duke, S. O. (1988) *Herbicides: Chemistry, Degradation and Mode of Action*, (Kearney, P. C. & Kaufman, D. D., eds.) pp 20-34, M. Dekker, New York.

4. Jaworski, E. (1972) *J. Agric. Food Chem.* *20*, 1195.

5. Rubin, J. l.; Gaines, C. G. and Jensen, R. A. (1982) *Plant Physiol.* *70*, 833-39.

6. For a review see Kishore, G. M. and Shah, D. M. (1988) *Annu. Rev. Biochem.* *57*, 627-663.

7. Smart, C. C.; Johanning, D.; Muller, G. and Amrhein, N. (1985) *J. Biol. Chem.* *260*, 16338-46. (b) Steinrucken, H. C.; Schulz, A.; Amrhein, N.; Porter, C. A. and Fraley, R. T. (1986) *Archiv. Biochem. Biophys.* *244*, 169-78.

8. Comai, L.; Facciotti, D.; Hiatt, W. R.; Thompson, G.; Rose, R. E. and Stalker D. (1985) *Nature* *317*, 741-44.

9. Haslam, E. (1974) *The Shikimate Pathway*, Wiley, New York.

10. Steinrucken, H. C. and Amrhein, N. (1984) *Eur. J. Biochem.* *143*, 351-57.

11. (a) Lewendon, A. and Coggins, J. R. (1983) *Biochem. J.* *213*, 187-91. (b) Lewendon, A. and Coggins, J. R. (1987) *Methods Enzymol.* *142*, 342-48.

12. (a) Ream, J. E.; Steinrucken, H. C.; Sikorski, J. A. and Porter, C. A. (1988) *Plant Physiol.* *87*, 232-38. (b) Mousdale, D. M. and Coggins, J. R. (1984) *Planta* *169*, 78-83. (c) Mousdale, D. M. and Coggins, J. R. (1987) *Methods Enzymol.* *142*, 348-54.

13. Boocock, M. R. and Coggins, J. R. (1983) *FEBS Lett.* *154*, 127-32.

14. (a) Anderson, K. S.; Sikorski, J. A. and Johnson, K. A. (1988) *Biochemistry* *27*, 1604-10. (b) Castellino, S.; Leo, G. C.; Sammons, R. D. and Sikorski, J. A. (1988) *Biochemistry*, *28*, 3856-68.

15. Rogers, S. G.; Brand, L. A.; Holder, F. B.; Sharps, E. S. and Brackin, M. J. (1983) *Appl. Environ. Microb.* *46*, 37-43.

16. Kresge, A. J. (1987) *Acc. Chem. Res.* *20*, 364-70.

17. Zemell, R. I. and Anwar, R. A. (1975) *J. Biol. Chem.* *250*, 4959-64.

18. Anton, D. L.; Hedstrom, L.; Fish, S. M. and Abeles, R. H. (1983) *Biochemistry* *22*, 5903-08.

19. Bondinell, W. E.; Vnek, J.; Knowles, P. F.; Sprecher, M. and Sprinson, D. B. (1971) *J. Biol. Chem.* *246*, 6191-96.

20. (a) Grimshaw, C. E.; Sogo, S. G.; Copley, S. D. and Knowles, J. R. (1984) *J. Am. Chem. Soc. 106*, 2699-2700. (b) Grimshaw, C. E.; Sogo, S. G. and Knowles, J. R. (1982) *J. Biol. Chem. 257*, 596-98.

21. (a) Lee, J. J.; Asano, Y.; Shieh, T.-L.; Spreafico, F.; Lee, K. and Floss, H. G. (1984) *J. Am. Chem. Soc. 106*, 3367-68. (b) Asano, Y.; Lee, J. J.; Shieh, T.-L.; Spreafico, F.; Kowal, C. and Floss, H. G. (1985) *J. Am. Chem. Soc. 107*, 4314-20.

22. Duncan, K.; Lewendon, A. and Coggins, J. R. (1984) *FEBS Lett. 165*, 121-27.

23. Anderson, K. S.; Sikorski, J. A. and Johnson, K. A. (1988) *Biochemistry 27*, 7395-7406.

24. (a) Anderson, K. S.; Sammons, R. D.; Leo, G. C.; Sikorski, J. A.; Benesi, A. J. and Johnson, K. A. (1990) *Biochemistry 29*, 1460-65. (b) Barlow, P. N.; Appleyard, R. J.; Wilson, B. J. O. and Evans, J. N. S. (1989) *Biochemistry 28*, 7985-91.

25. Pansegrau, P. D.; Anderson, K. S.; Jones, C.; Ream. J. E. and Sikorski, J. A. manuscript in preparation.

26. Leo, G. C.; Sikorski, J. A. and Sammons, R. D. (1990) *J. Am. Chem. Soc. 112*, 1653-54.

27. Anderson, K. S.; Sikorski, J. A.; Benesi, A. J. and Johnson, K. A. (1988) *J. Am. Chem. Soc. 110*, 6577-79.

28. Beau, J. M.; Schauer, R.; Haverkamp. J.; Kamerling, J. P.; Dorland, L. and Vliegenthart, J. F. G. (1984) *Eur. J. Biochem. 140*, 203-08.

29. Anderson, K. S. and Johnson, K. A. (1990) *J. Biol. Chem. 265*, 5567-72..

30. Alberg, D. G. and Bartlett, P. A. (1989) *J. Am. Chem. Soc. 111*, 2337-38.

31. Bartlett, P. A.; McLaren, K. L.; Alberg, D. G.; Fassler, A.; Nyfeler, R.; Lauhon, C. T. and Grissom, C. B. (1989) *Prospects for Amino Acid Biosynthesis Inhibitors in Crop Protection and Pharmaceutical Chemistry*, (Copping, L. G.; Dalziel, J. and Dodge, A. D., eds.) pp 155-70. British Crop Protection Council Mono. No. 42, The Lavenham Press Ltd., Lavenham, Suffolk, U.K.

32. Fersht, A. (1985) *Enzyme Structure and Mechanism* p 322, W. H. Freeman and Co., New York.

33. Huynh, Q. K.; Kishore, G. M. and Bild, G. S. (1988) *J. Biol. Chem. 263*, 735-39.

34. Padgette, S. R.; Huynh, Q.K.; Aykent, S.; Sammons, R. D.; Sikorski, J. A. and Kishore, G. M. (1988) *J. Biol. Chem. 263*, 1798-1802.

35. Padgette, S. R.; Smith. C. E.; Huynh, Q.K. and Kishore, G. M. (1988) *Arch. Biochem. Biophys. 266*, 254-62.

36. Modeling studies completed with version 3.2 of SYBYL Molecular Modeling Software, Tripos Associates, St. Louis, MO 63144.

37. Anderson, D. K.; Duewer, D. L. and Sikorski, J. A. unpublished results.

38. Sikorski, J. A. and Snyder, J. P. manuscript in preparation.

39. Ream, J. E.; Anderson, K. S.; Sammons, R. D. and Sikorski, J. A. manuscript in preparation.

40. Franz, J. E. (1978) *U.S. Patent 4,084,953* to Monsanto Company.

41. (a) Diehl, P. and Maier, L. (1988) *Phosphorous and Sulfur 39*, 159-64.
(b) Ream, J. E.; Anderson, K. S.; Andrew, S. A.; Phillion, D. P.; Knowles, W. S. and Sikorski, J. A. manuscript in preparation.

42. Stallings, W. C.; Abdel-Meguid, S. S.; Lim, L. W.; Shieh, H.-S.; Dayringer, H. E.; Leimgruber, N. K.; Stegeman, R. A. T.; Anderson, K. S.; Sikorski, J. A.; Padgette, S. R. and Kishore, G. M. (1990) *Science*, submitted.

MECHANISM OF ENYZMATIC PHOSPHOTRIESTER HYDROLYSIS

Steven R. Caldwell and Frank M. Raushel

Departments of Chemistry and Biochemistry & Biophysics
Texas A & M University
College Station, Texas 77840

Introduction. There are numerous reports from various sources of crude enzyme systems that appear to be capable of hydrolyzing organophosphotriesters (1-8). Unfortunately, none of these enzymes has ever thoroughly been characterized or purified to homogeneity. However, we have recently succeeded in the first isolation and characterization of a phosphotriesterase from *Pseudomonas diminuta* and this enzyme is the focus of this presentation.

The phosphotriesterase from *Pseudomonas diminuta* catalyzes the hydrolysis of typical phosphotriesters such as paraoxon to diethyl phosphate and p-nitrophenol as illustrated in Figure 1. The primary interest in enzyme catalyzed phosphotriesterase

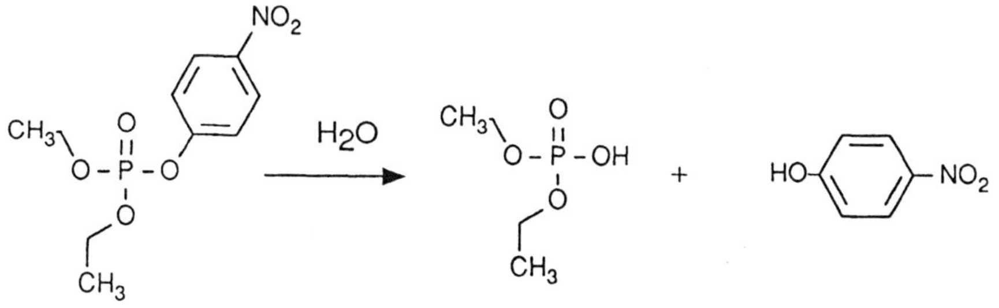

Figure 1. Hydrolysis of Paraoxon by the Phosphotriesterase.

hydrolysis stems from several experimental observations and results. For example, over 1000 organophosphate pesticides have been synthesized and approximately 3 x 10^7 kg of these toxic materials are consumed by the agricultural community in the U.S. annually (9). The consequence of this heavy usage is illustrated by the estimated 800 deaths per year and the thousands of reported pesticide poisonings attributed to exposure with these compounds (12). The lethal dose of paraoxon is 0.5 mg/kg. Recently, it has been shown

Chemical Aspects of Enzyme Biotechnology
Edited by T.O. Baldwin *et al.*, Plenum Press, New York, 1990

that a wide spectrum of these organophosphate pesticides are hydrolyzed by the enzyme from *Pseudomonas diminuta* (10). Many of those tested are among the more important and heavily utilized in the agricultural industry (11). The catalytic capability of the phosphotriesterase provides a promising method for detoxification of these pesticides.

Several chemical warfare weapons, namely sarin and soman, are also hydrolyzed by the phosphotriesterase at substantial rates. The lethal dose for these nerve toxins are about 0.01 mg/kg and the potential application of this enzyme is quite obvious and significantly important. The current estimated and reported stockpiles of sarin and soman are at least 10^4 tons in the United States. The Soviets are believed to have an equally large arsenal of these types of weapons (13).

In addition to the more practical aspects, it is interesting to note that no naturally occurring phosphotriester is known to exist. The first reported synthesis of a phosphotriester was in the 1937 by Schrader (14). The question thus arises concerning the function and the evolution of this enzyme. Is there a "natural" substrate for this enzyme or has the enzyme evolved in a very short period of time specifically for the hydrolysis of agricultural phosphotriesters?

Cloning, Overexpression, and Isolation. The capability of microorganisms to enzymatically hydrolyze phosphotriesters was observed in experiments performed by Munnecke and Hsieh with the soil microbe, *Pseudomonas diminuta* (15). Earlier work had also demonstrated that *Flavobacterium sp.* was able to degrade diazinon and parathion (16). Both *Pseudomonas diminuta* and *Flavobacterium sp.* (ATCC 27551) exhibited constitutive hydrolytic activity against parathion and the enzyme activity was later found to be associated with a plasmid borne gene in both organisms (17). The gene from each organism was isolated to a 1.3 kb *Pst*I restriction fragment and has been sequenced by two laboratories (18,19). The gene sequenced by the Serdar group at the Amgen Corporation indicates that the enzyme is synthesized as a proenzyme consisting of 365 amino acids. Apparently, 29 amino acids at the amino terminal end are subsequently removed in processing resulting in a protein with a molecular weight of about 36 kDa. It is presently believed that these 29 amino acids have the characteristics of a signal peptide. The predicted amino acid composition, based on the nucleotide sequence, is consistent with the observed amino acid content of the purified enzyme (10). Overexpression of the enzyme activity has been achieved using recombinant plasmids incorporated into *E. coli* vectors (17,19). The enzyme activity has also been significantly overexpressed using a baculovirus expression system developed in the laboratories of Summers (20). In this system, the gene expression is under the control of the polyhedron promoter. Cultured cells from *Spodoptera frugiperda* (sf9 cells) are infected with a recombinant baculovirus carrying the phosphotriesterase gene. Expression of the gene follows the infection of the cultured sf9 cells. Both systems provide an adequate source of enzyme and have been utilized to obtain significant amounts of homogeneous enzyme for biochemical studies.

Phosphotriesterase was first purified from *Spodoptera frugiperda* (sf9) cells which were infected with the recombinant baculovirus (10). The procedure employed four steps after the cells were sonicated and the typical results are shown in Table 1. The purification resulted in a 1500 fold purification of a homogeneous protein with a specific activity of 3200 units/mg for paraoxon hydrolysis. This scheme has allowed a very simple and straightforward purification of the phosphotriesterase.

Table 1. Purification of Phosphotriesterase.

STEP	UNITS	UNITS/MG	PURIFICATION
SUPERNATANT	5400	1.6	1
GREEN-A	7300	130	80
PHENYL SEPHAROSE	7900	1100	703
G-75	4100	3700	2400

Properties of the Purified Phosphotriesterase. The molecular weight of the purified protein was calculated at 39,000 gram/mole based on an estimate from a SDS-PAGE (10). This value was also confirmed using high resolution gel filtration chromatography. This value is very close to a predicted molecular weight of 37,826 based on the DNA sequence and suggests that the enzyme exists as a monomer in solution. Atomic adsorption analysis revealed the association of one zinc atom per molecule of phosphotriesterase (10). The kinetic parameters were measured for the homogeneous enzyme and a value of 2100 s^{-1} was obtained for V_{max}, while the value for V/K was $4 \times 10^7\ M^{-1}s^{-1}$ using paraoxon as a substrate. These rates are quite substantial and compare quite favorably with the V_{max} ($4.3 \times 10^3\ s^{-1}$) and V/K ($2.4 \times 10^8\ M^{-1}s^{-1}$) values for triose phosphate isomerase (21). The estimated rate enhancement produced by this enzyme relative to the rate of hydrolysis in water at pH 7.0 is 2.8×10^{11} (10).

Mechanism of Action. The enzymatic hydrolysis of paraoxon gives two products as shown in Figure 1. From a mechanistic viewpoint, the products can be obtained by a minimum of three different pathways. These limiting mechanisms are illustrated in Figure 2. First, the attack could occur by a nucleophilic substitution at the carbon atom of the phenyl ring with the formation of an tetrahedral intermediate followed by C-O bond cleavage to give the products. Secondly, the attack might occur at the phosphorus atom by the direct attack of an active site nucleophile to give a covalent enzyme intermediate and p-nitrophenol. This would follow with an attack by water at the phosphorus atom to give the second product diethyl phosphate. Lastly, in an S_N2 like process the reaction could proceed by the general base catalyzed attack of a water molecule directly at the phosphorus center with P-O bond cleavage to give two products in a single step.

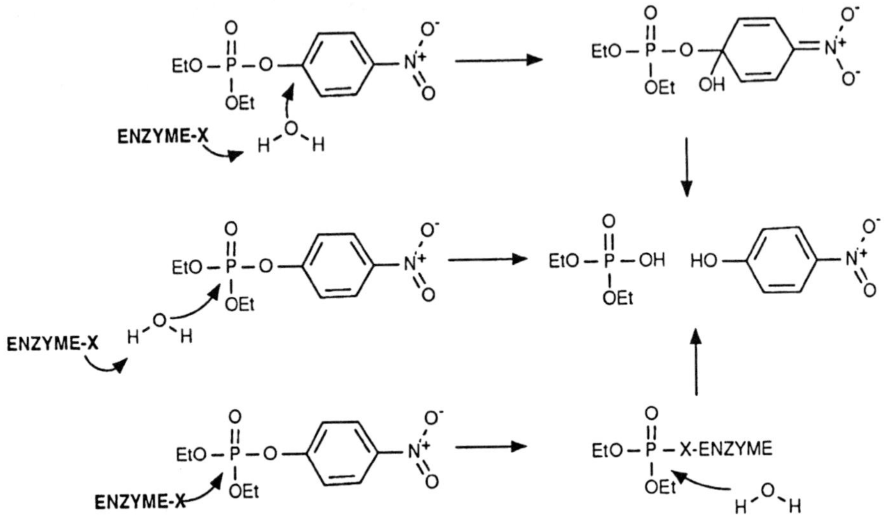

Figure 2. Possible Mechanisms for Phosphotriester Hydrolysis.

The mechanisms for P-O versus C-O bond cleavage were distinguished by conducting the enzymatic hydrolysis in oxygen-18 water and subsequent analysis by phosphorus-31 NMR spectroscopy (22). The analysis showed a 0.02 ppm upfield chemical shift in the phosphorus NMR spectrum of the diethylphosphate product which was expected with the incorporation of an oxygen-18 in the product, diethyl phosphate and not p-nitrophenol. These results are consistent only with the attack of the water at the phosphorus center and subsequent P-O bond cleavage, clearly eliminating the nucleophilic aromatic substitution mechanism. To distinguish between the covalent enzyme intermediate and the direct displacement mechanism required the determination of the net stereochemical outcome at the phosphorus center. The formation of a covalent phosphoenzyme intermediate suggests a two step mechanism, and thus, retention of configuration at the phosphorus. The single in-line displacement mechanism would consist of a stereochemical inversion at the phosphorus atom. Since paraoxon and diethyl phosphate are prochiral molecules the stereochemical assignments could not be accomplished with these compounds. Instead, the stereochemical analysis was performed by utilizing a chiral substrate, [O-ethyl O-(4-nitrophenyl)] phenyl phosphonothioate (EPN) (22). Each stereoisomer was independently synthesized from optically pure phenylphosphonothioic acid. It was found by phosphorus-31 NMR spectroscopy that only the (Sp) isomer of EPN was hydrolyzed by the enzyme to give only the (Sp) isomer of the product O-ethyl phenyl phosphonothioc acid as shown in Figure 3. Therefore, hydrolysis of EPN by the phosphotriesterase occurs with complete inversion of stereochemistry at the phosphorus center. These results eliminated the possible formation of an covalent phosphoenzyme intermediate and thus supported a single in-line displacement process at the phosphorus center with an activated water molecule.

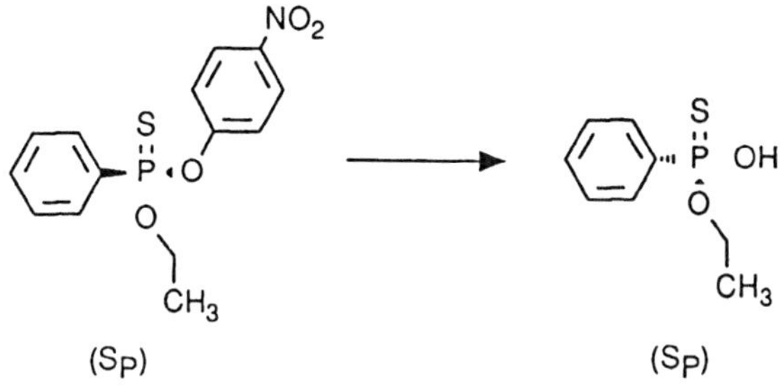

Figure 3. Hydrolysis of Chiral EPN.

Active Site Residues. It was observed from the pH rate profiles of the kinetic parameters, V_{max} and V/K, that a single unprotonated group was involved in catalysis (23). As shown in Figure 4, the V_{max} profile indicates that a species with a pK_a of 6.1 must be unprotonated when paraoxon is a substrate. Since paraoxon does not ionize in this pH range, these results are consistent with general base catalysis in the activation of the attacking water molecule. The identification of the active site base was attempted by measuring the pH dependence for paraoxon hydrolysis as a function of temperature and by the effect of organic solvent on the pK_a values (24). The results from the effect of temperature on the kinetic pK_a gave the thermodynamic values of 7.9 kcal/mole and -1.4 entropy units for enthalpy and entropy of ionization, respectively. These values are most consistent with the titration of a histidine residue (25). The effect of organic solvent suggested a cationic acid, also consistent with a histidine residue as the active site base.

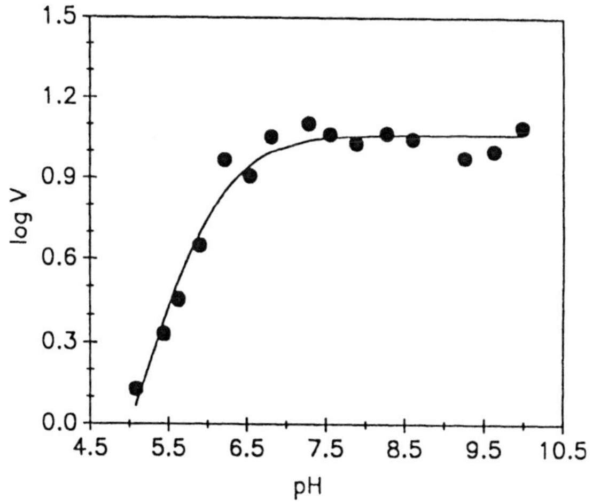

Figure 4. pH Rate Profile for Hydrolysis of Paraoxon.

Table 2. Survey of Chemical Modification Reagents.

Targeted Residue	Reagent	Inactivation Rate (h^{-1})
Cysteine	DTNB	<0.02
Cysteine	MMTS	<0.02
Aspartate, Glutamate	EDC	<0.07
Lysine	Pyridoxal	<0.06
Arginine	Butandione	<0.01
Histidine	Methylene Blue	0.72
Histidine	DEPC	1.92
Cysteine	Iodoacetamide	<0.01

The requirement for an essential histidine residue was confirmed using group specific chemical modification reagents for potential active site nucleophiles (24). The experimental survey is shown in Table 2 and reveals that only those reagents generally specific for histidine residues inactivate the enzyme. Diethyl pyrocarbonate (DEPC), a relatively specific reagent for the reaction with histidine, substantially inactivated the phosphotriesterase activity when only 2.9 of the 7 total histidine residues were modified. Diethyl-p-fluorophenyl phosphate, a competitive inhibitor of paraoxon, completely protected the enzyme from inactivation by DEPC and only two of the three histidines were modified, indicating that only one histidine was associated with the original inactivation event. The negligible effect by the other chemical modification reagents further supports a histidine as the sole species involved in catalysis. The inactivation of the phosphotriesterase by DEPC as a function of pH resulted in a pK_{inact} of 6.1, identical with the value obtained from the kinetic pH rate profile, providing additional evidence. Collectively, these data support a histidine as the active site base involved in catalysis but this evidence does not preclude the involvement of other essential residues such as metal ions.

The specific role of the zinc ion in the active site of phosphotriesterase has yet to be elucidated. A reasonable role for the zinc may involve polarization of the phosphoryl oxygen of the phosphotriester. This scenario would be analogous to carboxypeptidase, a zinc protease (26). Here, the zinc serves in polarizing the carbonyl oxygen, and thus facilitating the attack of a water molecule at the carbonyl carbon. The zinc further aids in stabilizing the developing negative charge on the carbonyl oxygen in the transition state. Another potential role for the zinc atom in phosphotriesterase may involve the stabilization of the phenolic leaving group, thereby assisting in catalysis (27).

Atomic absorption analysis of the holoenzyme detected a single zinc atom per molecule of phosphotriesterase (10). The removal of the zinc atom to form an apoenzyme was accomplished using o-phenanthroline, a metal chelator, and resulted in less than 0.1 zinc atom per enzyme molecule and concommitant loss in enzyme activity. Recent studies have shown that divalent metal ions, especially cobalt(II), when introduced into the growth media yield substantial increases in enzyme activity (19, 28).

Structure of the Transition State. In a determination of the substrate specificity for the phosphotriesterase, several paraoxon analogs were synthesized with varying substituents on the phenyl ring and then kinetically analyzed (23). This analysis revealed a linear structure activity relationship between the pK_a of the phenolic leaving group and the rate of chemical and enzymatic hydrolysis. This relationship can be defined by the Brφnsted equation, as illustrated in Equation 1

$$\log k_3 = (\beta \times pK_a) + \text{constant} \qquad (1)$$

and provides information concerning the transition state structure for the hydrolysis reaction. For the series of phosphotriesters examined, this relationship was expected to reveal the amount of P-O bond breakage in the transition state. Chemical hydrolysis of the phosphotriesters by KOH gave a linear plot over the entire pK_a range examined with a β value of -0.42, identical with a previous literature report (30). This result is consistent with a P-O bond order of about 1/3 in the transition state. A larger negative β value of -1.2 was obtained when the data for the enzymatic hydrolysis were analyzed (30). These data suggest a later transition state with a substantial amount of the negative charge associated with the leaving group. Enzymatic hydrolysis involves the addition of one substrate and the release of two products. If the simple kinetic model for enzymatic hydrolysis as illustrated in Equation 2 is assumed, then the following rate equations can be derived for the

$$E \underset{k_2}{\overset{k_1 A}{\rightleftharpoons}} EA \xrightarrow{k_3} EPQ \xrightarrow{k_5} E + \text{Products} \qquad (2)$$

kinetic parameters, V_{max} and V_{max}/K_m. With this model as a point of reference,

$$V_{max} = k_3 k_5 / (k_3 + k_5) \qquad (3)$$

$$V/K = k_1 k_3 / (k_2 + k_3) \qquad (4)$$

the Brφnsted plots obtained for the enzymatic hydrolysis data of several paraoxon analogs can be used to obtain values for k_1 through k_5 (30).

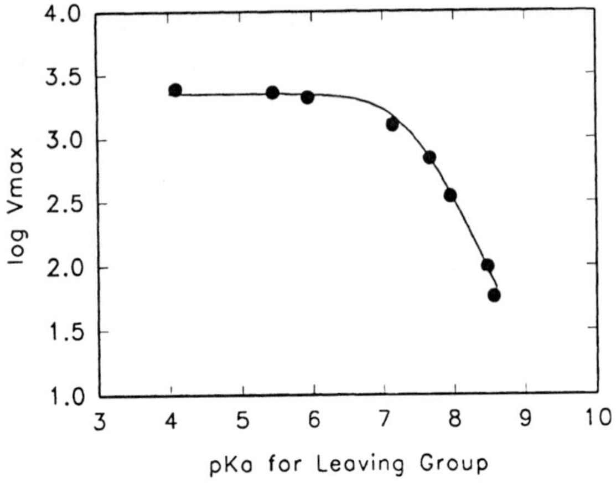

Figure 5. Brϕnsted Plot for log V_{max} Versus the pK_a of Leaving Group.

The Brϕnsted plot for log V_{max} versus the pK_a of the leaving group (Figure 5) is nonlinear for the pK_a range examined and thus there is a change in rate limiting step with these substrates. Thus, for those substrates with leaving group pK_a values ~7, the chemical step (k_3) is apparently rate limiting while the rate of hydrolysis is dictated by the magnitude of k_5 for those substrates with leaving group pK_a values below ~7. A fit of this plot to equation 1 and 3 provides the values for k_5 and β of 2200 s^{-1} and -1.2, respectively. A similar Brϕnsted plot is observed for the V/K data as shown in Figure 6.

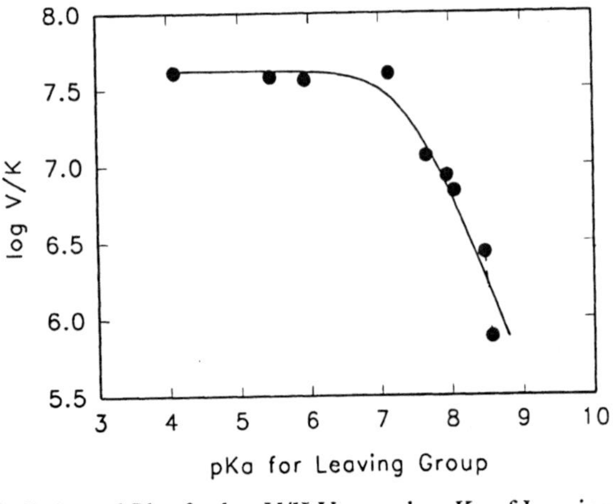

Figure 6. Brϕnsted Plot for log V/K Versus the pK_a of Leaving Group.

A fit of the data in this plot to equations 1 and 4 provides the value of k_1 and k_2 4.1 X $10^7\ M^{-1}s^{-1}$ and 2100 s^{-1}, respectively. Hydrolysis rates for those substrates in the plateau portion of the Brønsted plot are not affected by the pK_a of the leaving group, suggesting that another step in the enzymatic reaction is now more limiting than the chemical step. The nonlinearity of this plot and the relatively large value for k_1 suggests that diffusion of the substrate to active site may be limiting for V/K, since theoretically, second order rates of 10^7 to $10^9\ M^{-1}s^{-1}$ have been suggested for diffusion controlled enzymatic reactions. The measurement of diffusion controlled processes can be determined by varying the viscosity of the medium. In a diffusion limited system the rate of the reaction decreases proportionally with increasing viscosity of the medium. The effect of sucrose on the relative V/K is shown in Figure 7. A limiting slope value of 1.0 indicates a totally diffusion limited system. As predicted, those substrates whose leaving group effects did not affect hydrolysis rates are limited by the encounter of the substrate with the active site. The solid line in this plot is drawn using equation 5 and the calculated values obtained from the fits of the appropriate equations to Figures 5 and 6. The prediction is quite good and confirms the conclusion that for this group of substrates, V/K is limited by the chemical step for those substrates with leaving group pK_a's above ~7 and limited by diffusion for those substrates with leaving group pK_a values below ~7. These data show conclusively that paraoxon is poised such that its rate of enzymatic hydrolysis is limited predominately by diffusion.

<u>Substrate Specificity</u>. Although paraoxon remains one of the best substrates, other changes to the paraoxon molecule have been made in an effort to further define the substrate specificity (23). The nonbridging oxygen has been substituted with a sulfur atom with only minimal effects on V_{max}. When changes on the ester portion on the molecule were made, significant effects were observed. Diesters and monoesters are not hydrolyzed by the enzyme, and as the chain length of the ester increases on the phosphotriester both K_m and V_{max} decrease suggesting a hydrophobic binding pocket at the active site.

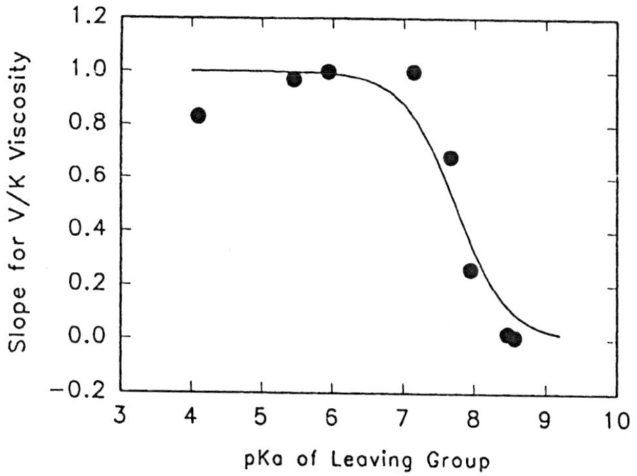

Figure 7. Effect of Solvent Viscosity on V/K.

Besides the hydrolysis of the P-O bond, phosphotriesterase can hydrolyze P-F bonds (31). The substrates of interest in this category are some of the chemical warfare agents, sarin, tabun, and soman. The P-F bond of sarin is hydrolyzed 1.5 X 10^6 times faster than the uncatalyzed reaction. Regardless, phosphotriesterase does exhibit the capability to hydrolyze these agents at a rate of 10^6 to 10^7 times greater than the uncatalyzed reaction at pH 7.0 and will perhaps be useful in degrading the enormous stockpiles of these compounds.

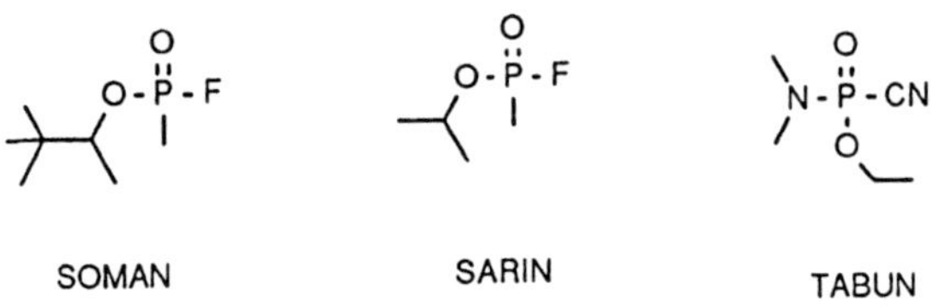

Phosphotriesterase has been shown capable of hydrolyzing commonly used pesticides (10). Since the gene for the enzyme could be incorporated and expressed in sf9 cells from the fall army worm, the possibility of infecting the larval stage of the fall armyworm was examined (24). Production of the enzyme in the baculovirus infected larva was successful up to 90 hours after infection. The time course for production of enzyme activity in the caterpillar shows the greatest expression about 90 hours after infection. In addition to the improved expression, the phosphotriesterase expressing catepillars exhibited a marked resistance to the insecticide, paraoxon. Figure 8 is a graphic representation of the enhanced resistance to paraoxon where the lethal dose increased 280-fold over the dosage for caterpillars not containing the phosphotriesterase gene. This model for the incorporation of the phosphotriesterase gene into an insect may have applications in the industrial sector where insecticidal resistance to commercially important insects such as the silk worm and the honeybee could be advantageous.

<u>Summary</u>. Phosphotriesterase activity can be expressed from a plasmid borne gene in a common soil microorganism, *Pseudomonas diminuta*. The gene for this enzyme has been cloned, sequenced, and overexpressed in bacterial and insect expression systems. Although no naturally occurring substrate for the enzyme has been identified, the phosphotriesterase exhibits a broad substrate specificity, hydrolyzing many organophosphate pesticides and chemical weapons. At the present, paraoxon appears to be the best substrate for the enzyme. From pH dependent studies and chemical modification experiments, there is an active site histidine important for catalysis and a zinc ion, is also required for activity. The size and the pK_a of the phenolic leaving group play an important role in the rate of substrate turnover. The natural role (if any) of this enzyme has not been determined and the hydrolysis of phosphotriesters may only be secondary to some other, but unknown, biological role. However, it is intriguing to speculate that evolutionary pressures due to pesticide exposure in the last fifty years on microorganisms may have

transformed a preexisting protease, phosphatase, or esterase into a highly efficient catalyst for the hydrolysis of phosphotriesters.

Acknowledgements: This work was supported in part by the Army Research Office, Robert A. Welch Foundation (A-840) and the Texas Advanced Technology Program. We would like to thank David Dumas, Bill Donarski, Vincent Lewis, George Omburo, Jennifer Newcomb, and James Wild for their efforts with this enzyme at Texas A&M University.

References

1. Mazur, A. (1946) *J. Biol. Chem.*, **164**, 271-289.
2. Main, A.R. (1960) *Biochem. J.*, **74**, 10-20.
3. Laveglia, J. and Dahm, P.A. (1977) *Ann. Rev. Entomol.*, **22**, 483-513.
4. Brown, K.A. (1980) *Soil Biol. Biochem.*, **12**, 105-112.
5. Baarsheris, W.H. and Heitland, H.S. (1986) *J. Agric. Food*, **34**, 707-709.
6. Nelson, L.M. (1982) *Soil Biol. Biochem.*, **14**, 219-222.
7. Hoskin, F.C.G. and Prusch, R.D. (1983) *Comp. Biochem Physiol.*, **75C**, 17-20.
8. Reiner, E., Aldridge, W.N., and Hoskin, F.C.G.(ed.), (1989) in *Enzymes Hydrolyzing Organophosphate Compounds*, John Wiley and Sons, New York.
9. Food & Agricultural Organization of the United Nations, Rome (1985) *FAO Production Yearbook* **37**, 292.
10. Dumas, D.P., Caldwell, S.C., Wild, J.R., and Raushel, F.M. (1989) *J. Biol. Chem.*, **264**, 19659-19665.
11. Brandt, E. United States Environmental Protection Agency, personal communication.
12. Feldman, J. (1983) *Subcommittee of Department Operation, Research and Foreign Agriculture,* U.S. House of Representatives, October 6, 1983, Government Printing Office, Washington, D.C.
13. Robinson, J.P.P. (1980)" *Stockholm International Peace Research Institute Chemical and Biological Warfare Studies 6"* in Chemical Weapons: Destruction and Conversion. Oxford University Press, Oxford and London. pp.9-56.
14. Schrader, G. (1952) *Angewandte Chemie und Chemie Ingenieur-Technik,* Verlag Chemie, Weinheim, No. 62.
15. Munnecke, D.M. and Hsieh, D.P.H. (1976) *Applied Environ. Microbiol.*, **31**, 63-69.
16. Sethunathan, N. and Yoshida, T. (1973) *Can. J. Microbiol.*, **19**, 873-875.
17. McDaniel, C.S., Harper, L.L., and Wild, J.R. (1988) *J. Bacteriol.*, **170**, 2306.
18. Harper, L.L., McDaniel, C.S., Miller, C.E., and Wild, J.R. (1988) *Applied and Environ. Microbiol.*, **54**, 2586-2589.

19. Serdar, C.M., Murdock, D.C., and Rohde, M.F. (1989) *Biotechnology*, **7**, 1151-1155.

20. Harper, L.L., Raushel, F.M., Luckow, V., Summers, M., and Wild, J.R. (1990) submitted to *Virology*.

21. Putmand, S.J., Coulson, F.W., Farley, I.R.T., Riddleston, B., and Knowles, J.R. (1972) *Biochem. J.*, **129**, 301.

22. Lewis, V.E., Donarski, W.J., Wild, J.R., and Raushel, F.M. (1988) *Biochemistry*, **27**, 1591-1597.

23. Donarski, W.J., Dumas, D.P., Heitmeyer, D.P., Lewis, V.E., and Raushel, F.M. (1989) *Biochemistry*, **28**, 4650-4655.

24. Dumas, D.P. (1989) PhD. Dissertation, Texas A & M University, College Station, Texas, *"The Purification and Characterization of the Phosphotriesterase from Pseudomonas diminuta"*.

25. Cleland, W. W. (1977) *Adv. Enzymol.* ***63***, 273.

26. Vallee, B.L., Riordan, J.F., and Coleman, J.E. (1963) *Proc. Natn. Acad. Sci. USA*, **49**, 109.

27. Herchlag, D. and Jencks, W.P. (1987) *J. Am. Chem. Soc.*, **109**, 4665-4674.

28. Omburo, G., Dept.of Chemistry, Texas A & M University, College Staion, Texas, unpublished results.

29. Khan, S.A. and Kirby, A.J. (1970) *J. Chem. Soc. (B)*, 1172-1182.

30. Caldwell, S.R., Newcomb, J., Schlect, K., and Raushel, F.M., Dept. of Chemistry, Texas A & M University, College Station, Texas, unpublished results.

31. Dumas, D.P., Durst, H.D., Landis, W.G., Raushel, F.M., and Wild, J.R. (1990) *Arch. Biochem. Biophys.*, **277**, 155-159.

Proline Isomerization and Protein Folding

Christy MacKinnon, Sudha Veeraraghavan, Isabelle Kreider, Michael J. Allen, John R. Liggins, and Barry T. Nall

Department of Biochemistry
University of Texas Health Science Center
San Antonio, Texas

Introduction

Properties of folding intermediates: expectations of the chemist

A reaction as complex as folding of a protein must involve a variety specific chemical processes. Hydrogen bonds and salt bridges are formed, and perhaps broken and interchanged. Solvent-induced hydrophobic associations may occur, or even a general hydrophobic collapse of the polypeptide to a less than fully ordered "globule". Polypeptide chains might become entangled with each other or with other chains and have to extricate themselves prior to further folding. Intricate shapes and structures may be constructed only to partially or fully unravel on transformation into other more stable forms. The objective of much experimental work on the process of folding has been to obtain direct evidence for some of these expectations. This has been a major challenge since many aspects of folding, rather than having the expected complexity, are, instead, well described by the simplest of chemical mechanisms: a two state process.

Experimental features of folding reactions

By a variety of physical and thermodynamic criteria the equilibrium properties of protein folding are in accord with the two state mechanism. This means that when folded and unfolded protein is in equilibrium, there are no partly folded species. That is, all of the protein molecules are either fully folded or fully unfolded. To the contrary, experiments which measure how fast an unfolded protein can refold (or a folded protein unfold) show considerable kinetic complexity proving the existence of a multitude of species, at least some of which must represent partially folded or misfolded species.

Reconciliation between expectation and experiment?

Two-state thermodynamic and multistate kinetic behavior are not necessarily in conflict. In many cases kinetic measurements apply to conditions quite different from those of the thermodynamic measurements. This is because measurements of equilibrium constants can only be made under conditions where detectable amounts of both folded and unfolded species are present. Kinetic measurements, however, can be made almost anywhere, and are often made far removed from the unfolding transition zone. Another aspect of the kinetic process of folding which is hardly apparent at equilibrium, is the role of proline imide bond isomerization in creating heterogeneous unfolded species in which otherwise unstructured polypeptides differ in the cis or trans states of proline imide bonds (Brandts et al., 1975).

Proline isomerization model

Amino acids are linked to one another by amide bonds in which, for unstructured polypeptides, the amide proton and carbonyl oxygen are almost always trans. Prolines, however, have an imide linkage and the carbonyl oxygen and the C_α proton can be either cis or trans to one another. The folded protein, of course, requires a unique proline conformer for each site in the tertiary structure in which a proline occurs. Both cis and trans types have been identified by X-ray crystallography although the trans form is much more common. On average the trans form of the proline imide bond is probably two to four times more common than the cis form for both folded proteins and unstructured polypeptides. The conversion between cis and trans forms of imide linkages is slow (time constants of 10-1000 s) and has a large activation enthalpy (20-25 kcal/mole). As first pointed out by Brandts (Brandts et al., 1975) these and other properties of proline imide bond isomerization are similar to those of the slow kinetic phases in protein folding reactions. A model emerged that explained much of the kinetic complexity of folding reactions: the presence of the slow kinetic phases in protein refolding:

1) Native proteins require specific isomeric states for each proline, cis or trans.
2) In the unfolded protein the conformational constraints on a proline are relieved and, from molecule to molecule, a particular proline occurs in both cis and trans forms.
3) For a protein molecule to refold quickly all prolines must be in the format required by the native tertiary structure.

4) Polypeptide chains with one or more prolines in a non-native format refold slowly as proline imide bonds isomerize prior to or along with refolding.

Prolines as kinetic traps for structural intermediates

For refolding conditions where native protein structure is very stable the slow folding process is not as simple as predicted by the simplest form of the proline isomerization model. This is because partially folded structures form which can strongly influence rates of chemically slow events like isomerization (Kim & Baldwin, 1980; Schmid & Baldwin, 1979). In some cases "native-like" structures form which are fully active enzymes (Cook et al., 1979). These native-like structures with incorrect proline isomers might be thought of as transient forms of "mutant" proteins.

Folding of Yeast Cytochrome *c*: a Model System

Advantages-unique features

Our own studies of folding have relied almost entirely on yeast cytochrome *c* as a model system. The nature of the relationship between (amino acid) sequence information and three-dimensional structure is a primary objective of studies of folding. How is a protein's tertiary structure encoded in its' amino acid sequence? Investigation of a member of the largest protein family of sequence variants seemed like the best place to look for an answer a related question: Why do different sequences fold to essentially the same tertiary structure? Moreover, when we started this work more than ten years ago, the two yeast isozymes, iso-1 and iso-2 cytochromes *c* , were the only small globular proteins for which a large number of mutant proteins had been isolated and characterized (Hampsey et al., 1986; Pielak et al., 1985; Sherman & Stewart, 1978).

Cytochrome *c* contains a covalently bound heme as a prosthetic group which, no doubt, plays a major role in folding and stability. The presence of a heme is a mixed blessing. It provides a very sensitive optical probe of folding and is located in the middle of the protein at the active site. In addition the heme allows estimation of the *in vivo* specific activity for normal and mutant forms of cytochrome *c* (Ernst et al., 1985; Schweingruber et al., 1979). The ability to estimate the amounts of cytochrome *c* in living cells makes this one of the better systems for attempts at relating experiments with purified mutant proteins to cellular function. One must, however, be on the look-out for features of folding of cytochrome *c* that are peculiar to heme containing proteins. Cytochrome *c* with a covalently attached co-factor is arguably one of the best systems for learning how a prosthetic group influences the folding process. It is perhaps surprising that, so far, the basic features of protein folding reactions seem to depend very little on the presence of prosthetic groups, disulfide bonds, or other special features of individual proteins.

Another important aspect of the yeast cytochrome *c* system is that the experimental groundwork needed for detailed physical characterization has been laid: 1) X-ray structures of twenty or so normal and mutant variants of oxidized or reduced yeast cytochromes *c* have been determined, some to an incredable 1.2 Å resolution (Leung et al., 1989; Louie & Brayer, 1989; Louie et al., 1988; Murphy & Brayer, 1990; Brayer,G.D., personal communication), 2) the 2D proton NMR spectrum of (horse) cytochrome *c* has been assigned (Feng et al., 1990; Feng et al., 1989; Wand et al., 1989) and extensive assignments have been made for yeast iso-1 (Pielak et al., 1988) and iso-2 cytochromes *c* (Nall, Feng, Englander, & Roder, unpublished). The X-ray structures provide the infomation needed to define the final folded states of the mutant proteins while the NMR assignments allow hydrogen-deuterium exchange to be used to monitor mutation-induced changes in folding and stability (Englander & Kallenbach, 1984; Roder et al., 1988).

Properties of folding

For accurate measurements of global protein stability it is necessary that unfolding transitions measured at equilibrium be reversible. A number of proteins, including iso-1 and iso-2 cytochromes *c* , exhibit highly reversible denaturant-induced unfolding transitions. Direct (mechanism independent) measures of protein stability can only be made by differential scanning microcalorimetry (DSC) which requires a reversible thermal-induced transition. Moreover, DSC is the most accurate measure of small differences in stability between mutant proteins. Fortunately, both iso-1[1] and iso-2 have highly reversible thermal transitions (Sturtevant & Nall, unpublished) with better than 96% of the unfolding enthalpy retained in a second scan of the same sample (Liggins, J. R., unpublished). There is, however, a slight concentration dependence of the transition midpoint: T_m decreases by a few degrees over an approximately ten-fold concentration range. This suggests preferential association of the unfolded protein (Sturtevant, J. S., personal communication), a complicating feature of the DSC transitions of cytochrome *c* and many other proteins.

---

[1]Naturally occuring iso-1-cytochrome *c* has a Cys at position 102. For reversible folding the reactive sulfhydral must be blocked by chemical reagents (Zuniga & Nall, 1983) or the Cys replaced by directed mutagenesis (Cutler, et al., 1987). The naturally occuring form of iso-2-cytochrome *c* has Ala at position 102.

The refolding kinetics of the yeast cytochromes *c* are very similar to those of the distantly related horse cytochrome *c* suggesting that the qualitative features of protein folding reactions are conserved within a protein family (Nall & Landers, 1981). Fast and slow phases are observed and both absorbance-detected phases give native enzyme as a product (Nall, 1983). The fast phase is relatively large with 70-80% of the amplitude for absorbance-detected folding and 80-90% for fluorescence-detected folding. Assuming that the proline model explains the existence of the slow folding phases, the large fast phase amplitude makes cytochrome *c* an excellent system for studies of the more poorly understood process of fast folding. Slightly different kinetics are found when different physical probes are used to monitor refolding rates. In particular, fluorescence-detected slow folding is about tenfold faster (τ=10-20 s) than absorbance-detected slow folding (τ=100-200 s) (Nall, 1983). Differences in the fluorescence vs. absorbance-detected rates of fast folding are observed too, but only at high and low pH (Nall et al., 1988). The small slow phase amplitudes taken together with the fact that all prolines are trans in the native protein suggests that the kinetic mechanism of slow folding should be relatively simple.

Kinetic properties of slow folding phases

For yeast cytochromes *c* the absorbance-detected slow phase is about ten-fold slower than fluorescence-detected slow folding. Refolding monitored by 695 nm absorbance, believed to be an indicator of native protein formation, shows only the slower ("absorbance-detected") of the two slow phases. Presumably, the fluorescence-detected slow phase monitors conversion of species with native-like absorbance but intermediate fluorescence to fully native species. Similarly, the absorbance-detected phase monitors conversion of species with native-like fluorescence but intermediate absorbance to fully native species. The properties of both slow phases are consistent with involvement of proline isomerization. The activation enthalpies are in the expected range for imide isomerization: $\Delta H^{\ddagger}$=21 kcal/mol for fluorescence and 27 kcal/mol for absorbance. In addition double jump experiments show that both absorbance and fluorescence-detected species are generated slowly in the unfolded protein with slightly different rates: τ_{DJ}(fluor)= 24 s, τ_{DJ}(Abs)=114 s (Osterhout & Nall, 1985).

Replacement of Conserved Proline Residues

Assignment of slow phases to specific prolines

Since many of the properties of the slow folding phases are those expected for reactions involving proline isomerization, considerable effort has been devoted to determining which prolines generate which slow phases. We have investigated slow folding of point mutants in which a proline is replaced by other amino acids. Iso-1 and iso-2 have prolines at identical locations with the exception of position (-1)[2] where a proline occurs in iso-2 but not in iso-1.

Since the slow folding behavior of iso-1 and iso-2 is essentially the same Pro(-1) does not appear to play a significant role in slow folding. Of the remaining prolines, those at positions 30, 71, and 76, are conserved among all members of the mitochondrial cytochrome *c* family, while Pro25 occurs at a location where non proline residues are also allowed.

Mutant proteins with replacements of Pro71 retain both the absorbance-detected and fluorescence-detected slow folding phases (Ramdas & Nall, 1986; White et al., 1987). Thus Pro71, which occurs between two short helical regions in the native protein, does not appear to block folding as detected by either fluorescence or absorbance probes. Another possibility is that local sequence constraints keep Pro71 in the native trans format, even in the unfolded protein. A mutant protein in which Pro25 is replaced by Gly also retains both slow phases, so Pro 25, found in a turn region, does not block refolding (MacKinnon, C., unpublished). Replacement of Pro76 leads to loss of the absorbance-detected phase while fluorescence-detected slow folding remains largely unchanged (Wood et al., 1988). So the absorbance-detected phase is assigned to (generated by) Pro76. Two different replacements of Pro30 have been attempted (Thr30 iso-2, Gly30 iso-2) but neither leads to a protein stable enough to study. If the fluorescence-detected slow phase is generated by a proline, Pro30 must be the critical residue since it is the only proline left. The critical question is whether or not the fluorescence-detected phase results from proline isomerization or from some other slow process, perhaps heme ligand exchange. Recent experiments (see below) show that fluorescence-detected slow folding involves proline isomerization, so we have tentatively assigned the fluorescence-detected slow phase to Pro30. Nevertheless, we still hope to show directly, in experiments underway, that the fluorescence-detected phase can be removed by site-directed mutagenesis of Pro30.

---

[2] The vertebrate cytochrome *c* numbering is used to denote amino acid positions in order to facilitate comparison between members of the cytochrome *c* family. Iso-1 has five additional amino-terminal residues, and iso-2 9 additional amino-terminal residues compared to vertebrate cytochromes *c*. Both iso-1 and iso-2 have one residue less on the carboxy terminus. Thus the vertebrate numbering of iso-1 and iso-2 starts at positions -5 and -9 respectively, and extends to

Simplified folding mechanism

The main result of replacing proline residues with other amino acids is that the kinetic mechanism of folding is greatly simplified: slow folding phases are eliminated. This allows focus on other, possibly more important aspects of the kinetics of folding: the fast folding phases. It seems reasonable that many of the physical interactions that modulate the slow folding kinetics may play a dominate role in fast folding. It is important to learn more about the fundamental physical processes limiting rates of fast folding processes. This is presently a poorly understood phenomena, but one which can be approached though folding studies of site-directed mutants. On the other hand real folding of real proteins almost always involves slow phases and thus, presumably, proline isomerization. For large proteins with many prolines, or small proteins requiring the more rare cis isomers, essentially all folding involves isomerization. Thus, it is also important to investigate further the interplay between the structure of intermediates kinetically trapped by non-native proline isomers and the resulting slow kinetic phases.

Shifts in conformational equilibria between distinct folded forms: proline replacement and alkaline conformational change

Proline replacement mutants have led to surprises with regard to folding to mutant or alternative folded conformations. Replacements of conserved prolines, at least those at positions 71 and 76, depress the pK of a pH-induced conformational change which occurs near pH 9 in normal cytochromes *c* but near pH 7 in many mutant proteins (Nall et al., 1989; Pearce et al., 1989; Ramdas & Nall, 1987). Thus, slightly above neutral pH the mutant proteins fold to mutant (or alkaline) conformations rather than native or native-like structures. Little is known about the alkaline conformational state of cytochrome *c* except that, unlike the native protein, it no longer contains Met as a heme ligand (Schechter & Saludjian, 1967). Surprisingly, the alkaline conformation of iso-2 is more stable towards denaturant-induced unfolding than the native protein (Osterhout et al., 1985) and, for horse cytochrome *c*, hydrogen-deuterium exchange studies suggest that the alkaline form has much but not all of the secondary structure of native cytochrome *c* (Dohne et al., 1989).

---

position 103 (Hampsey et al., 1986; Dickerson, 1972). For example, Pro-76 in the vertebrate numbering system corresponds to Pro-85 in the iso-2 numbering system.

Native-like intermediates are transient intermediates in folding to mutant conformations

Folding to the mutant (or alkaline) conformations occurs in a surprising manner: native or native-like species appear as transient intermediates in folding to the alkaline form (Nall, 1986). In the first (fast phase) step of folding a protein with the absorbance and reducibility of native cytochrome *c* is formed, but a slow conformational change follows leading to the mutant (alkaline) conformation. The second, slow conformational change can be induced in the absence of folding by pH jumps. So far this surprising feature of folding to a mutant conformation via a native-like state has been observed for both iso-1 and iso-2 (at high pH), for three different replacements of Pro71, and for a Pro76 to Gly76 mutant (Nall, 1986; Ramdas, 1987; Wood et al., 1988). This peculiar feature of folding to a non-native conformation shows that the folding process, at least some of the time, prefers to pass through specific states with structural features (heme ligation in the present case) differing from those of the final product of folding. This suggests the existence of unique species which direct folding, perhaps along a sequential pathway (Kim & Baldwin, 1982).

Proline isomerization *in vivo*

It is interesting to ask whether proline imide isomerization is important for folding in the cell. So far there is little direct experimental evidence one way or the other, although circumstantial evidence is beginning to build up. Creighton has suggested that isomerization-limited folding of large proteins with many prolines may be very slow (Creighton, 1978): half times of the order of ten minutes are estimated for a protein containing 20 prolines. Thus in some cases folding rates are slower than rates of protein synthesis and are comparable to bacterial doubling times. Moreover, unfolded proteins are expected to have cellular half-lives that are (on average) shorter than the slowest folding rates estimated by Creighton. So how do proteins fold in a cell? Is proline isomerization circumvented? Is isomerization catalyzed? There are four codons for proline, CCX where X=U,C,A, or G, so one possibility is that there are specific codons for cis and trans prolines. If so, the isomeric state of a proline needed by the folded protein might be specified during protein synthesis. As long as protein synthesis is a little faster than isomerization folding would proceed at a rapid (fast phase) rate. This possibility has been tested by inspecting proline codon preferences for proteins of known tertiary structure where the isomeric requirements of the native protein are known (Stickle, Rose, Nall, & Hardies, unpublished). The results show that there is no significant correlation between the use of particular codons and the isomeric state of prolines in folded proteins. Another possibility is that there are intracellular catalysts of isomerization and folding. This suggestion is strongly supported by the discovery of a peptidyl prolyl isomerase (PPI) (Fischer & Bang, 1985; Fischer et al., 1984; Fischer et al., 1989; Lang et al., 1987). Moreover, the enzyme has been shown to speed slow folding *in vitro*. To date there is no direct evidence that PPI acts as a folding catalyst *in vivo*.

Catalysis of Slow Folding by Peptidyl Prolyl Isomerase (PPI)

In preliminary experiments we have investigated catalysis of folding of yeast iso-2-cytochrome *c* by human PPI. Surprisingly there is little catalysis of either slow phase when folding experiments end at low denaturant concentrations far below the folding/unfolding transition zone. For folding ending at slightly higher guanidine hydrochloride concentrations efficient catalysis of the absorbance-detected phase (Pro76) is apparent. PPI catalysis of folding is very efficient and is apparent at substrate/catalyst ratios of 50 or higher. Mutant proteins are better substrates than normal iso-2, suggesting that global destabilization of structure by mutation or denaturant is important for efficient catalysis. Interestingly, the fluorescence-detected slow phase (probably Pro30) is catalyzed less efficiently than absorbance-detected slow folding. Nevertheless, essentially full catalysis of this phase is also possible but only at higher PPI concentrations. At present our working model is that structure in folding intermediates interferes with catalysis by PPI and that improved catalysis is achieved when structure in the protein substrate is weakened by mutation or denaturants (Veeraraghavan, S., unpublished).

Monoclonal Antibodies as Tools for Characterizing the Structure of Slow-Folding Intermediates and Conformational States of (Horse) Cytochrome *c*

In other work we have made use of monoclonal antibodies (MAbs) to characterize folding intermediates and other conformational states of horse cytochrome *c*. One of the MAbs binds at a peptide loop near proline 44 in horse cytochrome *c*. There are Trp residues in the vicinity of the MAb binding pocket. Since the heme is an efficient Trp fluorescence quencher, the bimolecular rate of Ab to cytochrome *c* binding can be measured by fluorescence quenching following stopped-flow mixing. Bimolecular rate constants for MAb association with oxidized vs. reduced cytochrome *c* are only slightly different, as might be expected for two forms of cytochrome *c* known to differ little in tertiary structure. The bimolecular rate for association with the alkaline conformation is, however, significantly slower than for either redox state of the native protein suggesting that the structure of the binding epitope on the alkaline protein is slightly perturbed.

An MAb can also be used to detect the rate of formation of an epitope during refolding (Blond-Elguindi & Goldberg, 1990; Murry-Brelier & Goldberg, 1988). For horse cytochrome *c* at low denaturant concentration, the antigenic site on the 40's loop appears with a first order rate that is almost twice that of the fluorescence-detected slow folding phase. For refolding at higher guanidine hydrochloride concentrations fluorescence-detected folding speeds up until, above 1.0 M guanidine hydrochloride, it is approximately equal to the rate of epitope formation (Allen, M., unpublished).

Conclusions

For the yeast cytochromes *c* Pro-76 generates the absorbance-detected slow folding phase while Pro-30 (probably) generates the fluorescence-detected species. The absorbance-detected phase can be removed by replacement of Pro76 with Gly. Both slow phases are catalyzed by PPI, although the absorbance-detected phase (Pro-76) is more susceptible to catalysis than the fluorescence-detected phase (Pro-30). Species present during slow folding are probably highly structured and contain some native-like elements. Mutations can affect preferences for alternative folded forms of protein and can alter the kinetics of protein folding. Folding is, at least in part, a highly directed process in which interactions not present in the final folded state are made transiently and then broken.

Acknowledgements

This work would not have been possible without the scientific contributions of colleagues and co-workers together with research grant support from NIH (GM32980), the Robert A. Welch Foundation (AQ0838), and the UTHSCSA University-Industry Cooperative Research Center.

References

Blond-Elguindi, S., & Goldberg, M. E. (1990) *Biochemistry 29*, 2409-2417.

Brandts, J. F., Halvorson, H. R., & Brennan, M. (1975) *Biochemistry 14*, 4953-4963.

Cook, K. H., Schmid, F. X., & Baldwin, R. L. (1979) *Proc. Natl. Acad. Sci. USA 76*, 6157-6161.

Creighton, T. E. (1978) *J Mol Biol 125*, 401-406.

Cutler, R. J., Pielak, G. J., Mauk, A. G., & Smith, M. (1987) *Protein Eng. 1*, 95-99.

Dickerson, R. E. (1972) *Sci. Am. 226*, 58-72.

Dohne, S. M., Elove, G. A., Roder, H., & Nall, B. T. (1989) *Biophys. J. 55*, 557a.

Englander, S. W., & Kallenbach, N. R. (1984) *Quart. Rev. Biophys. 16*, 531-625.

Ernst, J. F., Hampsey, D. M., Stewart, J. W., Rackovsky, S., Goldstein, D., & Sherman, F. (1985) *J. Biol. Chem. 260*, 13225-13236.

Feng, Y., Roder, H., & Englander, S. W. (1990) *Biophys. J. 57*, 15-22.

Feng, Y., Roder, H., Englander, S. W., Wand, A. J., & Di Stefano, D. L. (1989) *Biochemistry 28*, 195-203.

Fischer, G., & Bang, H. (1985) *Biochim. Biophys. Acta 828*, 39-42.

Fischer, G., Bang, H., & Mech, C. (1984) *Biomed. Biochim. Acta 43*, 1101-1111.

Fischer, G., Wittmann-Liebold, B., Lang, K., Kiefhaber, T., & Schmid, F. (1989) *Nature 337*, 476-478.

Hampsey, D. M., Das, G., & Sherman, F. (1986) *J. Biol. Chem. 261*, 3259-3271.

Kim, P. S., & Baldwin, R. L. (1980) *Biochemistry 19*, 6124-6129.

Kim, P. S., & Baldwin, R. L. (1982) *Annu. Rev Biochem. 51*, 459-489.

Lang, K., Schmid, F., & Fischer, G. (1987) *Nature 329*, 268-270.

Leung, C. J., Nall, B. T., & Brayer, G. D. (1989) *J. Mol. Biol. 206*, 783-785.

Louie, G. V., & Brayer, G. D. (1989) *J. Mol. Biol.* submitted.

Louie, G. V., Hutcheon, W. L. B., & Brayer, G. D. (1988) *J. Mol. Biol. 199*, 295-314.

Murphy, M. E. P., & Brayer, G. D. (1990) *J. Mol. Biol. submitted,*

Murry-Brelier, A., & Goldberg, M. E. (1988) *Biochemistry 27*, 7633-7640.

Nall, B. T. (1983) *Biochemistry 22*, 1423-1429.

Nall, B. T. (1986) *Biochemistry 25*, 2974-2978.

Nall, B. T., & Landers, T. A. (1981) *Biochemistry 20*, 5403-5411.

Nall, B. T., Osterhout, J. J., Jr., & Ramdas, L. (1988) *Biochemistry 27*, 7310-7314.

Nall, B. T., Zuniga, E. H., White, T. B., Wood, L. C., & Ramdas, L. (1989) *Biochemistry 28*, 9834-9839.

Osterhout, J. J., Jr., Muthukrishnan, K., & Nall, B. T. (1985) *Biochemistry 24*, 6680-6684.

Osterhout, J. J., Jr., & Nall, B. T. (1985) *Biochemistry 24*, 7999-8005.

Pearce, L. L., Gartner, A. L., Smith, M., & Mauk, A. G. (1989) *Biochemistry 28*, 3152-3156.

Pielak, G., Mauk, A. G., & Smith, M. (1985) *Nature 313*, 152-154.

Pielak, G. J., Boyd, J., Moore, G. R., & Williams, R. J. P. (1988) *Eur. J. Biochem. 177*, 167-177.

Ramdas, L. (1987) Ph.D. Thesis, University of Texas Health Science Center at Houston, TX.

Ramdas, L., & Nall, B. T. (1986) *Biochemistry 25*, 6959-6964.

Ramdas, L., & Nall, B. T. (1987) *J. Cell. Biochem. Supplement 11C*, 218.

Roder, H., Elove, G. A., & Englander, S. W. (1988) *Nature 335*, 700-704.

Schechter, E., & Saludjian, P. (1967) *Biopolymers 5*, 788-790.

Schmid, F. X., & Baldwin, R. L. (1979) *J. Mol. Biol. 135*, 199-215.

Schweingruber, M. E., Stewart, J. W., & Sherman, F. (1979) *J. Biol. Chem. 254*, 4132.

Sherman, F., & Stewart, J. W. (1978) in *Biochemistry and Genetics of Yeast* (Bacila, M., Horecker, B. L., & Stoppani, A. D. M., Ed.) pp. 273, Academic Press, New York.

Wand, A. J., Di Stefano, D. L., Feng, Y., Roder, H., & Englander, S. W. (1989) *Biochemistry 28*, 186-194.

White, T. B., Berget, P. B., & Nall, B. T. (1987) *Biochemistry 26*, 4358-4366.

Wood, L. C., White, T. B., Ramdas, L., & Nall, B. T. (1988) *Biochemistry 27*, 8562-8568.

Zuniga, E. H., & Nall, B. T. (1983) *Biochemistry 22*, 1430-1437.

INCREASING ENZYME STABILITY

C. Nick Pace

Biochemistry Department
Texas A&M University
College Station, Texas 77843

INTRODUCTION

Enzymes are active only when their polypeptide chain is folded into a unique globular conformation with a functional active site. In most naturally occurring enzymes, this globular conformation is only 5 to 10 kcal/mole more stable than unfolded, biologically inactive conformations (1). The conformational stability of a protein is defined as the free energy change for the reaction:

$$\text{folded (native)} \longleftrightarrow \text{unfolded (denatured)} \qquad (1)$$

under ambient conditions, such as water at 25°C. The most convenient methods of estimating the conformational stability of a protein are urea (or guanidine hydrochloride (GdnHCl)) unfolding curves and thermal unfolding curves. Estimates of the conformational stability based on urea unfolding curves are designated $\Delta G(H_2O)$, and estimates from thermal unfolding curves are designated $\Delta G(25^{\circ}C)$. In the first part of this article, we describe how to measure $\Delta G(H_2O)$ and $\Delta G(25^{\circ}C)$.

Our ability to construct enzymes to order ensures that they will be widely used in biotechnology. For most applications, the goal will be to construct the most stable enzyme that retains the desired biological activity. In general, it will be best to enhance the stability through changes in the amino acid sequence rather than by chemical modification of an enzyme after synthesis. For example, Groups at Genentech (2) and Genex (3) have been using this approach to increase the stability of subtilisin, and Degrado's group at Dupont has designed and constructed a protein based on a four-helix bundle structure with a conformational stability of 22.5 kcal/mole (4). In the last part of this article, we consider the most promising approaches being used at present to make enzymes more stable.

Chemical Aspects of Enzyme Biotechnology
Edited by T.O. Baldwin *et al.*, Plenum Press, New York, 1990

$\Delta G(H_2O)$ FROM UREA UNFOLDING CURVES

A urea denaturation curve for RNase T1 is shown in Fig. 1. (Practical information on measuring and analyzing a denaturation curve is given in Ref. 5.) For RNase T1 and many other globular proteins, the mechanism of unfolding has been shown to closely

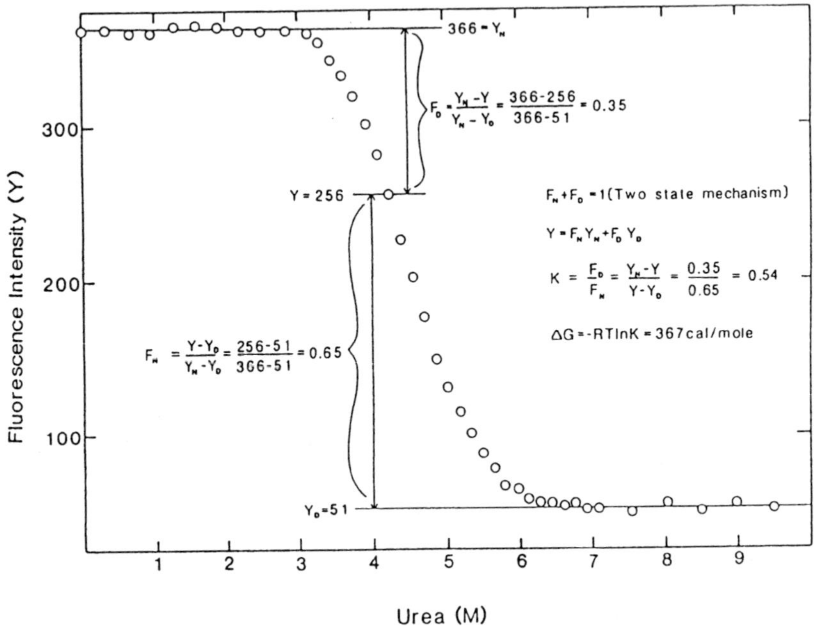

Fig. 1. Urea unfolding curve for RNase T1. The fluoresence intensity was measured at 320 nm after excitation at 278 nm on solutions at equilibrium containing 0.01 mg/ml RNase T1, pH 7, 30 mM MOPS buffer, 25°C. The analysis of the data based on a two-state folding mechanism (Eq. 1) is illustrated. The fluorescence intensity characteristic of native, Y_N, and denatured, Y_D, RNase T1 were obtained by a least-squares analysis of the pre- and post-transition baselines.

approach a two-state mechanism where the concentration of partially-folded molecules present at equilibrium is small enough to be neglected (6). By assuming a two-state folding mechanism (Eq. 1), the fraction of denatured protein, F_D, the fraction of native protein, F_N, the equilibrium constant, K, and the free energy change for unfolding, ΔG, can be calculated as a function of urea concentration as shown in Fig. 1.

ΔG is generally found to vary linearly with denaturant concentration, as shown in Fig. 2, and a least-squares ananlysis is used to fit the data from the transition region to the equation:

$$\Delta G = \Delta G(H_2O) - m\ [D] \qquad (2)$$

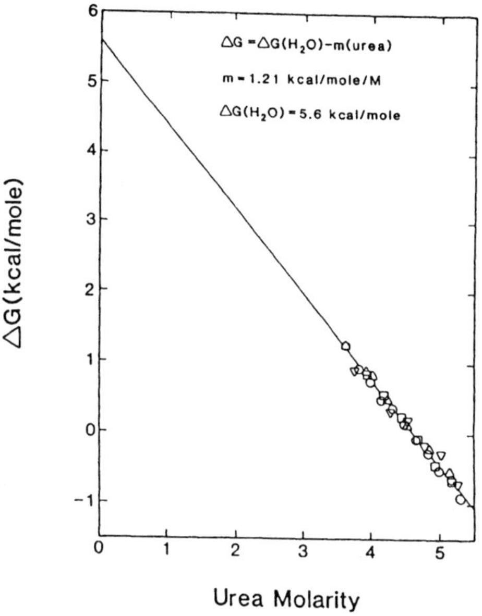

Fig. 2. ΔG as a function of urea molarity for the unfolding of RNase T1 in 30 mM MOPS buffer, pH 7, 25°C. ΔG values were calculated as explained in Fig. 1, and the solid line represents the equation and parameters given on the figure (Eq. 2).

where $\Delta G(H_2O)$ is the value of ΔG in the absence of denaturant, m is a measure of the dependence of ΔG on denaturant concentration, and [D] is the denaturant molarity. Values of $\Delta G(H_2O)$ and m obtained in this way are given on Fig. 2. Santoro and Bolen (7) have pointed out that this method of analysis underestimates the error in $\Delta G(H_2O)$ and m because no error is assumed for the pre- and post-transition baselines. As an alternative, they suggest using a nonlinear, least-squares program to fit the entire unfolding curve to the equation:

$$y = [(y_f + m_f [D]) + (y_u + m_u[D])(\exp -(\Delta G(H_2O)/RT - m[D]/RT))] / [1 + \exp -(\Delta G(H_2O)/RT - m[D]/RT)] \quad (3)$$

where y_f and m_f, and y_u and m_u are the slope and intercept of the pre- and post-transtion baselines, respectively, and [D], m and $\Delta G(H_2O)$ were defined above. Six parameters are best fit: y_f, m_f, y_u, and m_u, which characterize the pre- and post-transition baslines; and m and $\Delta G(H_2O)$ which characterize the transition region. The standard error in m, typically ± 150 cal/mole M, and $\Delta G(H_2O)$, typically ± 0.5 kcal/mole, is indeed substantially larger, and more reasonable, when this approach is used to analyze a urea or GdnHCl denaturation curve (8).

ΔG(25°C) FROM THERMAL UNFOLDING CURVES

Thermal unfolding curves for RNase T1 can be analyzed in exactly the same way as denaturant unfolding curves, but yield ΔG as a function of temperature rather than denaturant concentration, as shown in Fig. 3. These results can be used to determine the melting temperature, T_m, and the enthalpy change at T_m, ΔH_m, as described in the figure legend. The temperature dependence of ΔG is given by a form of the Gibbs-Helmholtz equation:

$$\Delta G(T) = \Delta H_m(1 - T/T_m) - \Delta C_p[T_m - T + T \ln (T/T_m)] \quad (4)$$

where $\Delta G(T)$ is ΔG at a temperature T, and ΔC_p is the change in heat capacity for the unfolding reaction. Thus, calculating ΔG(25°C) using results from a thermal unfolding curve requires a knowledge of ΔC_p.

Three methods have been used to measure ΔC_p. One simply requires a more detailed analysis of a thermal unfolding curve. If the equilibrium constant is measured as a function of temperature, the enthalpy change can be calculated using the van´t Hoff equation:

$$d(-R \ln K)/d(1/T) = \Delta H \quad (5)$$

For protein unfolding, ΔH is found to increase markedly with temperature because the heat capacity of the unfolded protein, $C_p(U)$, is greater than that of the folded protein, $C_p(F)$, and this is what gives rise to the large positive change in heat capacity for protein unfolding:

$$\Delta C_p = C_p(U) - C_p(F) \quad (6)$$

If ΔH can be measured as a function of temperature, ΔC_p can be calculated using the Kirchoff equation:

$$d(\Delta H)/d(T) = \Delta C_p \quad (7)$$

This approach has been successful in some cases (9,10), but could not be used with RNase T1 because ΔH is large and measurements

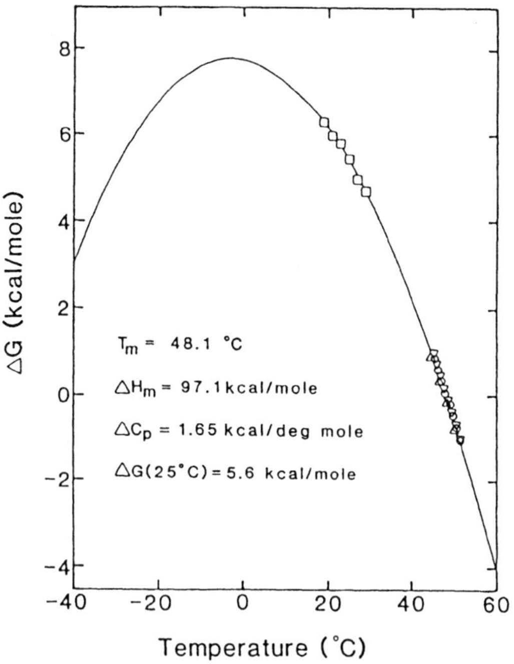

Fig. 3. ΔG as a function of temperature for the unfolding of RNase T1 in 30 mM MOPS buffer, pH 7. The ΔG values above 40°C were calculated from thermal unfolding curves. A least-squares analysis was used to determine T_m = T at ΔG = 0, and ΔH_m = (T_m in °K) X −(slope at T_m). The ΔG values below 30°C are $\Delta G(H_2O)$ values from urea unfolding curves determined at the corresponding temperatures. A best fit of these data to Eq. 4 was used to determine ΔC_p. The solid curve was calculated with Eq. 4 using the parameters given on the figure. (Adapted from Ref. 13.)

could be made over only a narrow temperature range, about 7°C. Note that this method involves taking the second derivative of the experimental data, so exceptionally good data are needed. A second method involves measuring T_m and ΔH_m as a function of pH, and obtaining ΔC_p from the slope of a plot of ΔH_m Vs. T_m. If proper precautions are taken and ΔH_m and ΔC_p do not depend on pH, an estimate of ΔC_p can be obtained (11,12). The method of choice is to determine ΔC_p using a differential scanning calorimeter (11).

Recently, we proposed a new method for estimating ΔC_p (13). A consideration of Eq. 4 shows that ΔG depends mainly on the first term at temperatures near T_m, but that the second term becomes increasingly important at temperatures below T_m. We suggest measuring $\Delta G(H_2O)$ at temperatures considerably below T_m using urea unfolding curves as described above. These values can be used directly in Eq. 4 along with T_m and ΔH_m to calculate ΔC_p. The six values of $\Delta G(H_2O)$ shown between 19 and 29°C in Fig. 3 yield ΔC_p values ranging from 1630 to 1870 with an average of 1720 ± 65 cal/mole deg. This method should be useful to researchers who do not have access to a calorimeter.

Becktel and Schellman (12) refer to plots such as the one shown in Fig. 3 as protein stability curves, and show that the temperature of maximum stability, T_S, can be calculated using:

$$T_S = T_m \exp(-\Delta H_m/T_m \Delta C_p) \qquad (8)$$

For RNase T1, T_S occurs near -5°C. The values of T_S generally fall between -10°C for a relatively hydrophillic protein like RNase A and 35°C for a relatively hydrophobic protein like β-lactoglobulin. Thus, the heat or cold induced unfolding of a protein like β-lactoglobulin can be observed directly (10). Several careful studies of cold denaturation have been published recently (14,15). The unusual temperature dependence of protein folding depends on ΔC_p and ΔC_p depends mainly on the ordering of water molecules around the nonpolar groups which are brought into contact with water when a protein unfolds, as do hydrophobic interactions (16-18).

CONSTRUCTING MORE STABLE PROTEINS

Adding disulfide bonds

Disulfide bonds make large contributions to the stability of some globular proteins. For example, RNase A unfolds completely when the four disulfide bonds are broken, and they are estimated to contribute about 19 kcal/mole to the conformational stability. Bovine pancreatic trypsin inhibitor (BPTI) is a very stable protein with a melting temperature around 95°C, but BPTI also unfolds when its three disulfide bonds are broken, and breaking the disulfide bonds between either residues 5 and 55, or 14 and 38 lowers the melting temperature by over 50°C (19). Chemical crosslinks between side chains can also be used to increase the stability as shown by the first two examples in Table 1. The second two examples show measured values of the decrease in conformational stability when one or two disulfide bonds are broken in an immunoglobulin light chain and ribonuclease T1.

This sort of evidence suggested that adding disulfide bonds to a protein by genetic engineering would be a good means of increasing their stability. The first results were discouraging. Seven different disulfide bonds were added to subtilisin, and none of the resulting proteins was significantly more stable than wild type (2). However, recent results with T4 lysozyme are more encouraging. As shown in Table 1, three different disulfide bonds have been added that increase the stability, and the protein containing all three of the disulfides has a melting temperature 23.4°C greater than the wild type protein (24). On the basis of these results, Matsumura et al. (31) make the following suggestions for engineering a disulfide bond that will increase stability: (1) "The size of the loop formed by the crosslink

Table 1. Measurements of increases in the conformational stability, $\Delta(\Delta G)$, and melting temperature, ΔT_m, of several globular proteins

Protein	Change	ΔT_m	$\Delta(\Delta G)$
		°C	kcal/mole
Adding crosslinks:			
Lysozyme[20]	35 to 108	29	5.2
RNase A[21]	7 to 41	25	4.9
C_L(lambda)[22]	26 to 86		4.7
RNase T1[23]	2 to 10 & 6 to 103	32	7.2
T4 Lyso[24]	3 to 97	5	---
	9 to 164	6	---
	21 to 142	11	---
	All Three	23	---
Adding salts:			
RNase T1[25]	+ 0.2 M $CaCl_2$	--	2.0
	+ 0.2 M Na_2HPO_4	--	4.4
α-Lact[26]	+ 0.1 mM $CaCl_2$	34	---
Single amino acid substitutions:			
Cro[27]	Tyr 26 ---> Cys	11	2.2
Cro[27]	Gln 16 ---> Leu	14	2.8
Cyt c[28]	Asn 57 ---> Ile	17	4.2
Barnase[29]	Ala 14 ---> Leu	12	4.8
Trp Sase[30]	Gly 49 ---> Ile	--	9.7

should be as large as possible."; (2) "---minimize the disruption or loss of interactions that stabilize the native structure."; and (3) "The strain energy introduced by the disulfide bond should be kept as low as possible." They suggest that placing the new disulfide in a flexible part of the molecule will aid in achieving the latter two goals (31).

Crosslinks increase the stability mainly by constraining the unfolded states and thereby decreasing their conformational entropy and increasing their free energy. In cases where strain in the folded state is minimal, the effect of a crosslink on the conformational entropy can be estimated with the following equation:

$$\Delta S_{conf} = -2.1 - (3/2)R \ln n \qquad (9)$$

where n is the number of residues in the loop forming the disulfide bond (23). This equation predicts that the stability will be increased by 3, 4, and 5 kcal/mole by adding loops of 15, 45, and 135 residues, respectively.

Adding cation and anion binding sites

It has long been known that compounds such as substrates and inhibitors that bind specifically to the native conformation of a enzyme will increase the stability. Assuming a single binding site on the native protein, the increased stability, $\Delta(\Delta G)$, can be calculated using:

$$\Delta(\Delta G) = \Delta G(L) - \Delta G(L=0) = RT \ln [1 + K_B(L)] \qquad (10)$$

where L is the free concentration of the molecule that binds, and K_B is the binding constant (32). Thus, the stability enhancement will depend on the binding constant and on the concentration of the compound that binds. For stabilizing an enzyme, it will be best to use a compound that can be added at moderate concentrations and whose binding does not interfere with the enzyme´s function. This is probably best achieved by adding sites on the surface of the protein that bind either cations or anions. For example, the conformational stability of RNase T1 is increased by 2 kcal/mole in the presence of 0.2 M $CaCl_2$. It was shown that this results primarily from the binding of a Ca^{2+} ion at a cation binding site with a very modest $K_B = 155\ M^{-1}$ (25). The Ca^{2+} ion is coordinated to two carboxylate oxygens from Asp 15 and to 5 water molecules (33). At the other extreme, T_m is increased by 33.8°C in the presence of 0.1 mM $CaCl_2$ for α-lactalbumin (26). The Ca^{2+} binding site also binds Na^+ ions and $K_B = 2.9 \times 10^9\ M^{-1}$ for Ca^{2+} ions and 1240 M^{-1} for Na^+ ions. In this case, the Ca^{2+} interacts with seven ligands, all oxygen: three carboxylates from Asp residues, two backbone carbonyls, and two water molecules (34). Using this general approach, the stability of subtilisin (35) and human lysozyme (36) have been increased by improving the binding of Ca^{2+} ions. The design of ion binding sites will be aided by the availability of detailed structural information on cation (34,37) and anion (38,39) binding sites.

Amino acid substitutions

Five single amino acid substitutions that result in large increases in the stability of four proteins are listed in Table 1. Yutani's group has measured the conformational stability of Trp synthase with every amino acid except Arg at Position 49 (30). The wild type enzyme has Glu at this position. The substitution listed in the table gives the stability difference between the least and most stable mutants. They concluded: "---the stability of the protein substituted at this position, which is buried in the interior of the molecule, tended to increase linearly with increasing hydrophobicity of the substituted residue, unless the volume of the substituted residue was over a certain limit." The substitution listed for barnase in also at a buried residue in the hydrophobic core of the molecule. In this case, the authors observe that the difference in free energy: "---exceeds by severalfold the values obtained from model experiments of the partitioning of relevant side chains between aqueous and nonpolar solvents." (29). The two substitutions listed for Cro and the one for cytochrome c are interesting because they are at largely exposed sites on the molecules and it is not clear how they increase the conformational stability. These results serve to illustrate that our understanding of how changes in the structure of a protein effect the stability is still very limited.

It is encouraging that large increases in the stability can be achieved by single amino acid substitutions. Several studies have shown that the effects of amino acid substitutions may be localized so that their effects on the stability are additive (3,40,41). For example, six amino acid substitutions have been made in subtilisin that each increase the stability by from 0.3 to 1.3 kcal/mole, and when they are all incorporated in the same molecule, the stability is increased by 3.8 kcal/mole (3).

On the basis of the results obtained to date, it seems likely that it will be possible to construct very stable enzymes when they are needed, and this should increase their usefullness in many biotechnological applications.

ACKNOWLEDGEMENTS

I am grateful to NIH (GM 37039), the Robert A. Welch Foundation (A 1060), and the Texas Agricultural Experiment Station for financial support.

REFERENCES

1. C.N. Pace, Conformational Stability of Globular Proteins, Trends Biochem. Sci. 15:14 (1990).
2. C. Mitchinson, and J.A. Wells, Protein Engineering of Disulfide Bonds in Subtilisin BPN', Biochemistry 28:4807 (1989).
3. M.W. Pantoliano, M. Whitlow, J.F. Wood, S.W. Dodd, K.D. Hardman, M.L. Rollence, and P.N. Bryan, Large Increases in General Stability for Subtilisin BPN' through Incremental Changes in the Free Energy of Unfolding, Biochemistry 28: 7205 (1989).
4. L. Regan, and W.F. DeGrado, Characterization of a Helical Protein Designed from First Principles, Science 241:976 (1988).

5. C.N. Pace, B.A. Shirley, and J.A. Thomson, Measuring the Conformational Stability of a Protein, in "Protein Structure: a practical approach" T.E. Creighton, ed., pp. 311, IRL Press, Oxford.
6. J.A. Thomson, B.A. Shirley, G.R. Grimsley, and C.N. Pace, Conformational Stability and Mechanism of Folding of Ribonuclease T1, J. Biol. Chem. 264:11614 (1989).
7. M.M. Santoro, and D.W. Bolen, Unfolding Free Energy Changes Determined by the Linear Extrapolation Method. 1. Unfolding of Phenylmethanesulfonyl α-Chymotryspsin Using Different Denaturants, Biochemistry 27:8063 (1988).
8. C.N. Pace, D.V. Laurents, and J.A. Thomson, PH Dependence of the Urea and Guanidine Hydrochloride Denaturation of Ribonuclease A and Ribonuclease T1, Biochemistry 29: in press.
9. W.M. Jackson, and J.F. Brandts, Thermodynamics of Protein Denaturation. A Calorimetric Study of the Reversible Denaturation of Chymostypsinogen and Conclusions Regarding the Accuracy of the Two-State Approximation, Biochemistry 9:2294 (1970).
10. C.N. Pace, and C. Tanford, Thermodynamics of the Unfolding of β-Lactoglobulin in Aqueous Urea Solutions between 5 and 55°C, Biochemistry 7:198 (1968).
11. P.L. Privalov, Stability of Proteins, Adv. Prot. Chem. 33:167 (1979).
12. W.J. Becktel, and J.A. Schellman, Protein Stability Curves, Biopolymers 26:1859 (1987).
13. C.N. Pace, and D.V. Laurents, A New Method for Determining the Heat Capacity Change for Protein Folding, Biochemistry 28:2520 (1989).
14. B. Chen., and J.A. Schellman, Low-Temperature Unfolding of a Mutant of Phage T4 Lysozyme. 1. Equilibrium Studies, Biochemistry 28:685 (1989).
15. P.L. Privalov, Y. Gricko, S.Y. Venyaminov, and V.P. Kutyshenko Cold Denaturation of Myoglobin, J. Mol. Biol. 190:487 (1986).
16. R.L. Baldwin, Temperature Dependence of the Hydrophobic Interaction, Proc. Nat. Acad. Sci. 83:8069 (1986).
17. P.L. Privalov, and S.J. Gill, Stability of Protein Structure and Hydrophobic Interaction, Adv. Prot. Chem. 39:191 (1988).
18. R.S. Spolar, J.-H. Ha, and T.M. Record, Hydrophobic Effect in Protein Folding and Other Noncovalent Processes Involving Proteins, Proc. Nat. Acad. Sci. 86:8382 (1989).
19. D.J. States, T.E. Creighton, C.M. Dobson, and M. Karplus, Conformations of Intermediates in the Folding of the Pancreatic Trypsin Inhibitor, J. Mol. Biol. 195:731 (1987).
20. R.E. Johnson, P. Adams, and J.A. Rupley, Thermodynamics of Protein Cross-Links, Biochemistry 17:1479 (1978).
21. S.H. Lin, Y. Konishi, M.E. Denton, and H.A. Scheraga, Influence of an Extrinsic Cross-Link on the Folding Pathway of Ribonuclease A. Conformational and Thermodynamic Analysis of Cross-Linked (Lysine 7 - Lysine 41) Ribonuclease A, Biochemistry 23:5504 (1984).
22. Y. Goto, M. Tsunenaga, Y. Kawata, and K. Hamaguchi, Conformation of the Constant Fragment of the Immunoglobulin Light Chain: Effect of Cleavage of the Polypeptide Chain and the Disulfide Bond, J. Biochem. 101:319 (1987).
23. C.N. Pace, G.R. Grimsley, J.A. Thomson, and B.J. Barnett, Conformational Stability of Ribonuclease T1 with Zero, One, and Two Intact Disulfide Bonds, J. Biol. Chem. 263: 11820 (1988).

24. M. Matsumura, G. Signor, and B.W. Matthews, Substantial Increase in Protein Stability from Multiple Disulfide Bonds, Nature 342:291 (1989).
25. C.N. Pace, and G.R. Grimsley, Ribonuclease T1 is Stabilized by Cation and Anion Binding, Biochemistry 27:3242 (1988).
26. M. Mitani, Y. Harushima, K. Kuwajima, M. Ikeguchi, M., and S. Sugai, Innocuous Character of EDTA as Metal-Ion Buffers in Studying Ca^{2+} Binding by α-Lactalbumin, J. Biol. Chem. 261:8824 (1986).
27. A.A. Pakula, and R.T. Sauer, Amino Acid Substitutions that Increase the Thermal Stability of the lambda Cro Protein, Proteins: Struc. Func. Gen. 5: 202 (1989).
28. G. Das, D.R. Hickey, D. McLendon, G. McLendon, and F. Sherman, Dramatic Thermostabilization of Yeast Iso-1-Cytochrome C by an Asparagine ---> Isoleucine Replacement at Position 57, Proc. Nat. Acad. Sci. 86: 496 (1989).
29. J.T. Kellis, K. Nyberg, and A.R. Fersht, Contribution of Hydrophobic Interactions to Protein Stability, Biochemistry 28:4914 (1988).
30. K. Yutani, K. Ogasahara, T. Tsujita, and Y. Sugino, Dependence of Conformational Stability on Hydrophobicity of the Amino Acid Residue in a Series of Variant Proteins Substituted at a Unique Position of Tryptophan Synthase α Subunit, Proc. Nat. Acad. Sci. 84:4441 (1987).
31. M. Matsumura, W.J. Becktel, M. Levitt, and B.W. Matthews, Stabilization of phage T4 Lysozyme by Engineered Disulfide Bonds, Proc. Nat. Acad. Sci. 86:6562 (1989).
32. C.N. Pace, and T. McGrath, Substrate Stabilization of Lysozyme to Thermal and Guanidine Hydrochloride Denaturation, J. Biol. Chem. 255:3862 (1980).
33. D. Kostrewa, H.-W. Choe, U. Heinemann, and W. Saenger, Crystal Structure of Guanosine-Free Ribonuclease T1 Complexed with Vanadate(V), Suggests Conformational Change upon Substrate Binding, Biochemistry 28:7592 (1989).
34. K.R. Acharya, D.I. Stuart, N.P.C. Walker, M. Lewis, and D.C. Phillips, Refined Structure of Baboon α-Lactalbumin at 1.7Å Resolution, J. Mol. Biol. 208:99 (1989).
35. M.W. Pantoliano, M. Whitlow, J.F Wood, M.L. Rollence, B.C. Finzel, G.L. Gilliand, T.L. Poulus, and P.N. Bryan, The Engineering of Binding Affinity at Metal Binding Sites for the Stabilization of Proteins: Subtilisin as a Test Case, Biochemistry 27: 8311 (1988).
36. R. Kuroki, Y. Taniyama, C. Seko, H. Nakmura, M. Kikuchi, and M. Ikehara, Design and Creation of a Ca^{2+} Binding Site in Human Lysozyme to Enhance Structural Stability, Proc. Nat. Acad. Sci. 86:6903 (1989).
37. A.L. Swain, R.H. Kretsinger, and E.L. Amma, Restrained Least Squares Refinement of Native (Calcium) and Cadmium-substituted Carp Parvalbumin Using X-ray Crystallographic Data at 1.6Å Resolution, J. Biol. Chem. 264:16620 (1989).
38. M. Fujinaga, T.J. Delbaere, G.D. Brayer, and M.N.G. James, Refined Structure of α-Lytic Protease at 1.7Å Resolution, J. Mol. Biol. 183:479 (1985).
39. J.W. Pflugrath, and F.A. Quiocho, Sulphate Sequestered in the Sulphate-Binding Protein of Salmonella Typhimurium is bound Solely by Hydrogen Bonds, Nature 314:257 (1985).
40. T. Alber, Mutational Effects on Protein Stability, Ann. Rev. Biochem. 58:765 (1989).
41. B.A. Shirley, P. Stanssens, J. Steyaert, and C.N. Pace, Conformational Stability and Activity of Ribonuclease T1 and Mutants: Gln 25 ---> Lys, Glu 58 ---> Ala, and the Double Mutant, J. Biol. Chem. 264:11621 (1989).

A STUDY OF SUBUNIT FOLDING AND DIMER ASSEMBLY *IN VIVO*

Thomas O. Baldwin

The Departments of Biochemistry and Biophysics and of Chemistry
Texas A&M University
College Station, Texas 77843

INTRODUCTION

Bacterial luciferase is a heterodimer, consisting of two nonidentical but homologous subunits, α and β (1). The enzyme has a single active center, that has been shown to be confined primarily, if not exclusively, to the α subunit (1, 2). The role of the β subunit is not known, but there is much support for the assertion that the β subunit is required for activity. The subunits apparently associate with high affinity; there have been no reports demonstrating dissociation of the complex of the wild-type luciferase under non-denaturing conditions.

We have undertaken a study of the folding and assembly of the subunits of bacterial luciferase (3, 4). We have found that the individual subunits, if allowed to fold independently, assume structures that do not associate to form the active heterodimeric structure. Based on these observations, we proposed that the two subunits must interact as intermediates on the folding pathway during the folding reaction, not following completion of the process of folding of each individual subunit. We have also discovered mutants of the luciferase enzyme that appear to be temperature sensitive in the folding reaction. A group of these mutants have in common deletions from the carboxyl terminus of the β subunit. These mutants are characterized by the inefficiency with which the subunits associate to form active heterodimeric luciferase, which, when formed, is as stable and as active as the wild-type enzyme. By site-directed mutagenesis, we have constructed a variant, βD313P, which has the same properties as the C-terminal deletion mutants. These studies show that the C-terminal region of the β subunit exerts little if any effect on the structure, stability, or activity of the native form of luciferase, but it appears to play a crucial role at some intermediate step in the process of folding and/or assembly of the heterodimer. The effect of several of the mutations suggests that it is not the precise amino acid sequence within the C-terminal region that is important, but rather some secondary structural element which is critical to the process of formation of the active luciferase. The phenotypic behavior of these mutants in folding *in vivo* and in refolding *in vitro* is the same, suggesting that folding in the cell is similar to folding *in vitro*.

METHODS

Plasmid Constructions and Bacterial Strains. Most of the plasmid constructions and bacterial strains used in this study have been described elsewhere. The plasmids pJH2 (α subunit, Amp^r), pJH5 (β subunit, Kan^r) and pJH3 (αβ, Amp^r) were all derived from pUC9 (3, 5). The plasmids pJS2, pJS12, and pJS23 were all derived from pTB7 (4, 5). The single site mutation of Asp to Pro at position 313 of the β subunit was constructed by Ke Wei in this laboratory by site-directed mutagenesis (6). The plasmid carrying the mutation, designated pDP313, is similar to pJH3. The plasmids were expressed in standard laboratory strains of *E. coli*, MC1009, LE392, and TB1, which have been described in this context elsewhere (3, 4, 7).

Growth Conditions. Cultures of cells harboring plasmids were grown with shaking at defined temperatures in Luria-Bertani broth supplemented with carbenicillin (100 μgm/ml) or, for cells harboring pJH5, kanamycin (30 μgm/ml).

Luciferase Subunit Complementation *In Vitro*. Luciferase subunits contained in lysates of recombinant *E. coli* were unfolded by incubation in 8 M urea and refolded by dilution (50-fold) into recovery buffer (0.5 M NaH_2PO_4/K_2HPO_4, 1 mM EDTA, 1 mM DTT, 0.2% bovine serum albumin, pH 7.0; 8). In some experiments, the amount of renaturation-competent subunit in a lysate was estimated by titration with the other subunit followed by refolding from urea, essentially as described (8).

Luciferase Assay. The activity of luciferase was determined by the flavin squirt assay, as described (9). The peak light intensity emitted was monitored with a calibrated (10) photomultiplier-photometer. One "light unit" with this photometer system was 7.58 x 10^9 q/sec.

RESULTS AND DISCUSSION

A common motif of polypeptide folding, especially for large polypeptides, is to fold into independent domains, such that, to the extent possible given the constraints of the covalent continuity of the peptide chain, the independent domains act as independent subunits (11). It is of interest to understand the impact of interactions *between* subunits, or between folding domains, on the structure of the folding unit. That is, does the amino acid sequence of a folding domain uniquely define the three dimensional structure of that unit, or do interactions between folding units contribute to the definition of the final structure? Furthermore, if such interactions do contribute to the definition of the final structure, it is of interest to have some feeling for the magnitude of the effects. It appeared to us that the luciferase enzyme might be a good model system with which to probe these and related questions. First, the enzyme is a heterodimer, such that the folding of each subunit can be studied independently without the complications associated with the second order dimerization process. Second, it appeared that the dimerization process could be studied independent of the folding process, such that the above questions might be approached. Finally, the activity of luciferase requires the heterodimeric structure, and the assay for the enzyme, by measurement of emitted light, is exquisitely sensitive, such that the formation of active luciferase in the living cell can be readily observed, allowing monitoring of the final step of folding *in vivo*.

Subunit Recombination *In Vivo* and *In Vitro*

To produce large amounts of the individual subunits, we constructed several vectors with the individual genes, *luxA* and *luxB*, under control of the *lac* promoter of the pUC9 plasmid. Using these plasmids, we have been able to produce large amounts of the individual subunits, each being absolutely free of the other.

Large quantities of α and β subunits accumulated in cells carrying the plasmids pJH2 and pJH5, respectively (3). In cells carrying both pJH2 and pJH5, both subunits accumulated, but, unlike cells carrying pJH3, the levels of the two subunits were different, presumably due to the different copy numbers of the two plasmids. We were surprised to find that upon mixing of the lysates from cells carrying pJH2 and pJH5, very little active luciferase was formed ($< 3 \times 10^8$ q/sec/mL), even with prolonged incubation (see Table I). However, lysates from cells carrying both pJH2 and pJH5 contained large amounts of active luciferase (1.63×10^{11} q/sec/mL), consistent with the level of expression of the subunit that accumulated at the lower level.

It seemed possible that the failure of the subunits produced independently to form active heterodimeric luciferase was a consequence of some covalent modification, such as proteolysis, that occurred in the free subunit, but not in the heterodimer. The lysates of cells carrying pJH2 and pJH5 were mixed, treated with 8 M urea, and then diluted into recovery buffer. Following incubation at 5° for 22 h, the activity of the recombined subunits had increased more than 100 times that in the undenatured control (see Table I) to approximately the level expected if 100% of the subunit present in each lysate had associated with the other subunit following mixing.

These observations were not expected. We had expected that the individual subunits would recombine readily to form the active

Table I. Complementation of Luciferase Subunits *In Vivo* and *In Vitro*.

Lysates[a]	Luciferase Activity ($q/sec/mL \times 10^{-10}$)	Fold Recovery Relative to Undenatured Control
pJH3[b]	48.0	
+ Urea	19.4	0.40
pJH2/pJH5[b]	16.3	
+ Urea	5.45	0.33
pJH2 + pJH5[c]	<0.03	
+ Urea	3.80	>127

[a]Cells carrying the plasmids indicated were grown to an $O.D._{600}$ of 1, harvested and lysed.

[b]Aliquots of the lysates of cells carrying pJH3, or both pJH2 and pJH5, were denatured in 8 M urea, diluted into recovery buffer and assayed following incubation for 22 h at 5°C. Control samples were not treated with urea.

[c]Aliquots of cells carrying pJH2 alone and of cells carrying pJH5 alone were mixed, lysed, and treated as in footnote *b*, above.

enzyme. Failure of the subunits to associate to form the active enzyme clearly demonstrated that the association step leading to the heterodimer does not involve the subunits as they would exist if both were allowed to fold to completion. Rather, the association between the two subunits must involve intermediate structures that exist during the folding of the independent subunits. If the subunits are allowed (or required, as in these experiments) to fold independently, one (or both) assume conformations that are experimentally irreversible without addition of a denaturant.

Gunsalus-Miguel et al. (8) found that the subunits of *V. harveyi* luciferase could be separated by chromotography on DEAE-Sephadex in 5 M urea and renatured independently. Mixing of the independently renatured subunits resulted in recovery of about 40% of the expected amount of active luciferase. The apparent contradiction between our results and those of Gunsalus-Miguel et al., which have been repeated in our laboratory (unpublished), suggests that the pathway of folding *in vivo* and *in vitro* might be different. An alternative explanation is that the subunits might not be completely unfolded in 5 M urea under conditions of chromatography, such that renaturation could allow refolding into a dimerization-competent structure. These two alternatives are not mutually exclusive. At present, no experimental data exist that permit further evaluation of the possible mechanisms. It does, however, appear clear that the structures of the subunits assumed *in vivo* are not in equilibrium with the structures of the subunits that can associate to form the active heterodimer.

	311				315					320				324	
WT	Val	Ile	Asp	Val	Val	Asn	Ala	Asn	Ile	Val	Lys	Tyr	His	Ser	
23	Val	Ile	Asp	Phe	Cys	Ala	Arg	Cys	Asp	Val	Phe	Leu	Leu	Asn	Arg
12	Val	Ile	Ala	Met	Trp										
2	Val	Ile	Gly												

FIGURE 1. Amino acid sequences of the β subunits of the luciferases encoded by the plasmids pTB7 (wild-type; W. T.), pJS23, pJS12 and pJS2. The C-terminal 14 residues of the wild-type sequence are shown, aligned with the same regions for the modified subunits.

Temperature Sensitive Folding Mutants

The coding region for the carboxyl terminal end of the β subunit of luciferase was truncated by deletion of specific fragments of DNA, resulting in a series of proteins with shortened and/or altered amino acid sequences within the C-terminal region. The first such protein to be constructed was encoded by the plasmid pJS2. Cells carrying this plasmid had the curious property of being dark if grown at 37° but of emitting detectable luminescence if grown at 20°, even after being shifted to 37°. Three of these proteins were analyzed in detail. The amino acid sequences of the C-termini of these proteins and of the wild-type subunit are presented in Figure 1.

Luminescence *in vivo* was monitored from cells carrying these plasmids during growth at 20° and at 37°. In addition, the relative accumulation of α subunit (estimated from Western blot analysis; 4) and the half-life of the luciferase activity *in vitro* at 37° were also determined. The results are presented in Table II.

Table II. Comparison of Relative Peak Luminescence *In Vivo* with α Subunit Accumulation and Enzyme Stability *In Vitro*.

Plasmid	Ratio[a] (37° Peak/20° Peak)	α Subunit[b] 20°	α Subunit[b] 37°	Half-Life *in vitro*[c] (Hours, 37°)
pTB7	0.20	100	100	8.5
pJS23	0.16	29	27	8.0
pJS12	0.02	76	42	5.8
pJS2	0.04	30	18	5.3

[a]The ratio of the peak luminescence measured from cells grown at 37° to the peak luminescence measured from cells grown at 20°. The absolute levels of luminescence at 20° were about the same for pTB7 and pJS12, but only about 0.25% of that level for pJS2. Cells carrying pJS23 attained about 30% of the wild-type level of luminescence at 20°, but other experiments showed that this was due to decreased mRNA stability in this construction (4).

[b]The relative levels of α subunit accumulation were determined from extracts of cells grown at each temperature. The level of α subunit in extracts of cells carrying the wild-type plasmid at 37° was about 1/2 that in cells carrying the wild-type plasmid at 20°.

[c]Centrifuged lysates in 0.2 M phosphate, 0.5 mM DTT, pH 7.0, of cells grown at 20° were incubated at 37°. Luciferase activity assays were performed over a 24 hour period.

These data show that the wild-type luciferase appears to be somewhat temperature sensitive in synthesis, folding and/or assembly, since about 5 times more active luciferase is produced in cells grown at 20° than at 37°. The enzyme encoded by pJS23 is comparable to the wild type in its ability to assemble *in vivo*. The decreased expression of pJS23 relative to the wild type appears to be the result of decreased mRNA stability in this construction (4). However, the enzyme encoded appears to be unaltered in its ability to fold into an active enzyme. The enzymes encoded by plasmids pJS12 and pJS2 appear to be very interesting. Both appear to be somewhat less stable than the wild type during prolonged incubation at 37°, but neither encodes an enzyme capable of folding into active luciferase during growth at 37°. Cells carrying pJS12 produce wild-type levels of luciferase when grown at 20°, while cells carrying pJS2 produce less than 1% of the wild-type level of active enzyme, even though the β subunits of the luciferases encoded by pJS12 and pJS 2 differ by only 2 amino acids in length (see Fig. 1).

Even though the apparent stability of the altered enzymes during incubation at 37° suggested that the truncations had not rendered them temperature sensitive, additional stability studies were carried out with pure enzymes. Luciferases were purified from cells carrying each of the 4 plasmids, the wild type and the 3 variants. The specific activity and reduced flavin binding affinity of the four enzymes were determined and found to be the same, within experimental error. The rates of thermal inactivation of the wild type and 3 mutant enzymes were determined at 45°, 50°, 55° and 60°; the enzymes encoded by pTB7 (wild type) and pJS23 had about the same apparent first order rates of inactivation, while the enzymes encoded by pJS12 and pJS2 were similar, and had rate constants for inactivation only slightly faster than the wild type (0.55 min^{-1}, compared with 0.40 min^{-1} and 0.32 min^{-1} for pJS23 and pTB7, respectively, at 60° in 0.50 M phosphate, pH 7.0). These results suggested that the differences in production of luciferase from the various plasmids were not due to differences in the thermal stability of the folded products. It is especially interesting to note that the enzyme encoded by pJS2, which accumulates even at 20° to less than 1% of the level of the wild type and pJS12 enzymes, has the same specific activity and is inactivated at the same rate at elevated temperatures.

An alternative possibility explaining the differences in accumulation of the various luciferases could be differences in protease lability. Wild-type luciferase has been shown to be exquisitely sensitive to proteolytic inactivation (12), and it is known that certain mutations actually enhance the sensitivity of luciferase to a variety of proteases (13). The apparent first order rates of inactivation of the luciferases by trypsin and by chymotrypsin were determined as described (14) and found to be essentially the same, demonstrating that the differences in accumulation of the enzymes were not caused by differences in protease lability.

The final possibility, that of temperature sensitivity in the actual folding/assembly process, was investigated by determining the extent of recovery of activity at 20° and at 37° following denaturation in 8 M urea. The results are presented in Table III. For ease of comparison with the data presented in Table II, the results are presented as a ratio of the maximum activity recovered at 37° to that at 20°. As can be clearly seen from these data, the efficiencies of the refolding/assembly processes from urea and the folding/assembly processes *in vivo* (Table II) have comparable temperature coefficients. The wild-type enzyme assembled with higher efficiency at 20° than at 37°, both *in vivo* and *in vitro*. Likewise, the truncated variants encoded by pJS12 and pJS2 showed a much greater efficiency of assembly at 20° than at 37°, both *in vivo* and *in vitro*. It is interesting to note that the absolute level of activity recovered at 20° for the enzyme encoded by pJS2 was significantly greater for renaturation from urea than for folding *in vivo*. Upon refolding from urea, the pJS2 enzyme gained about 10% of the wild-type activity, significantly greater than the ca. 0.25% *in vivo*, while the pJS12 enzyme regained about 75% of the wild-type level, the same as *in vivo*.

Table III. Comparison of Renaturation at 37° with Renaturation at 20°

Plasmid	Ratio[a] (37° Peak/20° Peak)
pTB7	0.34
pJS23	0.29
pJS12	0.09
pJS2	0.02

[a]Purified luciferases were denatured by dilution into urea-containing buffers. The enzymes were renatured by dilution (final enzyme concentration 0.2 μg/mL) into buffers containing 0.2% bovine serum albumin, 0.5 M phosphate, 1 mM EDTA, 1 mM DTT, pH 7.0. Following incubation at either 20° or 37°, the recovered activity was determined relative to an identical control that had been diluted into buffer without urea.

The results of these experiments suggest several interesting interpretations. First, it is clear that the precise amino acid sequence of the carboxyl terminal 10-12 residues of the β subunit has little or no effect on the structure, stability, folding or activity of the enzyme. Second, it is apparent that deletion of this region of the β subunit has no effect on the structure, stability or activity of the enzyme, but it does have a dramatic impact on the efficiency with which the active heterodimer assembles, both *in vivo* and *in vitro*. This interpretation suggests that the carboxyl terminal region of the β subunit must play some critical function *during* the folding process, perhaps in stabilizing some intermediate, but this function cannot rely directly on the amino acid sequence of the carboxyl terminus. It is possible that it is not the sequence *per se*, but rather the secondary structure which it assumes which is critical in the folding process. It is interesting to note that this is the last portion of the subunit to be released from the ribosome during synthesis *in vivo*, and it still has a critical function in the folding process.

To test these two interpretations, several point mutations were introduced (separately) into the 3' end of the *luxB* gene by site directed mutagenesis. These mutations were βD313P, βA317P, βV320P, and βH323P. The introduction of a prolyl residue was intended to disrupt secondary structure within the surrounding residues. Cells carrying plasmids with these constructions were grown at 20° and at 37°. Mutations βA317P, βV320P, and βH323P appeared to be without effect on the enzyme, while the mutant βD313P appeared to be very similar in all respects to the mutant pJS12 (see Table II). The enzyme βD313P was purified and shown to have the same specific activity as the wild-type enzyme. As with the other mutant enzymes, the βD313P enzyme was found to be as stable as the wild-type enzyme to thermal denaturation and proteolytic inactivation. Furthermore, as with the pJS12 mutant, the βD313P enzyme was temperature sensitive in refolding from urea.

These results support the conclusion that the secondary structure of a portion of the carboxyl terminus of the β subunit is critical for correct folding and assembly of the luciferase heterodimer. Substitution of prolyl residues at positions 317, 320 and 323 was without effect, but substitution of a prolyl residue at position 313 caused a dramatic effect on the efficiency of heterodimer assembly at 37°. The three-dimensional structure of the luciferase heterodimer is not known, so no insight regarding potential intermediate folding structures can be gained from inspection of the native structure. However, inspection of the three mutants, pJS12, pJS2, and βD313P, suggests several possibilities. The primary difference between pJS12 and pJS2 is that the enzyme encoded by pJS2 is 2 residues shorter than that encoded by pJS12. The enzyme encoded by pJS12 behaves like the wild type at 20°, while the pJS2 enzyme is seriously folding-deficient even at 20°. It is possible that the structure(s) critical for folding and assembly form from residues that precede the point of truncation in pJS2, and that the critical structure(s) become unstable as a result of the truncation. The effect of the substitution βD313P, then, could be similar to the effect of the truncation. It should be pointed out that other deletions were made in addition to those reported here (4). Deletion of as many as 16 residues did not further decrease the luciferase activity *in vivo* compared with pJS2.

It is difficult to imagine that such effects could be strictly thermodynamic. Rather, it would appear that these lesions must have kinetic effects on the folding process, such that under specific conditions, the correct folding pathway must become kinetically inaccessible. However, no measurements have been made of the kinetics of the folding/assembly process of the luciferase enzyme.

CONCLUSIONS

The experiments reported here demonstrate that the luciferase subunits fold independently to form structures that do not subsequently interact to form active heterodimers. The fact that successful dimerization requires that the two subunits fold together suggests that the encounter complex formed between the two subunits must involve intermediates on the folding pathway. Such an assembly mechanism, involving a second order binding reaction between intermediates, suggests that lesions or even growth conditions causing an alteration in the kinetics of folding of one or both of the two subunits might cause a dramatic alteration in the efficiency with which the two subunits assemble, simply by altering the concentrations of the interaction-competent species. Deletion of 11 (pJS2) or 9 (pJS12) residues from the carboxyl terminus of the β subunit, or substitution of a prolyl residue for the aspartyl residue at position 313 of the β subunit, had a dramatic effect on the efficiency of dimer assembly at 37°. These and other mutations demonstrate that the carboxyl terminus of the β subunit is not important in determining the structure, stability, or activity of the finally folded luciferase enzyme, but this region of the protein appears to play a crucial role in directing the assembly process, perhaps through some transient interaction that occurs during the process of folding and assembly of the active heterodimer.

ACKNOWLEDGEMENTS

This research was supported in part by grants from the National Science Foundation (NSF-DMB 87-16262) and the Robert A. Welch Foundation (A865). The manuscript was prepared while the author was a guest in the laboratory of Michel Goldberg, Institut Pasteur, Paris, and supported by a Fogarty Senior International Fellowship (F06 TW01606).

REFERENCES

1. M. M. Ziegler and T. O. Baldwin, Biochemistry of bacterial bioluminescence, in: "Current Topics in Bioenergetics," Vol. 12, pp. 65-113, Academic Press, New York (1981).
2. T. W. Cline and J. W. Hastings, Mutationally altered bacterial luciferase. Implications for subunit functions, Biochemistry 11: 3359-3370 (1972).
3. J. J. Waddle, T. C. Johnston and T. O. Baldwin, Polypeptide folding and dimerization in bacterial luciferase occur by a concerted mechanism *in vivo*, Biochemistry 26: 4917-4921 (1987).

4. J. Sugihara and T. O. Baldwin, Effects of 3' end deletions from the *Vibrio harveyi luxB* gene on luciferase subunit folding and enzyme assembly: Generation of temperature-sensitive polypeptide folding mutants, Biochemistry 27: 2872-2880 (1988).
5. T. O. Baldwin, T. Berends, T. A. Bunch, T. F. Holzman, S. K. Rausch, L. Shamansky, M. L. Treat and M. M. Ziegler, The cloning of the luciferase structural genes from *Vibrio harveyi* and the expression of bioluminescence in *Escherichia coli,* Biochemistry 23: 3663-3667 (1984).
6. K. Wei, Site specific mutagenesis study of the protein folding process of luciferase, M. S. thesis submitted to the Office of Graduate Studies, Texas A&M University, College Station, Texas (1990).
7. L. H. Chen and T. O. Baldwin, Random and site-directed mutagenesis of bacterial luciferase: Investigation of the aldehyde binding site, Biochemistry 28: 2684-2689 (1989).
8. A. Gunsalus-Miguel, E. A. Meighen, M. Ziegler-Nicoli, K. H. Nealson and J. W. Hastings, Purification and properties of bacterial luciferases, J. Biol. Chem. 247: 398-404 (1972).
9. J. W. Hastings, T. O. Baldwin and M. Ziegler-Nicoli, Bacterial luciferase: Assay, purification, and properties. in: "Methods in Enzymology," Vol. 57, pp. 135-152, M. Deluca, ed., Academic Press, New York (1978).
10. J. W. Hastings and G. Weber, Total quantum flux of isotropic sources, J. Opt. Soc. Amer. 53: 1410-1415 (1963).
11. D. B. Wetlaufer, Folding of protein fragments, in: "Advances in protein chemsitry", Vol. 34, pp. 61-92, Academic Press, New York (1981).
12. T. O. Baldwin, J. W. Hastings and P. L. Riley, Proteolytic inactivation of the luciferase from the luminous marine bacterium *Beneckea harveyi*, J. Biol. Chem. 253: 5551-5554 (1978).
13. T. F. Holzman, P. L. Riley and T. O. Baldwin, Inactivation of luciferase from the luminous marine bacterium *Beneckea harveyi* by proteases: Evidence for a protease labile region and properties of the protein following inactivation, Arch. Biochem. Biophys. 205: 554-563 (1980).
14. D. Njus, T. O. Baldwin and J. W. Hastings, A sensitive assay for proteolytic enzymes using bacterial luciferase as a substrate, Anal. Biochem. 61: 280-287 (1974).

CHARACTERIZATION OF A TRANSIENT INTERMEDIATE IN THE FOLDING OF DIHYDROFOLATE REDUCTASE

C. Robert Matthews

Department of Chemistry
Pennsylvania State University
University Park, Pennsylvania 16802

Edward P. Garvey

Division of Experimental Therapy
Burroughs-Wellcome Corporation
Research Triangle Park, North Carolina 27709

Jonathan Swank

Medical College of Pennsylvania
Philadelphia, Pennsylvania 19129

INTRODUCTION

The central dogma of molecular biology describes the flow of genetic information from DNA to RNA (transcription) and from RNA to protein (translation). Although the information is stored and expressed in a linear format, the final product spontaneously folds to a unique three dimensional structure which is required for function. The past three decades have seen tremendous advances in our understanding of the processes of transcription and translation, however, rather little is known about the details of how the amino sequence of a protein directs the rapid and efficient folding to the native conformation.

Perhaps this discrepancy can be understood in part by the fact that transcription and translation involve the formation of covalent chemical bonds which require enzymatic catalysis. Thus, the traditional biochemical approach of purification and characterization of the relevant enzymes has been successful in elucidating the elementary steps. In contrast, protein folding involves the formation of noncovalent bonds and proceeds in the absence of any added catalyst for many proteins (1). When this is coupled with the high cooperativity of the folding reaction (i.e., the absence of stable intermediates) observed in in vitro studies, one can appreciate why so much valiant effort over the last 30 years has produced relatively little insight into the mechanism (2).

The rapidly becoming classical technology of genetic engineering now provides access to materials which were unavailable even a few years ago. The opportunity to replace specific amino acids in a protein with others of the naturally occurring set permits one to evaluate their roles in the folding reaction. The recent review by Goldenberg (3) on genetic analyses of protein folding and stability nicely illustrates the power of this method.

We have applied this approach to the study of the folding of dihydrofolate reductase (DHFR) from E. coli, a relatively small (M_r 17,680) globular protein which undergoes reversible unfolding when exposed to high concentrations of urea or guanidine HCl. Previous studies in our laboratory on wild type DHFR have established a folding model which serves as a basis for the interpretation of the effects of mutations on the mechanism (4,5). This model suggests that a transient folding intermediate appears after a few hundred milliseconds before eventually proceeding to the native conformation. This species gives rise to a nonmonotonic change in the fluorescence intensity which was explained in terms of the burying of one or more tryptophan residues in a hydrophobic cluster.

In this paper we report the results of further studies on this folding intermediate, including a test of the hydrophobic cluster explanation and the assignment of this reaction to a specific tryptophan residue via genetic engineering. A more detailed discussion of the results and a description of the experimental methods can be found elsewhere (6).

RESULTS

In a previous study in our laboratory on the mechanism of folding of DHFR (4), it was found that the fluorescence intensity due to tryptophan residues does not change in a progressive way during refolding. Specifically, the intensity of the emission above 325 nm first increases above that of the unfolded form with a relaxation time of several hundred milliseconds. The intensity then decreases via a series of exponentials to a value for the native conformation which is less than of the unfolded form (Figure 1). This nonmonotonic change was interpreted in terms of a transient intermediate in which one or more tryptophan residues (out of five in E. coli DHFR) is sequestered in a hydrophobic environment and protected from solvent quenching mechanisms.

The possibility that only one tryptophan residue is involved in this intermediate was raised by the results of a mutagenic analysis of position 75 in strand D in DHFR (7). The replacement of Val 75 by either His or tryptophan dramatically reduces the amplitude of this fluorescence phase. Because the preceding residue is Trp 74, we speculated that it might be the primary contributor to this phase. When this tryptophan is replaced by phenylalamine the stability of DHFR is reduced by 1.2 kcal/mol, however, the mutant protein is active and undergoes a cooperative unfolding transition which is well described by a two state model (Figure 2). Examination of the transient fluorescence response in refolding for the W74F mutant shows that the intensity now decreases in a monotonic fashion (Figure 1). Recent folding studies on the W74L mutant show the same type of response (B. Finn, unpublished results). Thus, it appears that Trp 74 is responsible for the nonmonotonic change in fluorescence intensity observed for the wild type protein.

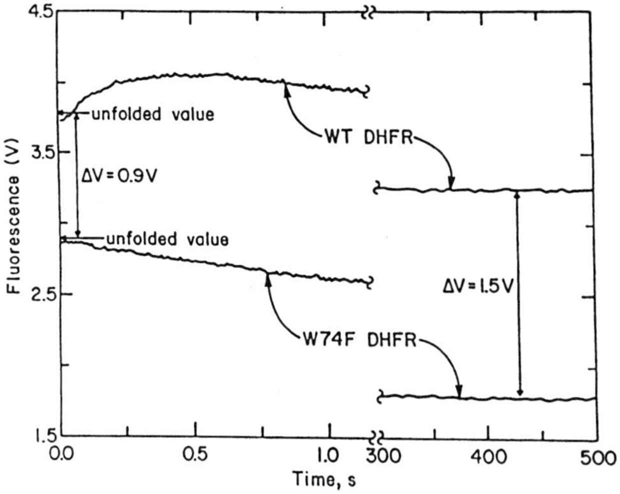

Figure 1. Time dependence of the fluorescence intensity observed for the refolding of the wild type DHFR (upper trace) and the W74F mutant DHFR (lower trace) at pH 7.8 and 15 C. The proteins were initially unfolded in 5.4 M urea and refolded into 1.0 M urea. The buffer contained 10 mM potassium phosphate, 0.2 mM K_2EDTA and 1 mM 2-mercaptoethanol. The samples were excited at 290 nm and the emission above 325 nm was detected on a stopped-flow spectrometer. This figure was taken from reference 6 with permission.

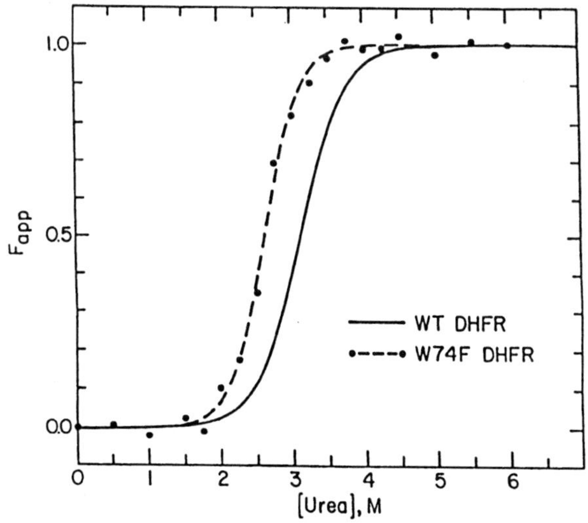

Figure 2. Apparent fraction of unfolded protein, F_{app}, as a function of the urea concentration for the wild type DHFR (solid line) and the W74F mutant protein (dashed line and data points) at pH 7.8 and 15 C. The calculation of F_{app} from the optical data is described elsewhere as is the free energy difference between native and unfolded forms (4). This figure was taken from reference 6 with permission.

This conclusion is a bit surprising when one examines the x-ray structure of DHFR (Figure 3 and reference 8). Trp 74 is indeed buried in a hydrophobic pocket in the native conformation, however, one of its nearest neighbors is Trp 47. Perhaps the loss in amplitude of the early fluorescence phase is actually due to Trp 47 either solely or in part and the replacement of Trp 74 with Phe (or Leu) quenches the amplitude. To test this hypothesis, we examined the behavior of a mutant in which Trp 47 was replaced by Tyr. The fluorescence intensity once again increases and then decreases as is the case for wild type DHFR (Figure 4). We also examined the mutant in which Trp 22 is replaced by histidine and observed the same behavior (Figure 4). In light of these additional results, we conclude that Trp 74 is indeed the source of the increase in fluorescence intensity and a marker for the presence of the associated intermediate.

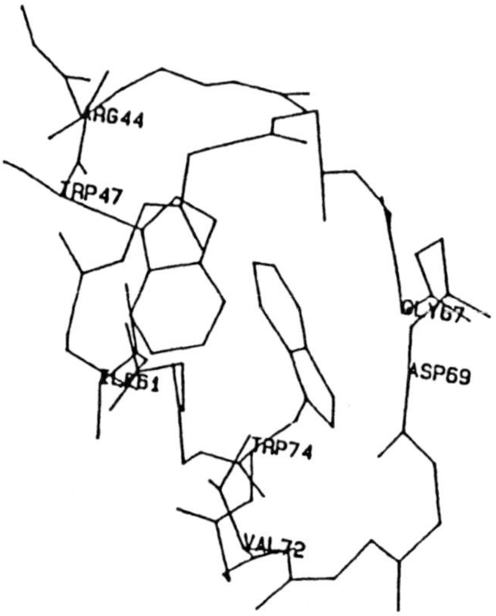

Figure 3. Representation of the local structure around Trp 74 in DHFR from E. coli (8). This figure was taken from reference 6 with permission.

To test the previous assumption that the increase in fluorescence intensity reflects the burying of Trp 74 in a hydrophobic pocket, we studied the effects of ionic quenching agents on the intensities of the various phases detected by fluorescence spectroscopy. In addition to the early phase which will now be designated as the tau 5 phase, four other slower phases appear during refolding. These phases are thought to lead from a set of intermediates to native or native-like forms which are capable of binding methotrexate, an inhibitor which binds at the active site (4). The relaxation times for these refolding phases range from 1 to several hundred seconds and all depend on the final urea concentration. These phases have been designated as tau 1 - tau 4, from the slowest to the fastest. Ionic quenching agents such as Cs^+ and I^- should be excluded from hydrophobic clusters and have a smaller effect on tryptophan residues which are buried in such clusters than on those with greater solvent exposure.

The effects of 0.2 M KI and 0.2 M CsCl on the relative intensities of the tau 5, tau 4 and tau 2 phases are shown in Figure 5; 0.2 M KCl

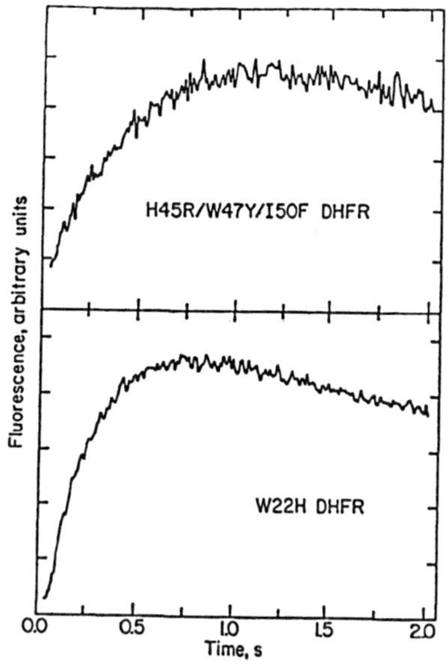

Figure 4. Time dependence of the fluorescence intensity during the refolding of the H45R/W47Y/I50F mutant DHFR (upper panel) and the W22H mutant DHFR (lower panel). The proteins were originally unfolded in 5.4 M urea and refolded into 1.0 M urea. Solution conditions were as described in Figure 1.

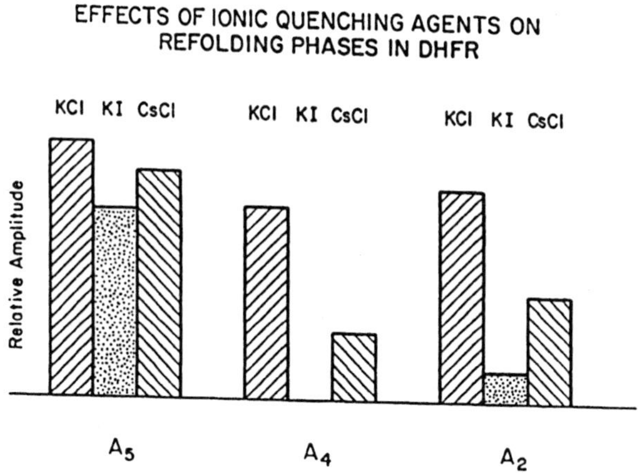

Figure 5. The relative amplitudes of the tau 5, tau 4 and tau 2 refolding phases in wild type DHFR in the presence of 0.2 M KCl (/////), 0.2 M KI (:::::) and 0.2 M CsCl (\\\\\). Solution conditions were as described in Figure 1.

serves as a control for ionic strength. Although I^- and Cs^+ cause a small decrease in the intensity of the tau 5 phase, they cause much more dramatic decreases in the intensities of both the tau 4 and tau 2 phases. The more potent effect of I^-, relative to Cs^+, may reflect local electrostatic properties near the tryptophan residues in the native and native-like intermediates. Thus we conclude that the tau 5 reaction indeed corresponds to the sequestering of Trp 74 in a hydrophobic cluster, early in the folding of DHFR.

DISCUSSION

The delineation of a protein folding mechanism requires detailed information on the structures of transient intermediates which appear on the folding pathway. One such intermediate in DHFR from E. coli involves the burying of one of the five tryptophan residues, Trp 74, in a hydrophobic cluster. This folding reaction occurs in the millisecond time scale and precedes the rate limiting steps which lead to the native conformation. Thus, one might suggest that the early appearance of this cluster plays a critical role in directing subsequent events.

A puzzling aspect of these results is why only Trp 74 contributes to the tau 5 phase detected by fluorescence spectroscopy. All five of the tryptophan residues are buried to one extent or another in hydrophobic clusters in the native conformation and several, including Trp 47, have a smaller solvent exposure than Trp 74. If the differential effects of the ionic quenching reagents can be taken as a monitor of solvent exposure in the intermediate, then one might propose that Trp 74 is in a nonnative environment which provides lesser exposure than is ultimately realized in the native conformation. Why Trp 74 is unique in this regard may be found be examination of the crystal of DHFR (8). Trp 74 is the only tryptophan whose peptide bond is involved in a beta strand; the remaining tryptophans are found in helices, loops or turns. If the formation of the hydrogen bonding network of the beta sheet is a critical part of the formation of the hydrophobic cluster then Trp 74 may be pulled into the cluster rather more tightly than tryptophan residues which dock on this sheet.

Obviously, more experiments are required to prove the validity of this hypothesis. What is clear at this stage is that we now have a spectroscopic probe for the formation of an important structure which appears early in the folding of DHFR. By making specific amino acid replacements at neighboring and distant positions in the three dimensional structure, we can identify other residues which also play a role in this species. Thus, we expect that genetic engineering will provide a means of mapping out the tertiary structure in a folding intermediate whose conformation was hither-to-fore inaccessible. The data obtained should enhance the possibility of understanding the mechanism by which the amino acid sequence of a protein determines its structure. We would then be in a position to extend the central dogma of molecular biology to include the folding of a polypeptide to its unique three dimensional structure.

REFERENCES

1. Fisher, G. and Schmid, F. X. (1990) Biochemistry 29, 2205-2212.
2. Kim, P. S. and Baldwin, R. L. (1982) Annu. Rev. Biochem. 51, 459-489.
3. Goldenberg, D. P. (1988) Annu. Rev. Biophys. Biophys. Chem. 17, 481-507.
4. Touchette, N. A., Perry, K. M. and Matthews, C. R. (1986) Biochemistry 25, 5445-5452.
5. Perry, K. M., Onuffer, J. J., Touchette, N. A., Herndon, C. S., Gittelman, M. S., Matthews, C. R., Chen, J.-T., Mayer, R. J., Taira, K., Benkovic, S. J., Howell, E. E. and Kraut, J. (1987) Biochemistry 26, 2674-2682.

6. Garvey, E. P. and Matthews, C. R. (1989) Proteins 6, 259-266.
7. Garvey, E. P. and Matthews, C. R. (1989) Biochemistry 28, 2083-2093.
8. Bolin, J. T., Filman, D. J., Matthews, D. A., Hamlin, R. C. and Kraut, J. (1982) J. Biol. Chem. 257, 13650-13662.

FROM BIOLOGICAL DIVERSITY TO STRUCTURE-FUNCTION ANALYSIS: PROTEIN ENGINEERING IN ASPARTATE TRANSCARBAMOYLASE

James R. Wild, Janet K. Grimsley, Karen M. Kedzie and Melinda E. Wales

Department of Biochemistry and Biophysics
Texas A&M University System
College Station, Texas 77843-2128

SUMMARY

Aspartate transcarbamoylase (ATCase, EC 2.1.3.2) is a common enzyme which catalyzes the first unique step in pyrimidine biosynthesis in divergent biological systems; however, it possesses tremendous architectural variety from one organism to another. For example, the *E. coli* ATCase holoenzyme is comprised of two catalytic trimers and three regulatory dimers, while the mammalian enzyme is part of a multifunctional protein aggregate encoding the preceding and subsequent enzymes in pyrimidine biosynthesis. Despite extreme differences in quaternary architecture and enzymatic organization, protein engineering studies have demonstrated the existence of highly conserved units of protein structure that impart specific functional characteristics.

1. The largest of these units are discrete polypeptides or superdomains within multifunctional proteins which have been shown to be uniquely involved in specific catalytic steps within the CAD or CA complexes of eukaryotic pyrimidine biosynthesis.
2. The catalytic polypeptides of various ATCases are organized into two discrete and separable binding domains for its substrates, carbamoyl phosphate and aspartate.
3. The regulatory polypeptides of the enteric bacterial enzymes also contain two discrete tertiary domains, the Allosteric Binding Domain and Cys^4 coordinated Zinc Domain involved in the protein:protein interface between the regulatory and catalytic polypeptides of the holoenzyme.
4. There are sub-domain structural units within the various polypeptides which have a coordinated impact on specific catalytic and regulatory functions in the enzyme.
5. Finally, it has been possible to ascribe some individual function to specific amino acids relative to ligand binding, zinc coordination, protein:protein interactions, and the structural reorganizations in the T-R transition of the enteric holoenzymes.

Comparisons of the catalytic sequences of various ATCases have revealed substantial conservation of primary and predicted secondary structures. Based upon the sequence alignment and the *E. coli* ATCase crystal structure, the hamster ATCase superdomain tertiary structure has been modeled with interactive computer graphics. The predicted conservation of structure/function relationships were verified experimentally through the construction of active catalytic bacterial/hamster chimeric enzymes. It has also been possible to verify the apparent structural homologies of ATCase and ornithine transcarbamoylase (OTCase, 2.1.3.3), the parallel enzyme from arginine biosynthesis, by exchange of the amino acid binding domains of the two enzymes. Extending these observations and protein engineering philosophies into the formation of hybrid and chimeric enzymes has resulted in the production of ATCases with altered catalytic and regulatory characteristics.

Chemical Aspects of Enzyme Biotechnology
Edited by T.O. Baldwin *et al.*, Plenum Press, New York, 1990

INTRODUCTION

Aspartate transcarbamoylase (carbamoylphosphate: L-aspartate carbamoyl-transferase, ATCase, EC 2.1.3.2) provides the first unique step in de novo pyrimidine biosynthesis and is associated with a variety of multimeric aggregates and multifunctional enzyme complexes in various organisms [1]. The enteric bacteria, including *E. coli*, possess a structurally complex, allosterically regulated enzyme composed of catalytic and regulatory polypeptides. In eukaryotic systems, ATCase is incorporated into multifunctional enzyme aggregates. For example, the "CAD" complex from animal systems is a trifunctional polypeptide complex of 800,000-1,200,000 daltons which includes ATCase, carbamoyl phosphate synthetase, (CPSase, EC 6.3.5.5) and dihydroorotase (DHOase, EC 3.5.2.3), the first three enzymes in pyrimidine biosynthesis.[2-5] Eukaryotic microorganisms possess a similar complex, but appear to have only two functional units ("CA")[6-9] although yeast has a translated, non-functional polypeptide sequence which is positionally correlated to "D".[8] The potential advantages of such multifunctional proteins include the coordinate regulation of related catalytic activities, oligomeric protein complex stability, intracellular localization, and efficiency of substrate channelling.

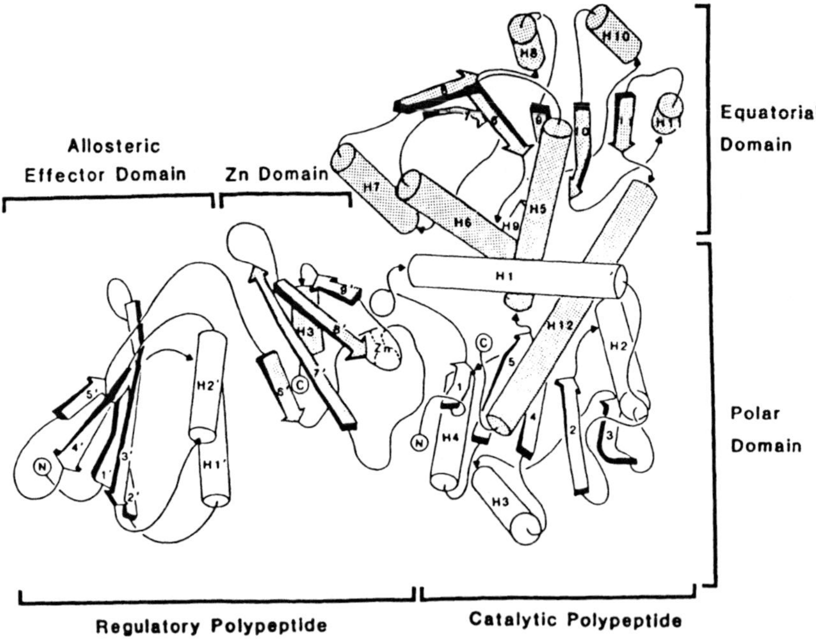

Figure 1. Schematic representation of the structural organization of the catalytic and regulatory polypeptide chains of ATCase from *E. coli* as determined by X-ray crystallography.[1,10]

The basis for the protein engineering studies described in this presentation involves the detailed structural analyses of the *E. coli* ATCase in various ligand associated states which have been determined to approximate 2.5 Å resolution by W.N. Lipscomb and his students and associates over the past twenty-five years.[10-14] The holoenzyme is a dodecamer, consisting of two catalytic trimers (c_3) and three regulatory dimers (r_2). Individual catalytic trimers are capable of unregulated catalysis with the two substrates, aspartate and carbamoyl phosphate. The catalytic polypeptide is divided into two tertiary structural domains with separate function (Fig. 1).

Table 1. Characterization of Native ATCases.

ATCase Type[a]	Examples of Organisms	Enzyme Regulation
Class A $(c\text{-}r)_2$	*Pseudomonas* *Rhizobacter*	UTP Inhibition
Class B1 (c_6r_6)	*Escherichia coli* *Salmonella typhimurium* *Shigella flexerni*	CTP Inhibition ATP Activation UTP/CTP Synergism[b]
Class B2	*Yersinia intermedia*	CTP Inhibition UTP Inhibition ATP Activation
Class B3	*Erwinia carnegiana* *Erwinia herbicola*	CTP s-Inhibition[c] No Activation UTP/CTP Synergism
Class B4	*Yersinia enterocolitica*	No Inhibition ATP Activation
Class B5	*Yersinia kristensenii* *Yersinia frederiksenii*	No Inhibition No Activation
Class B6	*Serratia marcescens*	No Inhibition CTP Activation ATP Activation UTP Counteracts CTP Activation[d]
Class B7	*Proteus vulgaris*	CTP Activation ATP Activation UTP/CTP Inhibition
Class C (c_3)	*Bacillus subtilis*	No allosteric regulation
Eukaryotic A (c_3) (CA or CDA?)	plants	UTP Inhibition
Eukaryotic CA (CD•A?)	*Saccharomycetes* *Neurospora*	UTP Inhibition - CPSase UTP Inhibition - ATCase
Eukaryotic CDA	*Drosophila* *Dictyostelium* mammals	UTP Inhibition - CPSase No Regulation of ATCase

[a] There are five general categories of ATCases: three bacterial types and two eukaryotic multienzyme complexes including CPSase ("CA") and CPSase plus DHOase ("CAD"). There are seven different species of class B enzymes based upon allosteric control patterns and catalytic parameters.
[b] UTP has no independent effect but serves to enhance or promote feed-back inhibition in synergism with CTP.
[c] s-Inhibition = slight inhibition (15-20%).
[d] No inhibition occurs with CTP or UTP, but UTP reduces CTP activation.

```
             1                                                      50
                        • •                                          ΔΔΔΔ
Escherichia MANPLYQKHI ISINDLSRDD LNLVLATAAK LKANPQPELL KHKVIASCFF EASTRTRLSF ETSMHRLGAS VVGFSDSANT
Serratia          H                E       G                                                  A GS
Salmonella
Proteus     .T               D E   EC  RV D  QQ NN       E                      AI            A GS
Bacillus    ......M  L TTMSE  TEE IKDL Q  QE   SGKTDNQ  TG FA NL    P     F    VAEKK  MN  L.NL GTS

                                  100                                                  150
             Δ  Δ  •••                  Δ•••• ••  ••  ••             •••Δ  Δ
Escherichia SLGKKGETLA DTI.SVISTYV DAIVMRHPQE GAARLATEFS GNVPVLNAGD GSNQHPTQTL LDLFTIQETQ GRLDNLHVAM
Serratia                  .                    . M S                  .                      S SI
Salmonella                .                               Q                                       I
Proteus                   . LR A       I            A A   DI          A                K         NI
Bacillus     V.Q     Y    RTLE IG   VC I  SED EY.YEELVSQ V I I       CG     S     M  Y EF NTFKG T SI

                                                         200
                   Δ        •                         •   •  •                           Δ Δ      ••
Escherichia VGDLKYGRTV H.SLTQALAK FDGNRFYFIA PDALAMPQYI LDMLDEKGIA WSLHSSIEEV MAEVDILYMT RVQKERLDPS
Serratia               .           E                A    K E  E Y S G         VP L
Salmonella              FALPRT     S                             M      G       D.
Proteus                 . A        T HL       KV    EH   HL E N VE Y Q ETLD    P L
Bacillus    H I HS .   AR NAEV TR L. A VL SG   ......SEW Q... ENTF GTYV.MD A  ES.S VVMLL  I N  HQSA

                          250                                                    300
              ••    •                        ΔΔ
Escherichia EYANV...KA QFVLRASDLH NAKANMKVLH PLPRVDEIAT DVDKTPHAWY FQQAGNGIFA RQALLALVLN RDLVL....
Serratia                   A . A G . L           I              Y              S .    V A A .....
Salmonella       ...        P. N G RQ            I   T                        A           SE S .....
Proteus          ...       TTA T HV D L I        I   T         Y Y         Y              KT .....
Bacillus    VSQEGYLN Y GLTVERAERM KRH IIMHPA  VN GV  DD SLVESEKSRI  K MK  V I  M VIQCA Q TNVKRGEAAY

              325

Escherichia ....
Serratia    ....
Salmonella  ....
Proteus     ....
Bacillus    VISH
```

Figure 2. Comparison of the deduced amino acid sequences of the catalytic polypeptides of ATCase from various enteric bacterial sources. (The triangles indicate residues implicated in substrate binding and the closed circles indicate residues implicated in d:r interactions. Modified from Wild and Wales.[1])

The "Carbamoyl Phosphate Binding Domain" (Polar Domain) is composed of the amino terminal 150 residues and the carboxy terminal helix (H-12) while the "Aspartate Binding Domain" (Equatorial Domain) is confined to amino acids 160-295. The three active sites in each trimer are formed at the interfaces between adjacent catalytic polypeptides, and the closure of the two domains of the catalytic polypeptide provides the necessary tight binding of substrate ligands for catalysis[10-12].

All of the ATCases from the enteric bacteria examined thus far possess the architecture of this dodecameric holoenzyme of *E. coli*, yet the regulatory characteristics of the enzymes can differ dramatically (Table 1). CTP is an inhibitor of the *E. coli* enzyme but has been shown to activate the ATCases of some enzymes and have no effect on others.[15-19] The enzymes from the Tribe 1 *Enterobacteriacae* are very similar to the *E. coli* enzyme with cooperative substrate kinetics, half-maximal aspartate requirements of 4.0 to 7.0 mM, allosteric inhibition by CTP and activation by ATP. In contrast, the enzyme of *Erwinia herbicola* (Tribe 4) has slight CTP inhibition and no ATP effect, while the *Serratia marcescens* (Tribe 2) and *Proteus vulgaris*(Tribe 3) enzymes[17] have half-maximal aspartate requirements of 20 mM to 30 mM and each is activated by both ATP and CTP. Furthermore, the ATCases from *Yersinia kristensenii* and *Yersinia frederiksenii* (Tribe 5) show no allosteric regulation even though the holoenzyme organization is apparently maintained.[18] The divergent regulatory characteristics of these enzymes in spite of their architectural similarities provide a unique opportunity for a comparative evaluation of the structural basis of these functional differences. The *pyrB* cistron of *E. coli* encodes a polypeptide of 311 amino acid residues initiated by an NH_2-terminal methionine that is not present in the catalytic polypeptide. The amino acid sequence of the *E. coli* catalytic polypeptide is compared with those of *S. marcescens*, *S. typhimurium*, and *P. vulgaris* (Figure 2)[1,19]. The c200-220 and the c250-265 regions contain the largest concentration of differences in the catalytic polypeptides. The c200-220 region appears to be involved in the intersubunit r:c interaction in the *E. coli* ATCase[10] while the c250-265 region is involved in domain closure[12]. Based on a Dayhoff analysis, most of the other differences represent conservative changes while maintaining size, charge, or hydrophobicity of the residues. With the exception of residue c267 in the *Bacillus subtilis* enzyme sequence (a class C ATCase),[20,21] all of the residues implicated in substrate binding are conserved throughout the available bacterial sequences. The ATCase from *Pseudomonas fluorescens*[22] is of great interest in that it has been reported to have regulatory and catalytic sites contained in a large (180,000 dalton) polypeptide which is functional as a dimer. In contrast, a recent abstract report by David Evans suggests that the enzyme may be a complex oligomer.[23] The pyrB gene has been cloned by several laboratories; however, no sequence or detailed biochemical analyses have been reported to date.

The genes encoding ATCase and OTCase (pyrBI and argI) are located within 3000 bases of one another at 96.5 minutes on the *E. coli* chromosome. Some strains of *E. coli* K-12 contain an additional OTCase gene, *argF*, located at 6.5 minutes; however, it is absent in other strains and other enteric bacteria. The gene products of *argI* and *argF* form true isoenzymes. Both of these OTCases are functional as trimeric catalytic subunits and share extensive amino acid sequence homology with ATCase, particularly in the amino-terminal regions.[24-26] These enzymes appear to have resulted from a functional differentiation in amino acid binding following the duplication of an ancestral carbamoyl transferase. The Carbamoyl Phosphate Binding Domains of these enzymes contain the amino acids proposed to be involved in the binding of the common substrate (CP) in homologous positions within the respective amino acid sequences; they share other amino acid sequence similarities as well. In addition, the predicted secondary structures of the *E. coli* ATCase and OTCase Carbamoyl Phosphate Binding Domains appear to be strongly conserved. As expected, the amino acid binding domains (Aspartate or Ornithine) do not have either primary sequence similarity or retention of secondary structures, except for the extreme carboxy-terminal H-12 region. This appears to reflect the differences in their amino acid specificies and suggests that the C-terminal, H-12 helix could retain a structural role in stabilizing the Amino Acid and Carbamoylphosphate Binding Domain interactions. In contrast, *Pseudomonas aeruginosa* possesses two entirely different OTCase genes which code for an anabolic OTCase (argF) and a catabolic OTCase (arcB).[27] These four OTCase genes are 50-55% homologous when conservative differences are considered. All of the transcarbamoylases studied so far have two highly conserved regions: positions 52-56 (Ser-Thr-Arg-Thr-Arg) and 134-137 (His-Pro-X-Gln) which contain six of the active site residues implicated in carbamoyl phosphate binding by X-ray crystallography of ATCase.

The early steps of pyrimidine biosynthesis in eukaryotic systems are provided by a contiguous arrangement of "superdomains" which provide enzymatic functions arranged as trimeric or hexameric complexes, $(CAD)_3$ or $(CAD)_6$ in metazoa from *Drosophila* to man[2-5,28] or "CA complexes" in yeast and *Neurospora*[7,8]. In the hamster CAD, the allosteric control appears to be associated with the CPSase superdomain[2] and there is no separate regulatory subunit. Several enzymatic studies have suggested that the CAD complex is organized into three distinct superdomains connected by protein bridges in the linear order NH_2-CPSase-DHOase-ATCase-COOH. Limited proteolysis has released an enzymatically active ATCase superdomain of 40 kilodaltons which aggregated into trimeric species from 140,000 to 1,100,000 daltons, corresponding to trimeric aggregates of 3, 6, 18, and perhaps higher multimers. An enzymatically active hamster ATCase superdomain has been isolated and manipulated from a CAD cDNA expression library.[28] The polypeptides aggregated into active trimers similar to the proteolyzed CAD protein[2,5]. A recent comparison of the deduced amino acid sequences of hamster and *E. coli* ATCases[29,30] revealed a 44% similarity with complete conservation of all 12 active site residues implicated from the *E. coli* ATCase crystal structure. In addition, the predicted secondary structure of the hamster ATCase suggested substantial conservation of structure between the two ATCases. Comparative genomic sequencing in the laboratory of Jeff Davidson have identified eight introns positioned through the ATCase superdomain of the chromosomal human gene; there is approximately a 98% sequence homology between the hamster and the human exon sequences (personal communication).

The independent ATCases and the ATCase-containing complexes occupy an important early step in de novo pyrimidine biosynthesis and most are characterized by either direct or indirect participation in allosterically effected, feed-back inhibition by pyrimidine nucleotides. The *B. subtilis* enzyme is not subject to heterotropic ligand effects,[20] while CPSase appears to be the primary target of UTP in the CAD multifunctional complexes[2-7,9], and both CPSase and ATCase are inhibited by UTP in CA microbial eukaryotic systems.[6-8] All of the ATCases from the enteric bacteria are comprised of discrete catalytic and regulatory polypeptides. A sequence comparison of the regulatory chains from the ATCases of the enteric bacteria reveal that sequence divergences are concentrated in three regions:

1) the hydrophobic Allosteric Domain::Zinc Domain interface within a single regulatory polypeptide (an r1:r1 monomeric interaction),

2) the r130s region which is involved in the r1:c1 interface between the Zinc Domain and the Aspartate Binding Domain, and

3) the carboxy-terminus of the regulatory polypeptides (the r1:c4 interface between the Zinc Domain and the Carbamoyl Phosphate Binding Domain).

There is a complete positional conservation of the amino acids implicated in the binding of CTP by X-ray crystallography in each of the sequenced regulatory polypeptides (Figure 3).

The architectural and functional variations in the ATCases from different biological systems (Table 1) provide an opportunity for the comparison of homology and divergence in these ATCases for the formulation of structure-function hypotheses. Due to the similarity of amino acid sequence (from 45 to 95% conserved positional identity) and the conservation of essential structures (virtually 100% for residues directly involved in substrate and allosteric ligand binding), it is possible to use the tools of molecular genetics, aided by computer graphics analysis, to exchange and evaluate specific functional roles for the various structural units. Through the exchange of specific protein subunits, tertiary structural domains, secondary structure units and site-directed mutagenesis, it is possible to alter the efficiency of catalysis, strengthen the allosteric control by nucleotide effectors, and begin to chart the molecular pathway of communication from the allosteric binding sites to the active sites of the enzyme.

```
              1                                                    50
                      Δ   Δ      ΔΔ                                       Δ   Δ
Escherichia   MTHDNKLQVE AIKRGTVIDH IPAQIGFKLL SLFKLTETDQ RITIGLNLPS GEMGRKDLIK IENTFL.SED
Serratia                    C                  T     A               N L              .T Q
Salmonella                  C           V                                             .T E
Proteus           H  K                  V   I     R          V F     ENL  K        V   T

                                                100
                                                           ......  . ....        .       .
                                  Δ     Δ Δ
Escherichia   QVDQLALYAP QATVNRIDNY EVVGKSRPSL PERIDNVLV. CPNSNCISHA EPVSSSFAVR KRANDIALKC
Serratia       AN   M    K             R LTL    DH  G  T.    G    RS    R   S K S GGEVH
Salmonella       N                   D              N    .                     K
Proteus        ANR  M          I D  Q  D MAL     LLED . P          N    Q   R K  L T VV  T

                           150
              .....      .
Escherichia   KYCEKEFSHN VVL.AN
Serratia      R      E Q    Q D
Salmonella             Y    .
Proteus              ERH A IR
```

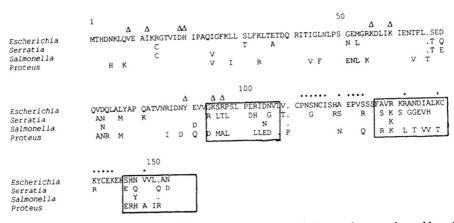

Figure 3. Comparison of the deduced amino acid sequences of the regulatory polypeptides of ATCases from various enteric bacterial sources. (The triangles above the sequence indicate residues implicated in binding the anti-conformation of CTP and the dots represent proposed c:r interactions.[14] Modifed from Wild and Wales.[1])

The following constructions are being used to decipher the molecular role for the various architectural units of ATCase whether existing as separate protein structures ("independent enzymes") or within large multifunctional complexes ("enzyme superdomains"):

1. The ATCase superdomains of both hamster and human CAD complexes have been separated from the other enzymatic structures by limited proteolysis to produce functional trimeric complexes.[2] These independent superdomains have been produced directly from expressed cDNA constructions in *E. coli.*[28]

2. It has been possible to exchange discrete tertiary domains from the ATCase superdomain of the hamster CAD complex (e.g. the Carbamoyl Phosphate Binding Domain or Aspartate Binding Domain) for the same domain from *E. coli* to form functional, but fragile, chimeric trimers.[29-31]

3. Due to the similarity of the Carbamoyl Phosphate Binding Domains, the Ornithine Binding Domain from the *E. coli* OTCase has been replaced by the Aspartate Binding Domain of ATCase to form a chimeric ATCase whose expression was controlled by arginine regulation.[32-33]

4. Hybrid ATCase holoenzymes have been formed from heterologous regulatory and catalytic subunits derived from different enteric bacteria. These hybrid enzymes have been shown to be stable and provide insight to the relative contributions of the two subunits to both regulatory and catalytic function of the enzyme.[19,34-36]

5. Hybrid enzymes have been formed from regulatory and catalytic polypeptides in which the four major tertiary domains of the enteric ATCases have been interchanged and used to form chimeric polypeptides.[32,37,38] The hybrid chimeric enzymes formed from these and other chimeric polypeptides in our laboratory (DeAndra Beck, Karen Foltermann) have proven to be stable; however, their regulatory characteristics are quite variable and require ongoing evaluation.

6. It is possible to exchange sub-domain protein structures (e.g. 8 to 37 amino acids) in order to evaluate the functional contribution of smaller units of the tertiary domains. It may be possible to correlate the importance of the position of these secondary and supersecondary structures in the *E. coli* enzyme with their functional contribution to enzymatic divergence in the enzymes from other enteric bacteria.[39-40]

7. Numerous site-directed mutational changes in our laboratory have been used to characterize the role of catalytic domain closure during the T-R conformational transition of ATCase, the stability and role of the zinc domain in transfer of the allosteric signals, and the ability of the allosteric binding site to discriminate nucleotide effector binding.[41-44] Additional site-directed modifications have been performed in the laboratories of R. Kantrowitz (Boston College), R. Cunin (CERIA, Brussels) and H. Schachman (U.C. Berkeley); however, they will not be discussed here.

These studies produce a variety of data that can be utilized to evaluate the complex structure-function relationships of these ATCases and have led to the definition of a common regulatory logic for the ATCases from varying biological sources.

DIVISION OF THE CAD COMPLEX INTO DISTINCT SUPERDOMAIN AGGREGATES

The CAD complex provides the central metabolic control for de novo pyrimidine biosynthesis in eukaryotic animal systems with CPSase targeted for allosteric inhibition by UTP and activation by phosphoribosyl 5'-pyrophosphate (PRPP). Since the first three catalytic activities and related regulatory functions are carried out by a single polypeptide chain, CAD must have a well defined domain structure. It has been proposed that there are as many as six distinct, functional responsibilities for catalysis or allosteric control which may involve discrete protein domains: the ATCase domain, the DHOase domain, the CPSase domain, a glutamine amidotransferase domain (GATase), and the allosteric regulatory sites.[2] It has been possible to physically separate functionally independent superdomains of the mammalian CAD complexes by proteolytic and genetic isolation.[2,5,28] The CPSase superdomain corresponds to the GATase, ATPase and CP synthetase functions as well as containing the allosteric regulatory control. The central portion of the protein complex can form functional DHOase dimers and the carboxy-terminal region of the complex provides multimeric aggregates for ATCase activity. Comparative DNA sequencing studies have revealed that the deduced amino acid sequences correspond to related enzymatic functions in polypeptides from microbial sources.[45] These independent polypeptide regions in CAD appear to be connected by amino acid linkers of 70-120 amino acids (Jeff Davidson, unpublished observations). Furthermore, it has been possible to remove portions of these linkers from various cDNA constructions without diminishing catalytic activity. In hamster, three of these domains have been characterized using limited protease digestion of the complex and the ATCase activity has been reported to be associated with a 120,000 dalton trimer (3 X 40,000). The *Drosophila melanogaster rudimentary* gene has been shown to encode four enzymatic domains which are linearly arranged in the order: NH_2-GATase-CPSase-DHOase-ATCase-COOH (a CDA complex); DNA hybridization studies between the *rudimentary* gene and hamster CAD sequences have demonstrated that these domains were similarly ordered (also CDA).[4] DNA sequencing studies with the yeast gene have identified a DHOase-like (D*) sequence within the *ura2* complex.[8] This D* region seems neither to encode a DHOase function nor possess extensive DHOase-homology, yet it is positioned within the gene to form a CD*A complex similar to the CAD complex of higher eukaryotes.

CHIMERIC ATCases FORMED BY INTERGENIC FUSION BETWEEN HAMSTER cDNA STRUCTURAL GENES AND THE *E. coli* GENE

It has been possible to construct two chimeric *E. coli*:hamster catalytic polypeptide fusions (chimeric ATCases) which have confirmed the functional homology of the Carbamoyl Phosphate and Aspartate Binding Domains of the two proteins.[29-31] Previous fusions of proteins with discrete functional domains have been successful with bacterial amino acid binding proteins, and between yeast and human yeast phosphoglycerate kinases;[1] however, the chimeric hamster:bacterial domain fusions represented a fusion between more distantly related proteins which possessed a 45% homology.[29,30] Although both fusions were active, both were also unstable as evidenced through Western blots; such chimeras might be sufficiently perturbed in structure to be proteolytically degraded in vivo as seen for other hybrid proteins. The *E. coli*-hamster (Ec-Carbamoyl Phosphate::hamster-Aspartate) chimera appeared to be somewhat more stable than the reverse fusion which contained the polar domain of the native hamster ATCase and the 7,000 dalton amino terminal linker. The linker may destabilize the trimeric association and Davidson et al. has found that deletions within the N-terminal linker of the ATCase produced from the initial cDNA expression plasmid (pAL4) dramatically increased enzymatic activity. The active site region of the hamster ATCase was evaluated by prediction modeling and appeared to be quite similar to that deduced from the PALA-liganded X-ray crystal structure of the *E. coli* enzyme. In addition to the twelve active site residues proposed to interact directly with PALA, other residues in *E. coli* ATCase have been implicated in catalysis based on experimental biochemical and genetic results and each could be appropriated placed by serial energy minimization. Furthermore, all of the residues could be fitted to the presumed active site geometry, supporting the importance of such indirect interactions between active site residues and neighboring amino acids. In fact, 102 of 103 enzyme:active site interactions were conserved between the two enzymes, and 28 of 42 secondary stage interactions with active site residues were identical in both enzymes. The other 14 secondary interactions, despite nonconservative substitutions of side chains, were conserved through alternate, but analogous, associations. These results extended the shell of the conserved catalytic core of the two ATCases well beyond the immediate active site. It should be noted that these interactions were not part of the initial modeling strategy but represent a validation of the model by analogy.[29-31]

From the apparent sequence homology between the hamster CAD ATCase and bacterial counterparts, it would appear that the CAD ATCase superdomain has evolved from a prokaryotic ancestor through a process of gene duplication, translocation and fusion with other enzymatic progenitors. It could be proposed that the ATCase genes have evolved more slowly than 80% of the measured proteins at a rate comparable to the glycolytic enzymes, both of which have crucial metabolic roles which would necessitate a slow rate of evolution.[29-31]

EXCHANGE OF ASPARTATE AND ORNITHINE BINDING DOMAINS IN CATALYTIC CHIMERIC ENZYMES

A chimeric polypeptide has been formed by fusing the appropriate structural regions of the ATCase and OTCase of *E. coli* in a hybrid gene which contained the *argI* promoter, the OTCase Carbamoyl Phosphate Binding Domain (up to ASPc162) and the Aspartate Binding Domain from ATCase (ASPc162 to carboxy-terminus). The resulting chimeric proteins aggregated to form functional ATCase trimers which completely lacked OTCase activity.[32,33] The successful formation of this active chimeric enzyme provided the first example of the direct exchange of genetic cassettes encoding functionally divergent protein domains which have been structurally defined based upon the X-ray crystallographic structure of ATCase and the presumed homology of the common Carbamoyl Phosphate Binding Domains in OTCase and ATCase. While these studies confirm the structure/function relationships in ATCase and OTCase based on primary sequence homology, they also support a model for the production of new biochemical function by the fusion of discrete genetic modules from ancestral genes.

Although the ATCase and OTCase catalytic trimers bear remarkable similarities and a close evolutionary relationship, the ATCase trimer will associate with regulatory dimers to form holoenzyme both in vivo and in vitro, but the OTCase trimer will not. Nonetheless, the chimeric polypeptide [Cotc-carbamoyl phosphate/atc-aspartate domain] assembled into functional catalytic trimers which formed stable holoenzyme with the *E. coli* regulatory subunit. There are a number of potential r:c contact sites which have been lost in the CP Binding Domain of OTCase and it had been assumed that the chimeric trimer would not associate with the regulatory subunit. Nevertheless, a relatively stable holoenzyme was formed.[32] This unexpected observation reinforced the potential role for the interaction between the Zinc Domain and the Aspartate Binding Domain in the r:c domain interaction and further supported the importance of the interaction between the r130s loop of the regulatory subunit with the c200s loop of the catalytic subunit. The chimeric hybrid holoenzyme had cooperative homotropic kinetics with aspartate and was inhibited by CTP; ATP had no effect on either the homotropic substrate saturation characteristics or on the activity of the enzyme.

FORMATION OF HYBRID ENZYMES FROM NATIVE CATALYTIC AND REGULATORY SUBUNITS FROM DIVERGENT BACTERIAL SOURCES

It is possible to form hybrid holoenzymes by mixing purified catalytic and regulatory subunits from varying bacterial sources.[19,34-36] The first such hybrid enzyme was formed in vitro from purified catalytic subunits of *Salmonella typhimurium* and regulatory subunits from *E. coli*; the two native enzymes were very similar and the results were somewhat inconclusive, although the allosteric responses to CTP were intermediate to the parental enzymes.[46] Since that time, this approach has been extended to form several hybrid ATCases from the native regulatory subunits of one bacterial enzyme and the catalytic subunits of another. The formation of over thirty hybrids has resulted in the following observations about the nature of the structure-function relationships of the subunits:

1) Stable hybrid enzymes can be formed both in vitro and in vivo from heterologous subunits expressed from different plasmids or assembled from separated subunits in vitro.
2) The hybrid enzymes demonstrate a pattern of allosteric control that is determined by the nature of the regulatory subunit.
3) Most of the hybrid enzymes formed from native subunits are catalytically efficient as indicated by low $S_{0.5}$ values of 3-8 mM aspartate and the maintenance of catalytic activity.
4) The r:c protein::protein interface (r130s::c200s) provides an important molecular interaction which affects the enzymatic characteristics expected in the T-R conformational transition of the enzyme.

One of the first hybrid enzymes formed (c_3-*Serratia marcescens*::r_2-*E. coli*) was catalytically paralyzed, and the aspartate concentration ($S_{0.5}$) required to promote catalysis increased from 5 mM to 125 mM aspartate.[35] The hybrid holoenzyme was a stable oligomer of 300 kD which possessed homotropic kinetic responses to aspartate and was subject to activation by ATP and inhibition by CTP. In contrast, the alternate construction (regulatory subunits from *S. marcescens* and catalytic subunits from other sources) were efficient for catalysis ($S_{0.5}$-ASP = 3-8mM) and the regulatory subunits determined the nature of the allosteric response (Table 2).

Table 2. The Regulatory Chain Dictates The Allosteric Response in Hybrids

catalytic subunit	regulatory subunit	CTP Effect	ATP Effect
Escherichia coli	*Escherichia coli*	-60%	+40%
	Serratia marcescens	+25%	+30%
	Salmonella typhimurium	-70%	+40%
Erwinia herbicola	*Erwinia herbicola*	-15%	---
	Escherichia coli	-50%	+30%
	Serratia marcescens	+45%	+30%
Serratia marcescens	*Serratia marcescens*	+35%	+50%
	Escherichia coli	-75%	+70%
	Proteus vulgaris	+45%	+50%
Proteus vulgaris	*Proteus vulgaris*	+30%	+110%
	Escherichia coli	-75%	+30%
	Serratia marcescens	+15%	+25%

[a] All enzymes are class B ATCases with catalytic trimer = 100-120 kD. The holoenzymes are 300-330 kD.
[b] Inhibition (-) or activation (+) in the presence of 2 mM nucleotide at the approximate $[S]_{0.5}$ for each enzyme in cell-free extracts.

The nature of the catalytic perturbation of the c_3-Sm::r_2-Ec hybrid enzyme was examined by forming a set of chimeric enzymes which interchanged the CP and ASP Binding Domains of the catalytic chains of *E. coli* and *Serratia marcescens* (c_3-Sm/Ec or c_3-Ec/Sm).[25] Comparison of the deduced amino acid sequences revealed that there were several altered r:c contacts between the CP Binding Domain of the catalytic polypeptide and the Zinc Binding Domain of the regulatory chain.[37,38] In spite of this, these studies have demonstrated that the catalytic paralysis seemed to be related to the interaction of the regulatory subunit and the Aspartate Binding Domain of the catalytic polypeptide (c_3-Ec/Sm::r_2-Ec) (Table 3). Subsequently, this observation was partially explained by the refinement of the native *E. coli* ATCase structure. Comparison of the unliganded T-state and the PALA-associated R-state revealed an important reorientation in the c200s loop (near H7, Figure 1).[10-13] In the less active T-state, ARGr130 formed a salt link with GLUc204 in a r1:c1 interaction which was broken as the Aspartate Binding Domain moved toward the Carbamoyl Phosphate Binding Domain during the T to R transition of the enzyme. In the R-state ARGr130 formed a new association with ASPc200, apparently to stabilize that region of the aspartate domain in concert with the domain closure at the active site. The effects of the loss of that salt link include the loss of the association between TRPc209 and ASPc203 in the c1 catalytic chain.

A similar approach has suggested that the Allosteric Binding Domain has the primary role in determining the pyrimidine nucleotide allosteric signal of the ATCase holoenzyme; however, the Zinc Binding Domain may have some interpretative role in transmitting the structural information to the catalytic subunits. When the allosteric binding domain of the *Proteus vulgaris* regulatory polypeptide (CTP-activating) was genetically fused with the zinc domain of the *E. coli* protein (CTP-inhibiting), the resulting chimeric regulatory dimer formed a stable holoenzyme with catalytic trimers from *Serratia marcescens* (K. Foltermann, W. Lagaly and J. Wild, in preparation). This enzyme possessed homotropic characteristics relative to aspartate and was inhibited by CTP. Several other such hybrid holoenzymes involving r_2-Pv/Sm chimeric polypeptides do not show either homotropic or heterotropic effects although relatively stable holoenzymes appear to be formed. These studies appear to suggest a role for the hydrophobic r1:r1 interface in transmitting both the homotropic and heterotropic communications. An alternative set of studies is being completed by D-A. Beck and J. Wild utilizing chimeric regulatory polypeptides formed from the Allosteric Binding Domains and Zinc Binding Domains from *Serratia marcescens* and *E. coli*. These studies indicate that the Allosteric Binding Domain generates a conformational signal which can be propagated through either the *E. coli* or the *S. marcescens* Zinc Binding Domains to homotrimeric catalytic subunits from both bacterial sources.

Table 3. Kinetic Characteristics of Chimeric ATCases[a].

	c_3-Ec/Sm	c_3-Sm/Ec	c_3-Ec/Sm:r_2-Ec	c_3-Sm/Ec:r_2-Ec
Enzyme	c_3	c_3	c_6r_6	c_6r_6
Kinetics	hyperbolic	hyperbolic	sigmoidal	sigmoidal
$[S]_{0.5}$[b]	16.5	7.5	125	11.5
Effector Responses[c]	none	none	ATP activation CTP inhibition	ATP activation CTP inhibition

[a] Protein nomenclature: c_3 catalytic trimer; r_2, regulatory dimer; c_6r_6, Holoenzyme;
Ec/Sm = Carbamoyl Phosphate Binding Domain from *E. coli*, Aspartate Binding Domain from *Serratia marcescens*; Sm/Ec is the reverse (modified from K.M. Kedzie, 1987[37])
[b] Concentration of aspartate (mM) required for half-maximal activity
[c] Effect of 2mM ATP or CTP on activity of the enzyme

THE EXCHANGE OF SUB-DOMAIN STRUCTURES WITHIN A TERTIARY DOMAIN

The nature of the catalytic paralysis resulting from the interaction between the Zinc Binding Domain of the *E. coli* regulatory subunit and the Aspartate Binding Domain of *Serratia marcescens* has been further explored by the formation of a chimeric catalytic subunit in which amino acids from the c_3-Ec polypeptide has been replaced by the analogous sequence from *S. marcescens*.[40] This c200s loop region (c196-228) is one of the most divergent areas of the deduced amino acid sequences of the catalytic polypeptide (Figure 2). In *S. marcescens* there are ten changes in this region; for example, ASPc200 is replaced by lysine, ASPc203 by glutamate, and TRPc209 by tyrosine. This sub-domain structure exchange permits the substitution of all ten amino acids within their original super-secondary structure as the crown of the Aspartate Binding Domain. The holoenzyme formed with the regulatory subunit from *E. coli* possessed partial paralysis and homotropic cooperativity in its aspartate requirement (60mM) with ATP activation and CTP inhibition; however, it was extremely sensitive to CTP/UTP synergism and the catalytic activity was virtually abolished in the presence of both feed-back inhibitors. This accentuation of the synergistic pyrimidine nucleotide effector response provides supportive evidence for the involvement of the r130s::c200s protein:protein interface for the transmission of the feed-back inhibition signal by pyrimidine nucleotides. Thus, the r1:c1 protein:protein interface appears to be primarily involved in propagating the pyrimidine allosteric signal.

SITE-DIRECTED MUTATIONAL CHANGES EFFECTING C:R INTERACTIONS AND ALLOSTERIC REGULATION

Site-directed changes of the amino acids in the *E. coli* catalytic polypeptide to their *Serratia marcescens* counterpart has better defined the roles of the c200s loop in the allosteric control of the catalytic activity of ATCase.[41,47] A lysine substitution at position 200c of the *E. coli* ATCase produced an enzyme with catalytic paralysis similar to that of the c_3-Sm::r_2-Ec hybrid ATCase and the c_3-Ec/Sm::r_2-Ec enzymes. Since the K_m for aspartate of the catalytic trimers are virtually identical for the native and chimeric trimers, the primary effect of this substitution appears to be at the c:r interface. This logical progression of studies from the comparative sequence analysis to hybrid/chimeric enzyme construction and site-directed analysis of individual residues has demonstrated an important role for the aspartate domain in c1:r1 bonding and communication. A communication that was initially suggested by the paralysis of the early c_3-Sm::r_2-Ec hybrid ATCase has been traced through domain exchange, sub-domain substitution and finally identified by site-directed mutagenesis.

Table 4. Site-directed modifications of the c200s loop of ATCase.

Site-directed Substitution	V_{max} [a]	$[S]_{0.5}$ [b]	N_{app} [c]	effect[d]
ASPc203-Glu/TRPc209-Tyr				
No effector	23.2	10.0	2.30	100
ATP	27.1	5.3	1.65	169
CTP	23.5	17.0	1.62	38
CTP/UTP	19.6	25.0	1.50	23
ASPc200-Lys				
No effector	5.83	70	1.51	100
ATP	6.72	31	1.80	190
CTP	4.21	150	1.59	35
CTP/UTP	4.75	200	1.70	20
GLUc204-Ala				
No effector	0.0086	10.5	1.90	100
ATP	0.0098	8.5	1.53	126
CTP	0.0098	15.0	2.52	58
CTP/UTP	0.0094	19.4	1.60	35

[a] Maximum velocity reported as ιmoles/hr/ιg
[b] Concentration of aspartate (mM) needed to achieve half-maximal velocity
[c] Hill coefficient estimated from the slope of v/v_{max}-v vs. log [S]
[d] The effect is reported as relative to the enzyme in the absence of effectors (100%)

A UNIFIED STRUCTURAL MODEL FOR THE SYNERGISTIC ROLE OF CTP AND UTP IN REGULATING ATCase ACTIVITY FROM THE ENTERIC BACTERIA

The ATCase from *E. coli* demonstrates homotropic cooperativity for both aspartate and carbamoyl phosphate and is subject to partial feedback inhibition by CTP which competes at the same allosteric binding site for activation by ATP.[48-50] It has been recently shown that the partial CTP-induced inhibition (approximately 60 percent) is enhanced by UTP to 95 percent inhibition even though UTP has no independent effect on catalytic efficiency.[51] In early studies of the direct binding of effectors to ATCase, a "half-site reactivity" was demonstrated for CTP binding by filter binding studies which indicated only 3 high affinity binding sites and two low affinity sites per holoenzyme in the presence of carbamoyl phosphate or the bisubstrate analogue PALA, N-phosphonacetyl-L-aspartate. Recent structural refinement has demonstrated an asymmetry across the twofold axis of molecular symmetry in the CTP-liganded enzyme such that CTP was bound to three regulatory sites associated with one of the catalytic trimers while the other three sites had an occupancy factor of less than 0.4.[14] In these studies it was noted that ARGc54 was oriented toward the substrate binding site in one half of the holoenzyme structure (that not binding CTP) and away from the active site in the CTP-associated half. In addition, there was another structural asymmetry which precluded ARGr130 from interacting with GLUc204 in the c:r (catalytic:regulatory polypeptide) interface; the phosphate groups of CTP in the crystal structure of a liganded holoenzyme were in a different conformation in the CTP saturated site due to a rotation about the oxygen-beta phosphate bond of the oxygen atom between the alpha and beta phosphates. This asymmetry across the 2-fold molecular axis may explain the "half-site reactivity" of the regulatory subunits in that the binding of CTP to one subunit of the regulatory dimer may impose a structural conformation on the second subunit of the same dimer that allows the binding of UTP not CTP. In this model, the binding of CTP would serve to provide for a primary inactivation of the adjacent catalytic site in the regulatory:catalytic polypeptide structure of the enzyme; however, it would not greatly affect the activity at the catalytic centers on the opposing catalytic trimers in the *E. coli* enzyme. The subsequent binding of UTP to the opposing allosteric sites could induce a reorientation of the active sites on the opposing catalytic trimer and inactivate the second set of active sites.

The bisubstrate analog PALA has been shown to promote the conformational transition from a relatively inefficient "T-state" conformation to an active "R-state". Thus the binding of substrate ligand molecules to some (1-3) of the six active sites would promote a structural transition for the entire enzyme (a secondary effect) which would make the unfilled active sites more efficient. It is still unclear as to how the binding of the heterotropic nucleotide ligands effects the T-R transition and what impact this secondary conformational change has on catalysis and subsequent ligand binding. The first step toward the development of a common model is provided by the observation that CTP may serve as a competitive inhibitor for ATP activation in varying ATCases from divergent bacterial sources.[52] CTP can inhibit ATP-activated ATCases, even if the CTP activates the enzyme by itself; CTP and ATP compete for a common or overlapping site in each of the ATCases examined. More recently, it has been noted that CTP and UTP synergistically inhibit the *P. vulgaris* enzyme even though CTP activates and UTP has no independent effect.[53]

The model of primary and secondary conformational effects can be readily adapted to those enzymes that appear to be activated by the presence of CTP but subsequently inactivated by the combination of CTP and UTP. In these cases, the binding of CTP could occur at three sites, similar to the *E. coli* enzyme, and exert primary inactivation on the three active sites associated with those regulatory sites. However, in the "non-*E. coli* enzymes" this primary effect could be accompanied by a secondary conformational change which results in the equivalent of a T-R transition of the entire holoenzyme. This conformation change could serve to provide the apparent activation of the enzyme at lower substrate concentations although only three sites (the non CTP-associated sites) are available for catalysis. By effecting both a primary and a secondary conformational change, the binding of CTP to the *Proteus vulgaris* or *S. marcescens* enzymes could inactivate the catalytic sites on one of the catalytic trimers while enhancing substrate access to the opposing trimer. Thus, CTP would appear to activate these ATCases. When UTP occupied the allosteric sites associated with the active catalytic subunit, the related active sites could lose their catalytic activity. Thus, CTP/UTP synergism could result in allosteric inhibition when each individual nucleotides had no effect (UTP) or actually appeared to activate (CTP) the enzyme on its own.

LITERATURE

1. J.R. Wild and M.E. Wales. Molecular Evolution and Genetic Engineering of Protein Domains Involving Aspartate Transcarbamoylase. Ann. Rev. Microbiol. 44:93-118.

2. D.R. Evans. CAD, a chimeric protein that initiates de novo pyrimidine biosynthesis in higher eukaryotes. In Multidomain proteins-structure and evolution. ed. D.G. Hardie, J.R. Coggins, pp. 283-331. New York: Elsevier. (1986) .
3. M.E. Jones. Pyrimidine nucleotide biosynthesis in animals: genes, enzymes, and regulation of UMP biosynthesis. Ann. Rev. Biochem. 49:253-79 (1980).
4. J.N. Freund and B.P. Jarry. 1987. The *rudimentary* gene of *Drosophila melanogaster* encodes four enzymatic functions. J. Mol. Biol. 193:1-13 (1987).
5. J.N. Davidson, P.C. Rumsby and J. Tamaren. Organization of a multifunctional protein in pyrimidine biosynthesis. J. Biol. Chem. 256:5220-5225.
6. G.A. O'Donovan and J. Neuhard. Pyrimidine metabolism in microorganisms. Bacteriol. Rev. 34:278-343 (1970).
7. A.J. Makoff and A. Radford. Genetics and biochemistry of carbamoyl phosphate biosynthesis and its utilization in the pyrimidine bisoynthetic pathway. Microbiol. Rev. 42:307-28.
8. J-L. Souciet, M. Nagy, M. LeGouar, F. LaCroute, S. Potier. Organization of the yeast *URA2* gene: identification of a defective dihydroorotase like domain in the multifunctional carbamoylphosphate synthetase aspartate transcarbamylase complex. Gene (Amst.) 79:59-70 (1989).
9. M. Faure, J.H. Camonis and M. Jacquet. Molecular characterization of a *Dictyostelium discoideum* gene encoding a multifunctional enzyme of the pyrimidine pathway. Eur. J. Biochem. 179:345-58 (1989).
10. K.L. Krause, K.W. Volz and W.N. Lipscomb, W.N. 2.5A structure of aspartate carbamoyl transferase complexed with the bisubstrate analog N-(phosphonacetyl)-L-aspartate. J. Mol. Biol. 193:527-53.75 (1987).
11. J.E. Gouaux and W.N. Lipscomb. The three dimensional structure of carbamyl phosphate and succinate bound to aspartate carbamoyltransferase. *Proc. Natl. Acad. Sci. USA* 85:4205-4208 (1988).
12. E.R. Kantrowitz and W.N. Lipscomb. *Escherichia coli* aspartate transcarbamoylase: the relation between structure and function. Science 241:669-74 (1988).
13. H.M. Ke. R.B. Honzatko and W.N. Lipscomb. Structure of unligated aspartate carbamoyltransferase of *Escherichia coli* at 2.6 Å resolution. Proc. Natl. Acad. Sci. USA 81:4037-40 (1984).
14. K.H. Kim, Z. Pan, R.B. Honzatko, H.M. Ke and W.N. Lipscomb. Structural asymmetry in the CTP-liganded form of aspartate carbamoyltransferase from *Escherichia coli*. J. Mol. Biol. 196:853-75 (1987).
15. M.R. Bethell and M.E. Jones. Molecular size and feedback-regulation characteristics of bacterial aspartate transcarbamoylases. Arch. Biochem. Biophys. 134:352-65 (1967).
16. J.R. Wild, W.L. Belser, W.L. and G.A. O'Donovan. Unique aspects of the regulation of the aspartate transcarbamoylase of *Serratia marcescens*. J. Bacteriol. 28:766-75 (1976).
17. J.R. Wild, K.F. Foltermann and G.A. O'Donovan. Regulatory divergence of aspartate transcarbamoylase within the *Enterobacteriaceae*. Arch. Biochem. Biophys. 201:506-17 (1980).
18. K.F. Foltermann, J.R. Wild, D.L. Zink and G.A. O'Donovan, G.A. Regulatory variance of aspartate transcarbamoylase among strains of *Yersinia entercolitica* and *Yersinia entercolitica*-like organisms. Curr. Microbiol. 6:43-47.71 (1981).
19. D-A. Beck, K.M. Kedzie and J.R. Wild. Comparison of the aspartate transcarbamoylases from *Serratia marcescens* and *Escherichia coli*. J. Biol. Chem. 264:16629-37 (1989).
20. J.S. Brabson and R.L. Switzer, R.L. Purification and properties of *Bacillus subtilis* aspartate transcarbamoylase. J. Biol. Chem. 250:8664-69 (1975).
21. C.G. Lerner and R.L. Switzer. Cloning and Structure of the *Bacillus subtilis* aspartate transcarbamoylase gene (*pyrB*). J. Biol. Chem. 261:11156-65 (1986).
22. L.B. Adair and M.E. Jones. Purification and characteristics of aspartate transcarbamoylase from *Pseudomonas fluroescens*. J. Biol. Chem. 247:2308-2315 (1972).
23. S.T. Berth and D.R. Evans. Characterization of the aspartate transcarbamoylase of *Pseudomonas fluorescens*. FASEB J. 4:A1836.
24. F. Van Vliet, R. Cunin, A. Jacobs, J. Piette, and D. Gigot. Evolutionary divergence of genes for ornithine and aspartate carbamoyltransferases-complete sequence and mode of regulation of the *Escheriahia coli argF* with *pyrB*. Nucl. Acids Res. 12:6277-89 (1984).
25. D.A. Bencini, J.E. Houghton, T.A. Hoover, K.F. Foltermann and J.R. Wild. The DNA sequence of *argI* from *Escherichia coli* K12. Nucleic Acids Res. 11:8509-18 (1983).
26. J.E. Houghton, D.A. Bencini, G.A.. O'Donovan and J.R. Wild. Protein differentiation: a comparison of aspartate transcarbamoylase and ornithine transcarbamoylase from *Escherichia coli* K-12. Proc. Natl. Acad. Sci. USA 81:4864-68 (1984).
27. Y. Itoh, L. Soldati, V. Stalon, P. Falmagne, and Y. Terawaki. Anabolic ornithine carbamoyltransferase of *Pseudomonas aeurginosa*: nucleotide sequence and transcriptional control of the *argF* gene. J. Bacteriol. 170:2725-34 (1988).

28. J.A. Maley and J.N. Davidson, J.N. The aspartate transcarbamoylase domain of a mammalian multifunctional protein expressed as an independent enzyme in *Escherichia coli*. Mol. Gen. Genet. 213:278-84 (1988).
29. J.G. Major. Structural modelling of the hamster CAD aspartate transcarbamoylase and comparisons to the *Escherichia coli* ATCase trimer. PhD Dissertation, Texas A&M Univ. College Station. 122 pp (1989).
30. J.G. Major, M.E. Wales, J.E. Houghton, J.A. Maley, J.N. Davidson and J.R. Wild. Molecular evolution of enzyme structure: construction of a hybrid hamster/*Escherichia coli* aspartate transcarbamoylase. J. Mol. Evol. 28:442-450 (1989).
31. J.G. Major, R. Radhakrishnan, J. Villano, E.F. Meyer and J.R. Wild. Structural modeling of the hamster CAD aspartate transcarbamoylase superdomain and comparisons to the *Escherichia coli* ATCase trimer. (submitted for publication).
32. J.E. Houghton. Structural and functional comparisons between OTCase and ATCase leading to the formation and characterization of an active domain fusion between *argI* (OTCase) and *pyrB* (ATCase). PhD Dissertation, Texas A&M University, College Station. 140 pp (1986).
33. J.E. Houghton, G.A. O'Donovan and J.R. Wild. Reconstruction of an enzyme by domain substitution effectively switches substrate specificity. Nature 338:172-74 (1989).
34. K.F. Foltermann, M.S. Shanley, and J.R. Wild. Assembly of the aspartate transcarbamoylase holoenzyme from transcriptionally independent catalytic and regulatory cistrons. J. Bacteriol. 157:891-98 (1984).
35. M.S. Shanley, K.F. Foltermann, G.A. O'Donovan and J.R. Wild. Properties of hybrid aspartate transcarbamoylase with native subunits from divergent bacteria. J. Biol. Chem. 259:12672-77 (1984).
36. K.F. Foltermann, D-A. Beck and J.R. Wild. In vivo formation of hybrid aspartate transcarbamoylases from native subunits of divergent members of the family Enterobacteriaceae. J. Bacteriol. 167:285-90 (1986).
37. K.M. Kedzie. Characterization of the *pyrB* gene of *Serratia marcescens* and hybrid gene formation with the *pyrB* gene of *Escherichia coli*, leading to the production of chimeric ATCase. Ph.D Dissertation, Texas A&M Univ., College Station. 158 pp (1987).
38. K.M. Kedzie, G.A. O'Donovan and J.R. Wild. In preparation.
39. R. Cunin, A. Jacobs, D. Charlier, M. Crabeel, and G. Hérve. Structure-function relationship in allosteric aspartate carbamoyltransferase from *Escherichia coli* I. Primary structure of a *pyrI* gene encoding a modified regulatory subunit. J. Mol. Biol. 186:707-13 (1985).
40. M.E. Wales, J. Moehlman and J.R. Wild. Characterization of a exchange fusion aspartate transcarbamoylase involving the 200s loop of the catalytic subunit which effects the T-R transition. (submitted for publication).
41. J.K. Grimsley. Role of the 200s loop in the heterotropic and homotropic response of *Escherichia coli* aspartate transcarbamoylase. PhD Dissertation. Texas A&M University, College Station, Texas. 99pp (1989).
42. T.S. Corder and J.R. Wild. Discrimination between nucleotide effector responses of aspartate transcarbamoylase due to a single site substitution in the allosteric binding site. J. Biol. Chem. 264:7425-7430 (1989).
43. M.E. Wales. T.A. Hoover and J.R. Wild. Site-specific substitutions of the Tyr-165 residue in the catalytic chain of aspartate transcarbamoylase promotes a T-state preference in the holoenzyme. J. Biol. Chem. 263:6109-14 (1988).
44. M.E. Wales and J.R. Wild. Role of the 240s loop interactions in stabilizing catalytic site domain closure in defining kinetic parameters of aspartate transcarbamoylase. (submitted for publication).
45. H. Nyunoya, K.E. Broglie and C.J. Lusty, The gene coding for carbamoyl-phosphate synthetase I was formed by fusion of an ancestral glutaminase gene and a synthetase gene. J. Biol. Chem. 259:9790-98 (1985).
46. G.A. O'Donovan, H. Holoubek and J.C. Gerhart. Regulatory properties of intergeneric hybrids of aspartate transcarbamoylase. Nature New Biol. 238:264-266 (1972).
47. J.K. Grimsley, M.E. Wales and J.R. Wild. Role of the 200s loop of the catalytic polypeptide in transmitting the homotropic and heterotropic responses of aspartate transcarbamoylase. (submitted for publication)
48. J.C. Gerhart and A.B. Pardee. The enzymology of control by feedback inhibition. J. Biol. Chem. 237:891-96 (1962).
49. J. Monod, J. Wyman and J-P. Changeux. On the nature of allosteric transition: A plausible model. J. Mol. Biol. 12:88-118 (1965).
50. H.K. Schachman. Can a simple model account for the allosteric transition of aspartate transcarbamoylase? J. Biol. Chem. 263:18583-18586 (1988).

51. J.R. Wild, S.J. Loughrey and T.C. Corder. In the presence of CTP, UTP becomes an allosteric inhibitor of aspartate transcarbamoylase. Proc. Natl. Acad. Sci. USA. 86:52-56 (1989).
52. J.R. Wild, J.L. Johnson and S.J. Loughrey. ATP-liganded form of aspartate transcarbamoylase, the logical regulatory target for allosteric control in divergent bacterial systems. J. Bacteriol. 28:766-75 (1988).
53. K.F. Foltermann, W. Lagaly and J.R. Wild. The aspartate transcarbamoylase from *Proteus vulgaris*: feedback inhibition requires CTP/UTP synergism. (submitted for publication).

CATALYTIC ANTIBODIES:

PERSPECTIVES AND PROSPECTS

Donald Hilvert

Departments of Chemistry and Molecular Biology
Research Institute of Scripps Clinic
10666 North Torrey Pines Road
La Jolla, California 92037

INTRODUCTION

Enzymes make complex life possible. With very few exceptions, each of the tens of thousands of chemical reactions that sustain living systems takes place quickly and smoothly through the action of a specific enzyme. The high rates and selectivities of enzymes make them ideal catalysts for *in vitro* processes, as well, and they are being utilized increasingly in research, industry, and medicine.

Nevertheless, for specific reactions of interest, enzymes are not always available. The development of effective strategies for preparing protein catalysts with tailored activities and specificities is therefore an important scientific goal. Clearly, if scientists could create new enzymes from scratch in the laboratory, many of the difficult, inefficient and expensive reactions that are still required today in the production of pharmaceuticals, industrial chemicals, and synthetic materials of all kinds would become straightforward.

One exciting approach to *de novo* enzyme design, the preparation of catalytic antibodies, takes advantage of the selectivity and diversity of the mammalian immune system to create families of protein receptor molecules with tailored catalytic activities. Progress in this rapidly developing field is described below in the context of ongoing studies in the author's laboratory. Both the limitations of this new technology and its prospects for the future are considered.

ANTIBODIES AND THE IMMUNE SYSTEM

The immune system is unquestionably the most prolific source of high affinity receptor molecules known.[1] To protect the body against disease-causing microorganisms and toxins, it can create a virtually limitless number of unique antibody molecules that bind foreign substances (antigens) tightly and specifically. By current estimates, the number of different receptors that can be created is greater than *100 million*, each with its own distinctive binding site. Scientists have learned to exploit this protein generating system to provide antibody molecules on demand with tailored affinities for specific ligands. A wide range of materials can be used as antigens, including natural and man-made compounds, charged and neutral species, small organic molecules, inorganic complexes, and large polymers. Their ease of preparation and intrinsic molecular properties have made antibodies enormously important as diagnostic agents, as components of drug delivery systems, and as agents for affinity purification of diverse materials.

Linus Pauling recognized the similarities between antibodies and enzymes nearly forty years ago.[2] While immunoglobulins do not normally catalyze reactions, they are protein molecules that bind ligands with high specificity and affinity. Dissociation constants for typical antibody-antigen complexes fall in the range 10^{-4} to 10^{-14} M, and antigen binding apparently involves the same factors that are important for ligand binding to enzymes. These include hydrophobic interactions, hydrogen bonding, ion-pairing, and dispersion forces.[1] Structural studies reveal that the basic sizes and shapes of the binding pockets of enzymes and antibodies are similar, as well.[1] Given the spectacular diversity of the immune system, therefore, it was not unreasonable to expect that antibodies might be found, or engineered, that possess intrinsic catalytic activity.

Chemical reactions proceed through the formation of high energy, short-lived transition states that are intermediate in structure between substrate and product. Enzymes work because they possess active sites that are complementary in shape and charge to the ephemeral transition states of the reactions they catalyze. Antibodies, on the other hand, normally leave their ligands chemically unchanged because their active sites are complementary to a ground state species rather than to the transition state of a reaction. In order to prepare an antibody that catalyzes a chemical transformation, it was therefore necessary to develop stable analogs of the transition state of particular reactions for use as haptens.[3] Over the past two decades a number of such compounds have been prepared as high affinity inhibitors of enzymes[4] and many of these materials are now being used to elicit catalytic antibodies. When the transition state analog design is a good one, a significant fraction of the immune response should possess the desired catalytic activity. Indeed, this approach has been used successfully in the last few years to create immunoglobulins that promote a wide variety of chemical transformations, from hydrolytic reactions to photochemical processes.

In my laboratory at Scripps we have used transition state analogs to create antibody catalysts for two important carbon-carbon bond forming reactions. These are the bimolecular Diels-Alder reaction and a unimolecular Claisen rearrangement. Both are examples of concerted pericyclic processes of enormous importance to theoreticians and synthetic chemists alike. Inasmuch as the probability of inducing an effective constellation of catalytic groups (e.g., general acids, general bases, nucleophiles) in the binding site of an antibody during immunization is low, concerted reactions that do not require chemical catalysis are especially appropriate targets for antibody catalysis. Pericyclic reactions, like Diels-Alder cycloadditions and sigmatropic rearrangements, do not require chemical catalysis but should be sensitive to proximity and strain, the principal catalytic effects antibodies are likely to impart. The development of effective protein catalysts for such processes is likely to enhance our understanding of how enzymes work and may lead to practical applications in the area of organic synthesis.

DIELS-ALDER CYCLOADDITIONS

The Diels-Alder reaction involves concerted addition of a conjugated diene to an olefin to give a cyclohexene derivative.[5] It is one of the most important transformations available to the organic chemist for the construction of carbon-carbon bonds in complex natural products and other molecules. Although of great value in the laboratory, Diels-Alder reactions are unknown in biological systems. Nature has apparently elected other strategies for joining carbon atoms together. Nevertheless, concerted cycloadditions are eminently suited to catalysis by approximation.[6] The bimolecular version of the Diels-Alder reaction has a substantial entropic barrier (typically in the range of -30 to -40 e.u.). By binding the two substrates together in the active site of antibody, however, it should be possible to "pay for" the large losses in translational and rotational entropy that are incurred in the course of the reaction and thus achieve large rate accelerations.[6]

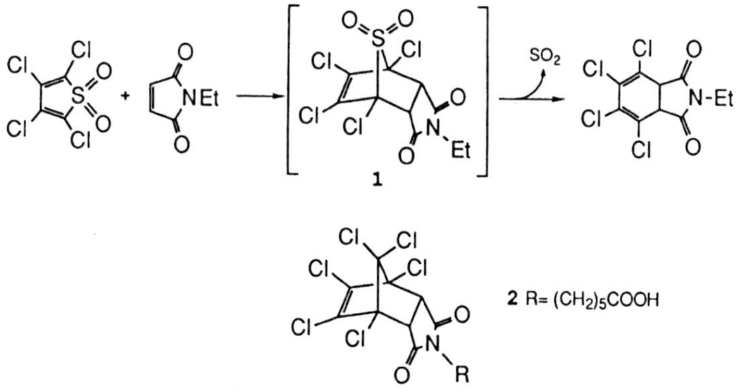

Scheme 1

Indeed, we recently succeeded in a catalyzing the bimolecular Diels-Alder reaction depicted in Scheme 1 with a chemically modified antibody.[7] Tetrachlorothiophene dioxide and N-ethylmaleimide undergo [4 + 2] cycloaddition to give a bicyclic adduct **1** that rapidly eliminates sulfur dioxide yielding dihydro-(N-ethyl)-tetrachlorophthalimide as the final product. The latter is further oxidized under the reaction conditions to tetrachlorophthalimide. The Diels-Alder reaction has a highly ordered transition state structure that resembles products more than substrates,[8] and we chose **2** as our hapten for eliciting the catalytic antibody. This material mimics the initially formed product of the cycloaddition, the unstable bicyclic intermediate **1**, and hence, the transition state structures for both the initial cycloaddition and the subsequent cycloreversion. We expected that antibodies generated against this compound would have the correct shape for bringing the two substrate molecules together in the proper orientation for reaction. In addition, because **2** has a very different shape than the final products of the reaction (phthalimide and SO_2), complications due to product inhibition were expected to be minimized.

A number of antibodies raised against **2** substantially accelerated the reaction shown in Scheme 1. The best of these, 1E9, was investigated in some detail. Catalysis was severely inhibited by the transition state analog, confirming that the reaction takes place in the induced binding pocket. The catalyst was also shown to undergo multiple (> 50) turnovers without significant product inhibition, vindicating our choice of hapten. From competition ELISA experiments we estimate that the K_d for product is greater than 3 mM, while the hapten binds with sub-μM affinity. Kinetic analysis of 1E9 revealed it to be quite efficient, with an effective molarity of at least 110 M per binding site. The effective molarity represents the concentration of substrate that would be needed in the uncatalyzed bimolecular reaction in order to achieve the same rate as occurs in the enzyme ternary complex. The true effective molarity for 1F7 must be substantially higher than 110 M, but low solubility of tetrachlorothiophene dioxide in aqueous buffer prevented its determination. In fact, the observed effective molarity is five orders of magnitude larger than the maximum concentration of tetrachlorothiophene dioxide obtainable under the reaction conditions. This rate acceleration is likely to be a consequence of a substantially smaller entropic barrier for the antibody catalyzed reaction compared to the uncatalyzed process, and we are currently determining the thermodynamic parameters for 1E9 to verify this expectation. Indeed, we anticipate that Diels-Alderase antibodies will be invaluable tools for enhancing our understanding of catalysis by approximation.

The antibody accelerated Diels-Alder reaction provides an excellent demonstration of the feasibility of exploiting the immune system to catalyze important non-physiological processes. Given the ubiquity of Diels-Alder reactions in organic synthesis, extension of our design strategy to other [4 + 2] cycloadditions is important. For example, those reactions that occur poorly at ambient temperatures and pressure, or with low regio- and enantioselectivity, are especially attractive targets for this methodology. It may be necessary to develop improved second-generation transition state analogs and new strategies for overcoming product inhibition in order to obtain successful antibody catalysts for such processes. The investment would, however, be worthwhile as significant practical applications are likely to result. In particular, use of antibodies to influence the regioselectivity of Diels-Alder reactions, and to induce asymmetry into the products obtained, has great promise.

A number of pericyclic processes, in addition to Diels-Alder reactions, have been utilized by organic chemists to construct carbon-carbon bonds in a selective fashion. Sigmatropic reactions, like the Cope and Claisen rearrangements, have been especially valuable in this regard.[9] As in the case of cycloadditions, they are good candidates for catalysis by antibodies because they do not require chemical catalysis but should be sensitive to strain and proximity effects.

The conversion of (-)-chorismate into prephenate (Scheme 2) is an example of a biologically relevant Claisen rearrangement. This [3,3]-sigmatropic process is the key step in the biosynthesis of aromatic amino acids in plants and lower organisms and is catalyzed by the enzyme chorismate mutase. Although the detailed mechanism of action of the enzyme remains to be clarified, the structure of the transition state, **3**, has been inferred from extensive inhibitor studies[10] and elegant stereochemical experiments.[11] Compound **4** was synthesized by Bartlett and coworkers as a mimic of this putative transition state species.[12] It is currently the best inhibitor of chorismate mutase known, binding two orders of magnitude more tightly to the enzyme than does chorismate itself.

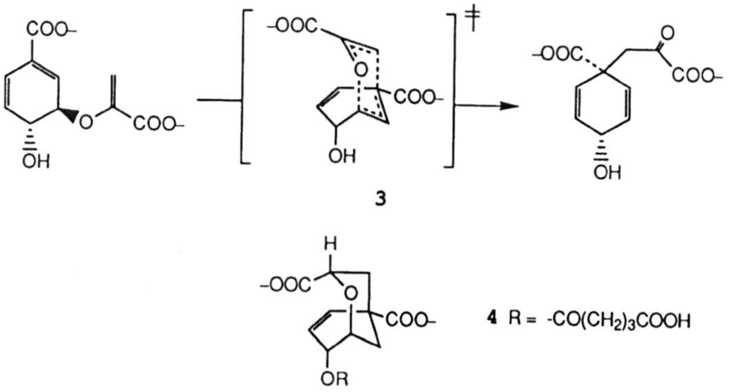

Scheme 2

We prepared 45 high affinity monoclonal antibodies against **4** and purified them to homogeneity.[13] Two of the isolated immunoglobulins catalyzed the targeted reaction with a rate acceleration of more than two orders of magnitude compared to the uncatalyzed process. Another antibody elicited independently against **4** by Schultz and coworkers exhibited a 10,000-fold rate acceleration, only two orders of magnitude lower than that achieved by the natural enzyme isolated from *Aerobacter aerogenes*![14] These results are significant, as they demonstrate that an imperfect transition state analog can be used to obtain catalytic antibodies having relatively high chemical efficiency. In addition, they underscore the importance of screening the immune system widely to obtain catalysts with the desired specific activity. We currently sample only a small fraction of the total diversity inherent in a given immune response. With improved screening techniques the probability of identifying more efficient, or particularly rare, antibody catalysts will increase.

Several lines of evidence indicate that antibody catalysis of the chorismate rearrangement is not artifactual: i) saturation kinetics were observed, indicating that the immunoglobulin and substrate must combine to form a Michaelis complex prior to the rearrangement event; ii) the catalyzed reaction was specifically inhibited by the transition state analog; and iii) not all antibodies that bind the transition analog tightly are catalysts. Thermodynamic study[13] of our chorismate mutase antibody revealed that the observed 200-fold rate enhancement is due entirely to a 5.7 kcal/mol lowering of the enthalpy of activation. The entropy of activation for the antibody reaction was actually 12 e.u. less favorable than for the spontaneous thermal rearrangement. These findings are consistent with the notion that the antibody uses strain energy to facilitate the reaction.

Since antibodies are inherently chiral objects, it is not surprising that they are able to influence the stereochemical outcome of the reactions they catalyze. In fact, our chorismate mutase antibody exhibits high enantioselectivity, accepting only the (-)-isomer of chorismate as a substrate.[15] The level of discrimination was greater than 90:1, allowing us to carry out a kinetic resolution of racemic chorismate. Because the efficacy of many drug molecules depends on the availability of a single enantiomer in highly pure form, the superb stereospecificity of catalytic antibodies, similar to that of natural enzymes, will be of great practical value.

We are currently dissecting the active site of our chorismate mutase antibody with a combination of chemical and genetic techniques. Attempts are also being made to obtain crystals suitable for x-ray diffraction studies. Information obtained in the course of these investigations may allow us to deduce the fundamental relationship between protein structure and function and will likely shed light on the specific chemical mechanism(s) by which a protein can accelerate sigmatropic rearrangements. More generally, we expect that continued and detailed work on this system will enhance our basic understanding of ligand-receptor interactions and catalysis.

LIMITATIONS AND FUTURE PROSPECTS

The development of catalytic antibodies represents a considerable advance in the field of enzyme engineering. This technology provides researchers with the means to create entirely new protein catalysts for manifold applications in chemistry and biology. In addition to the pericyclic processes described above,[7,13,14] ester and amide hydrolyses,[16,17] acyl transfer to amines and alcohols,[18] photodimerization and photo-cleavage processes,[19,20] redox reactions,[21] and a β-elimination[22] have been successfully catalyzed by immunoglobulins. The list continues to grow rapidly, and it is likely to be only a matter of time before antibodies are generated that can synthesize or modify proteins, polysaccharides and oligonucleotides in a site specific fashion, or that effect regio- and stereoselective aldol condensations, polyene cyclizations, etc. In principle, any chemical transformation is amenable to this new technology, so long as a suitable transition state analog can be devised and the reaction itself is compatible with the aqueous milieu and the protein environment of the immunoglobulin active site.

The production of truly *practical* immunoglobulin catalysts is, however, still limited by a number of considerations. Although large rate accelerations have been achieved with some catalytic antibodies, even the best cases fall several orders of magnitude short of the enhancements obtained with analogous enzymes. This fact is not surprising given that stable compounds will never mimic transition state species perfectly. Antibodies complementary to such imperfect analogs cannot be expected to provide optimal stabilization of true transition states, excect by chance. Product inhibition is another important problem with catalytic antibodies (and also with many enzymes). If the immunogen resembles the product of the targeted reaction too closely, severe inhibition will prevent effective turnover of the catalyst. This complication can be especially severe in the case of bimolecular reactions and strategies will need to be devised, as in the case of the Diels-Alder reaction discussed above, to circumvent this problem. Another difficulty involves generation of an optimal constellation of catalytic groups within the antibody binding site during the immunization event. Unlike the pericyclic processes considered here, many chemical transformations require general acid, general base or nucleophilic catalysis in order to proceed at an appreciable rate. Such catalytic residues will be especially important in energetically more demanding reactions than those examined to date. Taking advantage of charge complementarity between antibody and hapten, Schultz and coworkers were able to induce a catalytically active carboxylate in an immunoglobulin combining site.[22] However, the generality of this approach and its extension to systems requiring that two or three such groups act in concert must still be demonstrated. Consequently, identification of first generation immunoglobulins with more than modest activity for many of the more interesting, and difficult, chemical transformations remains a serious challenge.

Because of the enumerated limitations, and as discussed above, the successful identification of a catalytic antibody depends in large part on our ability to access the enormous diversity of the immune system. Only a fraction of any given immune response contains antibodies with the desired specific activity or regio- and stereo-selectivity. Consequently, efficient screening of large numbers of immunoglobulins must always be an important experimental consideration. Standard methods of making monoclonal antibodies are tedious, time consuming, and expensive. Generally, after months of effort, only a relatively small number of monoclonals (typically less than 50) are obtained having the desired specificity. Thus, only a very small fraction of the total diversity inherent in a given immune response is actually available for study with conventional protocols.

A recent advance, pioneered by the Lerner group at Scripps, makes a much larger fraction of the immunological diversity system available for screening. Using the polymerase chain reaction and lambda phage technology these researchers are able to produce very large combinatorial libraries of Fab proteins in *E. coli*.[23] Monoclonal Fab fragments can be prepared against individual transition state analogs in 2 weeks with these techniques rather than several months, and in a form that is suitable for genetic manipulation. Since 10^6 to 10^7 different antibodies can be examined in a day, this new methodology should greatly facilitate the production of catalytic antibodies. Much effort must now be focused on developing rapid and powerful assays for screening these large numbers of molecules for the desired activities.

In addition to extensive screening of antibody libraries for better abzymes, a number of strategies can be envisaged for improving the chemical efficiency of first generation antibody catalysts. For example, specific catalytic groups can be introduced into an antibody combining site via site-selective chemical reaction. The substrate itself can be utilized to deliver the masked functionality to the binding pocket in roughly the correct orientation, in analogy to standard affinity labeling practices.[24] If properly engineered, the resulting "semisynthetic" antibodies will combine the chemical reactivity of the prosthetic group with the intrinsic binding specificity of the protein template. Recent progress in expressing antibodies in microorganisms such as *E. coli*[25] and yeast[26] makes a molecular biological approach possible too. Recombinant DNA technology is currently being used to reengineer the affinity and specificity of individual antibody molecules and to introduce general acids, general bases or nucleophiles into their combining sites.[27]

The rational design of improved antibodies by site-directed mutagenesis, using either chemical or recombinant methods, will be most successful when a great deal is known about the structure of the immunoglobulin active site. However, for most catalytic antibodies detailed structural information will be unavailable, at least initially. For this reason, schemes that exploit random mutagenesis and classical genetic selection, which do not presuppose any knowledge of the active site, may ultimately be more valuable for improving the catalytic properties of immunoglobulins. Such an approach would mimic in a microorganism what the mouse immune system does naturally. However, in a microorganism it will be possible to select directly for function, rather than simply for tight binding to a transition state analog. The chorismate mutase system is ideal for testing these ideas, and we have successfully expressed our chorismate mutase antibody as a functional Fab protein inside a yeast cell that lacks natural chorismate mutase activity (unpublished results, K. Bowdish, Y. Tang, J. Hicks, D. Hilvert). These cells are now being grown under conditions such that only those will survive that produce an improved version of the antibody. In conjunction with the phage technology, selection methods will provide a particularly powerful tool for identifying antibody catalysts that rival the efficiency of enzymes.

In short, catalytic antibody technology goes a long way toward providing researchers with protein catalysts having tailored activities and specificities, even for reactions with no physiological counterpart. The use of chemistry, molecular biology and genetics to improve readily available, first-generation immunoglobulins will greatly enhance the power and utility of this approach to catalyst design and will make highly efficient catalyst molecules generally available for use in medicine and industry. Clearly, continued work on this exciting frontier of molecular engineering will bring us closer to our goal of being able to design enzyme-like catalysts for virtually any chemical transformation.

ACKNOWLEDGMENT

This work was supported in part by grants from the American Cancer Society (JFRA-45195) and the National Institutes of Health (GM-38273).

REFERENCES

1. E. A. Kabat, "Structural Concepts in Immunology and Immunochemistry," Holt, Reinhart and Winston, New York, 1976; D. Pressman and A. Grossberg, "The Structural Basis of Antibody Specificity," Benjamin, New York, 1968; A. Nisonoff, J. Hopper and S. Spring, "The Antibody Molecule," Academic Press, New York, (1975).
2. L. Pauling, Chemical achievement and hope for the future, Amer. Sci. 36:51 (1948).
3. W. P. Jencks, "Catalysis in Chemistry and Enzymology," McGraw Hill, New York, p.288 (1969).
4. R. Wolfenden, Transition state analog inhibitors and enzyme catalysis, Ann. Rev. Biophys. Bioeng. 5:271 (1976); G. E. Lienhard, Enzymatic catalysis and transition-state theory, Science 180:149 (1973).
5. J. Sauer, Diels-Alder reactions: New preparative aspects, Angew. Chem., Int. Ed. Engl. 5:211 (1966); Diels-Alder reactions: The reaction mechanism, ibid. 6:16 (1967); A. Wassermann, "Diels-Alder Reactions", Elsevier Publ. Co.; Amsterdam, (1965).
6. M. I. Page and W. P. Jencks, Entropic contributions to rate accelerations in enzymic and intramolecular reactions and the chelate effect, Proc. Nat. Acad. Sci. USA 68:1678 (1971); W. P. Jencks, Binding energy, specificity, and enzymic catalysis: The Circe effect, Adv. Enzymol. 43:219 (1975).
7. D. Hilvert, K. W. Hill, K. D. Nared, and M.-T. M. Auditor, Antibody catalysis of a Diels-Alder reaction, J. Am. Chem. Soc. 111:9261 (1989).
8. F. K. Brown and K. N. Houk, The STO-3G transition structure of the Diels-Alder reaction, Tetrahedron Lett. 25:4609 (1984).
9. F. E. Ziegler, The thermal aliphatic Claisen rearrangement, Chem. Rev. 88:1423 (1988); R. P. Lutz, Catalysis of the Cope and Claisen rearrangements, Chem. Rev. 84:205 (1984).
10. P. R. Andrews, E. N. Cain, E. Rizzardo, and G. D. Smith, Rearrangement of chorismate to prephenate. Use of chorismate mutase inhibitors to define the transition state structure, Biochemistry 16:4848 (1977); H. S.-I. Chao and G. A. Berchtold, Inhibition of chorismate mutase activity of chorismate mutase-prephenate dehydrogenase from *Aerobacter aerogenes*, Biochemistry 21:2778 (1982).

11. S. G. Sogo, T. S. Widlanski, J. H. Hoare, C. E. Grimshaw, G. A. Berchthold, and J. R. Knowles, Stereochemistry of the rearrangement of chorismate to prephenate: Chorismate mutase involves a chair transition state, J. Am. Chem. Soc. 106:2701 (1984).
12. P. A. Bartlett and C. R. Johnson, An inhibitor of chorismate mutase resembling the transition-state conformation, J. Am. Chem. Soc. 107:7792 (1985).
13. D. Hilvert, S. H. Carpenter, K. D. Nared, and M.-T. M. Auditor, Catalysis of concerted reactions by antibodies: The Claisen rearrangement, Proc. Natl. Acad. Sci. USA 85:4953 (1988).
14. D. Y. Jackson, J. W. Jacobs, R. Sugasawara, S. H. Reich, P. A. Bartlett, and P. G. Schultz, An antibody-catalyzed Claisen rearrangement, J. Am. Chem. Soc. 110:4841 (1988).
15. D. Hilvert and K. D. Nared, Stereospecific Claisen rearrangement catalyzed by an antibody, J. Am. Chem. Soc. 110:5593 (1988).
16. A. Tramontano, K. D. Janda, and R. A. Lerner, Catalytic antibodies, Science 234:1566 (1986); S. J. Pollack, J. W. Jacobs, and P. G. Schultz, Selective chemical catalysis by an antibody, Science 234:1570 (1986); A. Tramontano, A. A. Ammann, and R. A. Lerner, Antibody catalysis approaching the activity of enzymes, J. Am. Chem. Soc. 110:2282 (1988); K. D. Janda, S. J. Benkovic, and R. A. Lerner, Catalytic antibodies with lipase activity and R or S substrate selectivity, Science 244:437 (1989).
17. K. D. Janda, D. Schloeder, S. J. Benkovic, and R. A. Lerner, Induction of an antibody that catalyzes the hydrolysis of an amide bond, Science 241:1188 (1988); B. L. Iverson, and R. A. Lerner, Sequence-specific peptide cleavage catalyzed by an antibody, Science. 243:1184 (1989); S. Paul, D. J. Volle, C. M. Beach, D. J. Johnson, M. J. Powell, R. J. Massey, Catalytic hydrolysis of vasoactive intestinal peptide by human autoantibody, Science. 244:1158 (1989).
18. S. J. Benkovic, A. D. Napper, and R. A. Lerner, Catalysis of a stereospecific bimolecular amide synthesis by an antibody, Proc. Natl. Acad. Sci. USA 85:5355 (1988)
19. A. Balan, B. P. Doctor, B. S. Green, M. Torten, and H. Ziffer, Antibody combining sites as templates for selective organic chemical reactions, J. Chem. Soc., Chem. Commun. 106 (1988).
20. A. G. Cochran, R. Sugasawara, and P. G. Schultz, Photosensitized cleavage of a thymine dimer by an antibody, J. Am. Chem. Soc. 110:7888 (1988).
21. K. M. Shokat, C.J. Leumann, R. Sugasawara, and P. G. Schultz, An antibody-mediated redox reaction, Angew. Chem. Int. Ed. Engl. 27:1172 (1988); N. Janjic and A. Tramontano, Antibody-catalyzed redox reaction, J. Am. Chem. Soc. 111:9109 (1989).
22. K. M. Shokat, C. J. Leumann, R. Sugasawara, and P. G. Schultz, A new strategy for the generation of catalytic antibodies, Nature (London) 338:269 (1989).
23. W. D. Huse, L. Sastry, S. A. Iverson, A. S. Kang, M. Alting-Mees, D. R. Burton, S. J. Benkovic, and R. A. Lerner, Generation of a large combinatorial library of the immunoglobulin repertoire in phage lambda, Science 246:1275 (1989).

24. S. J. Pollack, G. R. Nakayama, and P. G. Schultz, Introduction of nucleophiles and spectroscopic probes into antibody combining sites, Science 242:1038 (1988).
25. A. Skerra and A. Plückthun, Assembly of a functional immunoglobulin F_v fragment in *Escherichia coli*, Science 240:1038 (1988); M. Better, C. P. Chang, R. R. Robinson, and A. H. Horwitz, *Escherichia coli* secretion of an active chimeric antibody fragment, Science 240:1041 (1988).
26. A. H. Horwitz, C. P. Chang, M. Better, K. E. Hellstrom, and R. R. Robinson, Secretion of functional antibody and Fab fragment from yeast cells, Proc. Natl. Acad. Sci. USA 85:8678 (1988); J. R. Carlson, A new means of inducibly inactivating a cellular protein, Mol. Cell. Biol. 8:2638 (1988).
27. S. Roberts, J. C. Cheetham, and A. R. Rees, Generation of an antibody with enhanced affinity and specificity for its antigen by protein engineering, Nature (London) 328:731 (1987); E. Baldwin, and P. G. Schultz, Generation of a catalytic antibody by site-directed mutagenesis, Science 245:1104 (1989).

POTENT AND SELECTIVE OXYTOCIN ANTAGONISTS OBTAINED BY CHEMICAL MODIFICATION OF A *STREPTOMYCES SILVENSIS* DERIVED CYCLIC HEXAPEPTIDE AND BY TOTAL SYNTHESIS

M. G. Bock*, R. M. DiPardo, P. D. Williams, R. D. Tung, J. M. Erb, N. P. Gould, W. L. Whitter, D. S. Perlow, G. F. Lundell, R. G. Ball[#], D. J. Pettibone[+], B. V. Clineschmidt[+], D. F. Veber, and R. M. Freidinger

Departments of Medicinal Chemistry, Biophysical Chemistry[#], and New Lead Pharmacology[+], Merck, Sharp & Dohme Research Laboratories, West Point, Pennsylvania 19486 and Rahway, New Jersey 07065

Oxytocin (OT) is a neurohypophyseal hormone which has an important function in parturition.[1] There is considerable evidence that the uterotonic action of OT and its stimulation of uterine prostaglandin release combine to initiate labor.[2,3] Additionally, OT mediates the postpartum function of contracting the mammary myoepithelium to elicit milk letdown[4] and has also recently been implicated as a key element in preterm labor.[5,6] Attempts to further delineate these OT connected events have provided the impetus to discover agents which interact selectively and competitively at the OT receptor. Such compounds are invaluable in determining the physiological and pathophysiological role of OT and its close structurally related hormone, arginine vasopressin (AVP). Additional interest derives from the prospect of their use as novel therapeutic agents.

Since the breakthrough synthesis of OT three decades ago, considerable research has been devoted to the design of antagonists of this peptide hormone.[7-9] The approach for achieving this objective has generally been limited to modifications of the native hormone structure itself.[3,5,10-13] One noteable example is the development of the competitive OT antagonist, 1-deamino-2-D-Tyr-(OEt)-4-Thr-8-Orn-oxytocin.[14,15] Initial clinical evidence suggests that this peptide is efficacious in the inhibition of uterine contractions in premature labor in humans.[6,16]

An alternative for finding antagonist leads is the use of receptor based screening. With such an approach we recently discovered the cyclic hexapeptide OT antagonist **1** (Figure 1), unrelated to OT in structure and derived from a microbial source.[17] While the OT/AVP antagonist properties of this compound are attractive (Table 1), its poor aqueous solubility limits its utility, especially for intravenous (i.v.) administration. Further, some improvement in potency and receptor selectivity would be desirable. Nonetheless, the novel structure of this compound provides a unique starting point for optimizing its pharmacological profile via chemical functionalization and to apply these findings to the design and de novo synthesis of structural analogues. Our initial results bearing on these issues comprise the subject of the following discussion.

On inspection of the gross structure of **1**, several sites amenable for chemical modification are indicated. Among these, we focused primarily on the piperazic acid (Piz) residues 4 and 5 and on the peptide backbone (especially between residues 2 and 3). Indications that this approach would prove fruitful were obtained during the very preliminary stages of our program. Thus, removal of the hydroxyl group of the N-hydroxyisoleucine in **1** with titanium trichloride in water-methanol provided **2**.[18] This dehydroxylation step caused a significant reduction in the inherent non-OT/AVP receptor-dependent contractile activity of **1**, thereby allowing this analogue and subsequent derivatives (vide infra) to be studied for OT agonist or antagonist activity in functional assays (e.g. rat isolated uterus).[17] Concomitantly, the OT receptor affinity of **2** was increased 5-fold relative to **1** and the OT vs AVP selectivity was improved approximately 3-fold. While an improvement in receptor binding potency and selectivity was noted for **2**, the compound was surprisingly chemically labile. Prolonged exposure of **2** to air resulted in the partial oxidation of the Piz^4 and Piz^5 residues to afford a mixture of **2** and **3**. This transformation could be driven to completion by briefly exposing **2** to tert-butylhypochlorite in pyridine. Compound **3** (L-365,209)[17], which exhibited a 20-fold increase in OT receptor affinity and a 2.5 to 5-fold increase in OT vs AVP selectivity relative to **1**, thus provided the structural basis for our subsequent studies.

One rationale for the improved OT receptor affinity of **3** provides that changes in the conformation of the Piz^4 and Piz^5 residues, brought about by the introduction of unsaturation, in turn influence the bioactive conformation of the 18-membered cycle. Extensive ^1H NMR investigations of **3**[19] confirm that its conformation differs, at least in solution, from the X-ray crystal structure[20] obtained for **1**. In order to gauge the affect of the envelope (Cß-exo) conformation of the L-Piz^5 residue on the conformation of the 18-member cycle and correspondingly on the receptor binding affinity of **3**, the latter compound was regioselectively oxidizing with DDQ to afford the more planar diene **4**. This compound was 3-fold less potent than **3** vs the

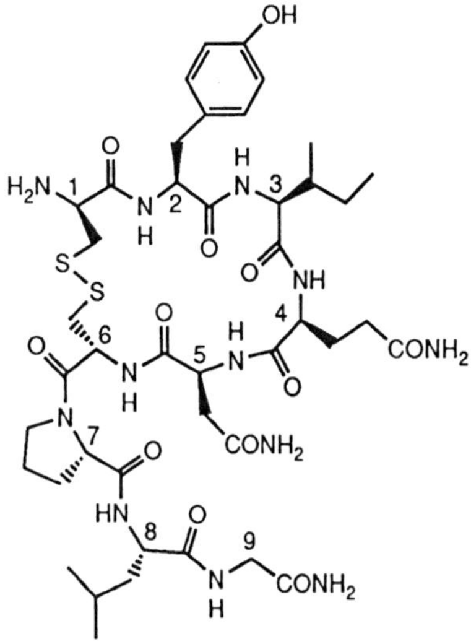

Oxytocin

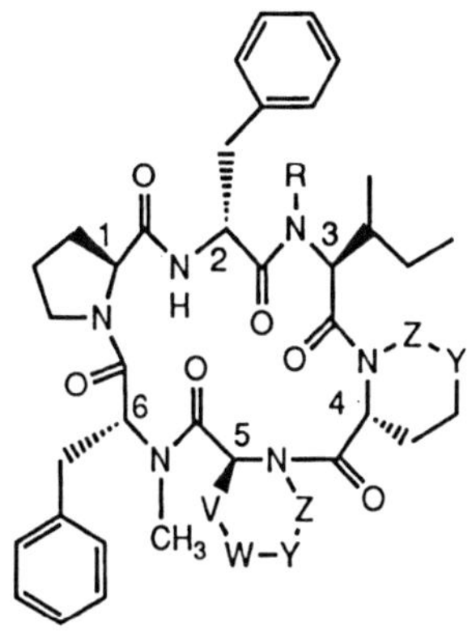

(1) R = OH; V-W = CH_2-CH_2, Y-Z = CH_2-NH.
(2) R = H; V-W = CH_2-CH_2, Y-Z = CH_2-NH.
(3) R =H; V-W = CH_2-CH_2, Y-Z = CH=N.
(4) R = H; V-W = CH=CH, Y-Z = CH=N.
(6) R =H; V-W = CH_2-CH_2, Y-Z = CH_2-NCH_3.

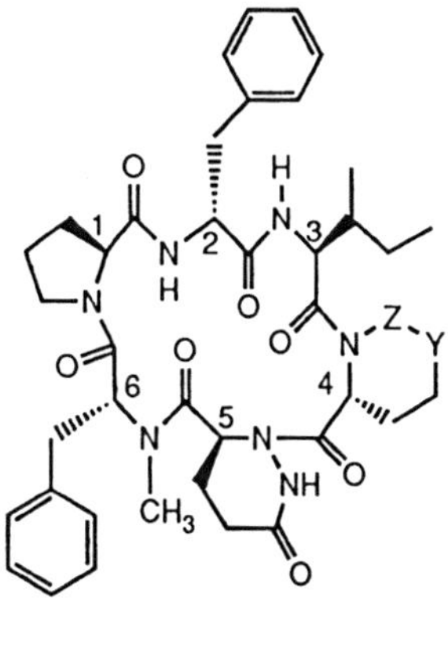

(5) Y-Z = CH=N.
(7) Y-Z = CH_2-NH.

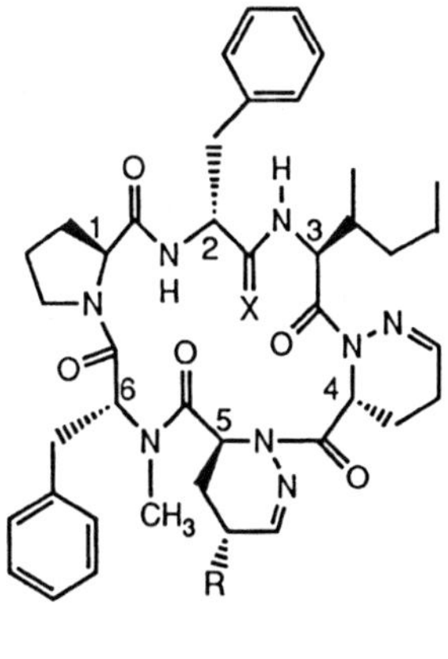

(8) X = O, R = $CH_2N(CH_3)_2$.
(9) X = S, R = H.
(10) X = H_2, R = H.

Figure 1

Table 1. Affinities of *Streptomyces silvensis* derived cyclic hexapeptide **1** and analogues for OT and AVP receptors in the rat.

	K_i (nM)[a]		
		[^{3}H] AVP	
Compd	[^{3}H]OT(uterus)	V_1(liver)	V_2(kidney medulla)
OT	0.89	61±7	88±32
1	150±23	2,200±260	3,400±420
2	30±3.5	1,500±130	1,500±130
3	7.3±0.58	730±180	540±30
4	8.1(1)	390(2)	580(2)
5	9.6±1.5	1,600(2)	6,800(2)
6	2,500(1)	7,400(1)	3,100(1)
7	89(1)	3,900(1)	7,800(1)
8	3.4±0.46	550±83	1,100±84
9	1.1±0.15	140±28	70±5.6
10	>30,000(1)	>30,000(1)	~30,000(1)

[a] Details of the assay methodology are contained in Ref. 17. K_i values are the group mean ± SE for three to six determinations (unless otherwise noted in parentheses) and were determined[35] from the IC_{50} values generated using five to eight concentrations of compound in triplicate.

OT receptor and displayed similarly modest differences in affinity for the AVP receptors. Alternatively, chemoselective and regioselective oxidation of **3** with alkaline hydrogen peroxide afforded **5**. Since the carbon atom adjacent to the Piz^5 ß-nitrogen in both **3** and **5** is sp^2 hybridized, it may be assumed that the conformations of both Piz^5 residues are similar. The OT receptor binding affinities of **3** and **5** are in concert with this analysis. However, any loss in binding affinity of **5**, resulting from conformational differences vs **3**, may be compensated for by increased van der Waals interactions and/or hydrogen bond donor-acceptor capabilities of the Piz^5 residue in **5** not available to Piz^5 in **3**. The enhanced OT vs AVP selectivity of **5** may also be a reflection of this extended capability. It is noteworthy that reductive alkylation of the ß-nitrogen of both Piz residues in **2** with methyl groups yielded compound **6** which is 260 to 340 fold less potent than **5** and **3**, respectively. Interestingly, compound **2** could not be induced to react with iodomethane or dimethyl sulfate reflecting the weak nucleophilicity of the Piz ß-nitrogens. A substantial reduction in binding affinity for the OT and AVP receptors was obtained when the Piz^4 residue in **5** was reduced ($NaCNBH_3$), (cf **7**). Again, it may be argued that conformational changes induced by the saturated Piz^4 residue and/or its modified hydrogen bond donor-acceptor capabilities are responsible for this result. This and similar observations[21] lead to the conclusions that an unsaturated (envelope) Piz^4 residue is a potency enhancing feature in this series of cyclic hexapeptides and that the Piz^5 residue is more amenable to functional group interchange, without loss in receptor binding affinity, than the Piz^4 residue. A noteworthy example of the latter point is the chemoselective transformation of **3** to **8** on exposure to Eschenmoser's salt and camphorsulfonic acid in THF. The dimethylaminomethyl derivative **8** is equipotent with **3** and shows enhanced OT vs AVP selectivity. Moreover, **8** (as its acetate salt) is 40-fold more soluble in aqueous media than **3** (cf 2.7 mg/mL vs 0.068 mg/mL at pH 7.3, respectively).

The consequence of changes in the peptide backbone of **3** was also explored. The introduction of pseudopeptide linkages in place of one or more amide bonds is a means of modifying peptide properties. To date, few peptide bond replacements have been reported for cyclic peptides, and these have been prepared by total synthesis.[22] By exploiting the presence of only two secondary amides in **3**, we succeeded in directly and selectively introducing pseudopeptide linkages into **3**. Reaction of **3** with the Lawesson reagent in toluene provided the potent and selective OT receptor ligand **9** as a crystalline solid.[23] Its structure was established by single crystal X-ray diffraction analysis[24,25] and is displayed in stereoview in Figure 2. The 18-membered ring of **9** is markedly non-planar due to a turn at the Pro^1 residue which appears to be stabilized by a hydrogen bond of 2.93A between N1 and O2 (Figure 1 numbering). There is, however, no evidence for a classic ß-turn. This backbone conformation results

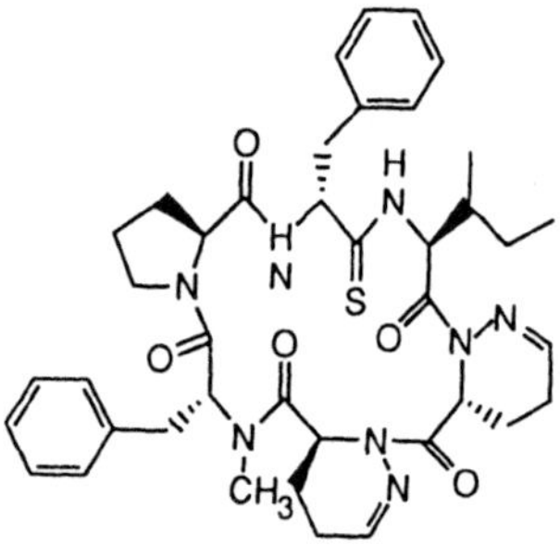

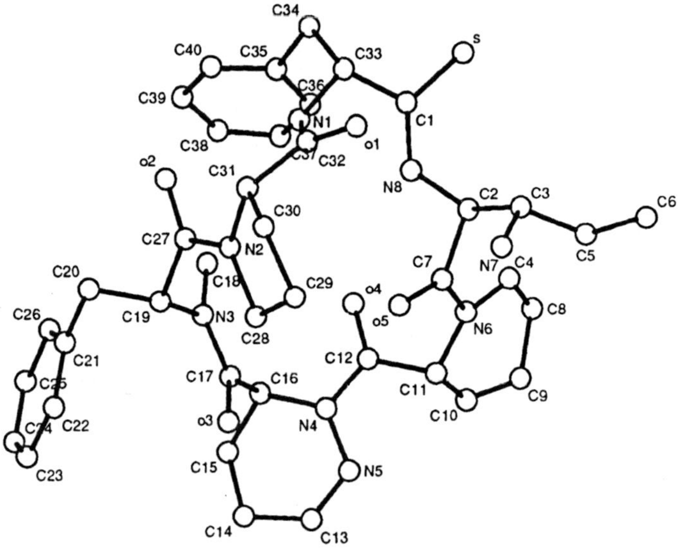

Figure 2. Computer generated drawing of 9 from X-ray coordinates with hydrogens omitted for clarity.

in the thioamide sulphur atom pointing away from the ring while the phenyl ring of the adjacent Phe^2 residue is positioned over the central space of the macrocycle in close proximity to the methyl group of N-methyl Phe^6. Interestingly, no upfield chemical shift for the N-methyl group of **9** was observed by ^1H NMR, relative to **3**, indicating that the positioning of the phenyl ring is likely a function of crystal packing.

The successful introduction into **3** of a single thioamide moiety afforded the opportunity of selectively obtaining the PheΨ[CH_2NH]Ile reduced peptide analogues. This modification has been useful for obtaining enzyme inhibitors[26] and hormone antagonists.[27,28] Further, it was anticipated that this change would augment the aqueous solubility of the products. Desulfurization of **9** with Raney-nickel gave the expected **10**. In contrast to **9**, the OT and AVP receptor affinities of **10** were drastically reduced.

While chemical transformations carried out directly on the *Streptomyces silvensis* derived OT antagonist **1** can lead to analogues with improved OT receptor binding potency, OT/AVP selectivity, and aqueous solubility, the reliance of this approach on the core structure of the natural product limits the analogue design possibilities. Additional structural changes not readily imposed on the lead compounds might be expected to result in agents displaying even more favorable pharmacological profiles. Thus, our strategy for developing OT antagonists encompassed chemical probing of the lead structure **3**, as well as application of empirical observations to the design of analogues more readily accessible via total synthesis.

The preparation of cyclic hexapeptides related to **3** by total synthesis presents significant synthetic challenges because of the presence of four contiguous secondary amino acids which include the unusual hydrazone-containing dehydropiperacic acids (Δ-Piz). The most difficult problems involve the preparation and incorporation of D- and L-Δ-Piz. The general synthetic route outlined in Scheme 1 was developed for synthesis of these analogues. The use of base-labile N-Fmoc protection was dictated by the acid lability of many of the intermediates.[29] Particular care is required in couplings to the dipeptide ester which demand sufficient activation of the acyl component to minimize diketopiperazine formation. Couplings at pipecolic acid are known to be particularly sluggish[30] and in this series required the use of acid chlorides as the acylating agent. Analogues **11** and **12**, (Table 2) were prepared in solution using a linear coupling approach. It has proven possible to adapt the acid chloride approach to the solid phase method. Accordingly, the remaining peptides shown in Table 2 were prepared on solid phase. In all cases, L-Pro was the carboxy terminal amino acid of the linear hexapeptide intermediate, consistent with previous studies which

Scheme I. Synthesis of Cyclic Hexapeptides

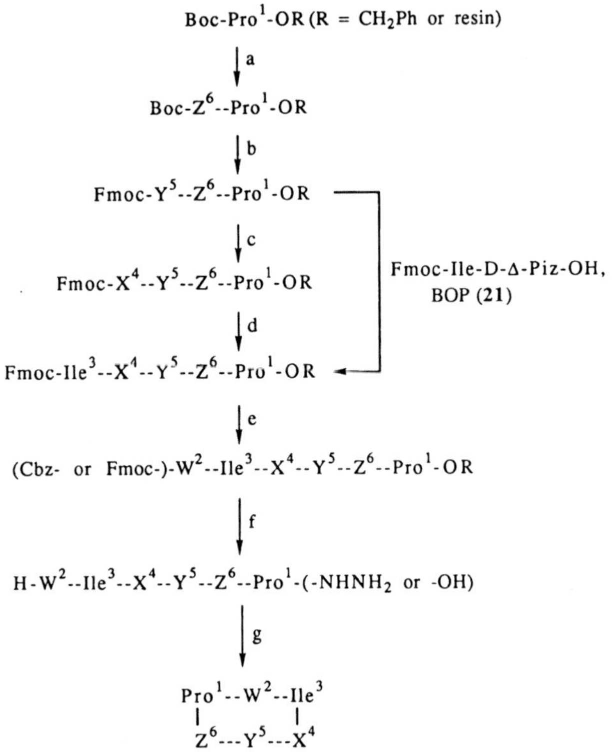

(a) HCl or TFA; Boc-Z^6-OH, BOP (b)HCl; Fmoc-Y^5-Cl or Fmoc-(N^{δ}-Cbz)Ppz-Cl (c) piperidine or Et_2NH; Fmoc-X^4-Cl (d) piperidine or Et_2NH; Fmoc-Ile-Cl or Fmoc-Ile-Cl/AgCN/toluene/-80° (**12,14**) (e) piperidine or Et_2NH; Fmoc-W^2-OH, BOP or Fmoc-D-Phe^2-Cl or Cbz-D-Phe^2-Cl (**12**) (f) piperidine or Et_2NH; NH_2NH_2 or H_2, $Pd(OH)_2$ (g) i-AmONO, thenpH 8 or DPPA, $NaHCO_3$, DMF; compounds **17, 21-24** were obtained by hydrogenolysis[H_2, $Pd(OH)_2$] to remove the N^{δ}-Cbz group on Ppz^5; compound **14** was obtained by hydrogenolysis to remove the N^{β}-Cbz group on D-Piz^4, followed by oxidation to D-Δ-Piz^4 with t-BuOCl. Compound **13** was obtained on solid phase using Boc protection, TFA deblocking, and symmetrical anhydride couplings. The linear hexapeptide was cleaved from the resin with HF and cyclized with DPPA.

predicted efficient cyclization at such a juncture.[31] Piperazic[32,33] and piperazine[34] carboxylic acids were synthesized and resolved according to literature procedures. The absolute configuration of the latter amino acid was established by chemical means by correlation with L-serine. Macrocyclizations were performed by either azide or DPPA procedures on the linear hexapeptides. Protecting groups were removed employing standard protocols and the final products were purified to homogeneity by chromatographic means.

Replacement of both dehydropiperazic acids in **3** with the commercially available pipecolic acids (Pip) was an early target of our synthetic studies. This substitution gave **11** and resulted in a 10-fold loss in oxytocin receptor affinity. Removal of the N-methyl group from N-Me-D-Phe in **11** produced a further 2-fold loss (**12**). The D-Pro^4, L-Pro^5 double substitution analogue **13** was reduced in binding affinity by another order of magnitude and supports the hypothesis (vide supra) that the larger six-membered rings are important for confomational reasons. The L-Pip^5 analogue **14** proved to be equipotent to **3** and, in agreement with the structure-activity profile generated by the semi-synthetic analogues (vide supra), demonstrates the importance of the D-Δ-Piz^4 residue for achieving good receptor binding affinity. Concurrently, the potency and selectivity displayed by **14** underscores the previous observation that the L-Δ-Piz^5 in **3** is alterable. However, in order to replace both Δ-Piz residues in **3** without detriment to receptor binding affinity, compensatory potency enhancing modifications are required. In this regard we pursued simplifications and potency enhancing modifications in other parts of the cyclic peptide system. A variety of cyclic and acyclic amino acids in place of L-Pro^1 uniformly led to loss of potency. However, replacement of N-Me-D-Phe^6 with D-His resulted in a 2-fold gain in receptor binding affinity (**15**). An additional benefit derived from this substitution was the concomitant enhancement of aqueous solubility which D-His^6 imparted on compound **15**. Incorporation of piperazine-2-carboxylic acid (Ppz) in the 5-position of the cyclic hexapeptides also had a modest potency, as well as solubility enhancing effect (cf. **12** and **17**). Increasing the size of the 2-position aromatic group was found to produce an enhancement in receptor binding affinity as well. For example, substituting D-Phe^2 in **11** with D-Trp yielded **16** which bound the OT receptor more avidly by one order of magnitude. In this regard, D-2-naphthylalanine proved to be the optimum substitution for D-Phe^2 (cf. **18** and **20**). Combination of the above potency and solubility enhancing features produced the group of analogues **21 - 24**. Some of the most potent compounds cited here have been characterized as functional oxytocin antagonists in a rat uterus preparation and have sufficient aqueous solubility for i.v. administration. Consequently, they may have utility as research tools and in certain therapeutic applications.

Table 2. Inhibition of binding of ^{3}H-oxytocin to rat uterine receptors and ^{3}H-arginine vasopressin to rat liver (V_1) and kindney medulla (V_2) receptors by synthetic cyclic hexapeptides.

```
Pro¹--W²--Ile³
|          |
Z⁶---Y⁵---X⁴
```

Compound	W	X	Y	Z	Ki (nM)[a] OT	Ki (nM)[a] V_1	Ki (nM)[a] V_2
11	D-Phe	D-Pip	L-Pip	N-Me-D-Phe	83(2)	890(1)	1600(1)
12	D-Phe	D-Pip	L-Pip	D-Phe	140±0.18	1600(1)	4500(1)
13	D-Phe	D-Pro	L-Pro	N-Me-D-Phe	1400(1)	9300(1)	33000(1)
14	D-Phe	D-Δ-Piz	L-Pip	N-Me-D-Phe	5.9±1.9	970(1)	480(2)
15	D-Phe	D-Pip	L-Pip	D-His	36±3.5	2700(1)	6800(1)
16	D-Trp	D-Pip	L-Pip	N-Me-D-Phe	8.1±0.81	3500±330	260±22
17	D-Phe	D-Pip	L-Ppz	D-Phe	91±8.7	3000(1)	4300(1)
18	D-Trp	D-Pip	L-Pip	D-His	7.8±0.66	2200±250	1800±79
19	D-1-Nal	D-Pip	L-Pip	D-His	3.3(2)	1000(1)	1000(1)
20	D-2-Nal	D-Pip	L-Pip	D-His	1.6±0.11	760±100	320±25
21	D-Trp	D-Δ-Piz	L-Ppz	N-Me-D-Phe	1.1±0.38	2000(2)	61(2)
22	D-2-Nal	D-Pip	L-Ppz	N-Me-D-Phe	2.3±1.0	4400(2)	120(2)
23	D-Trp	D-Pip	L-Ppz	N-Me-D-Phe	4.2±0.39	9500±550	240±19
24	D-2-Nal	D-Pip	L-Ppz	D-His	0.74±0.11	1400(2)	240(2)

In summary, the selective chemical transformations carried out on the *Streptomyces silvensis* derived OT antagonist **1** have led to several analogues with improved OT receptor binding potency, OT/AVP selectivity, and/or aqueous solubility. Our results illustrate how subtle structural modifications can have dramatic effects on the ability of these compounds to bind to the OT receptor, presumably through effects on conformation and/or hydrogen bonding potential. Application of these observations to the design of cyclic hexapeptides related to the natural product lead demonstrates that high levels of OT receptor affinity can be realized by incorporation of certain amino acids at the 2- and 4-positions. Aqueous solubility can be increased substantially by introducing amino acids containing basic groups at the 5- and 6-positions.

Acknowledgement It is a pleasure to acknowledge the contributions of Dr. J. P. Springer (X-ray), Dr. S. M. Pitzenberger (^{1}H NMR), and Dr. M. J. Kaufman (solubility studies). We are indebted to Dr. M. A. Goetz for providing the fermentation product **1**, Dr. B. E. Evans for stimulating discussions, and Dr. P. S. Anderson for encouragement and support.

REFERENCES

1. Pritchard, J. A.; MacDonald, P. C.; Gant, N. F.; in *Williams Obstetrics*; Appleton-Century-Crofts:Norwalk, 1985, ed.17, 295.
2. Fuchs, A.; Fuchs, F.; Husslein, P.; Soloff, M. S.; Fernstrom, M. J. *Science (Washington, D.C.)* (1982) **215**, 1396.
3. Chan, W. Y.; Powell, A. M.; Hruby, V. J. *Endocrinology* (1982) **111**(1) 48.
4. Soloff, M. S.; Alexandrova, M.; Fernstrom, M. J. *Science (Washington, D.C.)* (1979) **204**, 1313.
5. Chan, W. Y.; Hruby, V. J.; Rockway, T. W.; Hlavacek, J. *J. Exp. Ther. Pharmacol.*, (1986) **239**, 84.
6. Akerlund, M.; Stromberg, P.; Hauksson, A.; Andersen, L. F.; Lyndrup, J.; Trojnar, J.; Melin, P. Br. *J. Obstet. Gynaecol.* (1987) **94**, 1040.
7. Berde, B.; Boissonnas, R. A. in *Neurohypophysial Hormones and Similar Polypeptides*, Berde, B., Ed.; Springer Verlag:Berlin; 1968; p 802.
8. Sawyer, W. H.; Manning, M. in Oxytocin: Clinical and Laboratory Studies, Amico, J. A.; Robinson, A. G., Eds.; Elsevier Science Publishers B. V. (Biomedical Division):Amsterdam; 1985; p 423.
9. Manning, M.; Sawyer, W. H. *J. Lab. Clin. Med.* (1989) **114**(6), 617.
10. Manning, M.; Sawyer, W. H. *Trends Neurosci.* (1984) **7**, 6.
11. Chan, W. Y.; Nestor, J. J.; Ferger, M. F.; du Vigneaud, V. *Proc. Soc. Exp. Biol. Med.* (1974) **146**, 364.

12. Kruszynski, M.; Lammek, B.; Manning, M.; Seto, J.; Haldar, J.; Sawyer, W. H. *J. Med. Chem.* (1980) **23**, 364.
13. Chan, W. Y.; Fear, R.; du Vigneaud, V. *Endocrinology* (1967) **81**, 1267.
14. Melin, P.; Trojnar, J.; Johansson, B.; Vilhardt, H.; Akerlund, M. *J. Endocrinol.* (1986) **111**, 125.
15. Hahn, D. W.; Demarest, K. T.; Ericson, R.; Homm, R. E.; Capetola, R. J.; McGuire, J. L. *Am. J. Obstet. Gynecol.* (1987) **157**, 977.
16. Andersen, L. F.; Lyndrup, J.; Akerlund, M.; Melin, P. *Am. J. Perinat.* (1989) **6**(2), 196.
17. Pettibone, J. D.; Clineschmidt, B. V.; Anderson, P. S.; Freidinger, R. M.; Lundell, G. F.; Koupal, L. R.; Schwartz, C. D.; Williamson, J. M.; Goetz, M. A.; Hensens, O. D.; Liesch, J. M.; Springer, J. P. *Endocrinology* (1989) **125**(1), 217.
18. Mattingly, P. G.; Miller, M. J. *J. Org. Chem.* (1980) **45**, 410.
19. Pitzenberger, S. M.; unpublished results.
20. Springer, J. P.; personal communication.
21. To be reported in the full account.
22. Sherman, D. B.; Spatola, A. F. *J. Amer. Chem. Soc.* (1990) **112**, 433.
23. Lajoie, G.; Lepine, F.; Maziak, L.; Belleau, G. *Tet. Lett.* (1983) **24**(36), 3815
24. Crystals for X-ray diffraction studies formed from ethanol in space group P21 with cell constants of a=10.413, b=17.225, c=11.199 and b=97.77R. The final unweighted refinement residual is 0.067.
25. Ball, R. G. *Acta Crystallogr.* (1990), to be submitted.
26. Szelke, M.; Leckie, B.; Hallett, A.; Jones, D. M.; Sueiras, J.; Atrash, B.; Lever, A. F. *Nature (London)* (1982) **299**, 555.
27. Rodriguez, M.; Bali, J.-P.; Magous, R.; Castro, B.; Martinez, J. *Int. J. Peptide Protein Res.* (1989) **27**, 293.
28. Coy, D. H.; Heinz-Erian, P.; Jiang, N.-Y.; Sasaki, Y.; Taylor, J.; Moreau, J.-P.; Wolfrey, W. T.; Gardner, J. D.; Jensen, R. T. *J. Biol. Chem.* (1988) **263**(11), 5056.
29. Anteunis, M. J. O.; Van Der Auwera, C. *Int. J. Pept. Prot. Res.* (1988) **31**, 301.
30. Nutt, R. F.; Holly, F. W.; Homnick, C.; Hirschmann, R.; Veber, D. F. *J. Med. Chem.* (1981) **24**, 692.
31. Brady, S. F.; Varga, S. L.; Freidinger, R. M.; Schwenk, D. A.; Mendlowski, M.; Holly, F. W.; Veber, D. F. *J. Org. Chem.* (1979) **44**, 3101.
32. Hassall, C. H.; Johnson, W. H.; Theobold, C. H. *J. C. S. Perkin I* (1979) 1451.
33. Caldwell, C. G.; manuscript in preparation.
34. Felder, E.; Maffei, S.; Pietra, S.; Pitre, D. *Helv. Chim. Acta* (1960) **43**, 888.
35. Cheng, Y.-C.; Prusoff, W. H. *Biochem. Pharmacol.* (1973) **22**, 3099.

DESIGN OF PEPTIDE LIGANDS THAT INTERACT WITH SPECIFIC MEMBRANE RECEPTORS

Victor J. Hruby, T. Matsunaga, F. Al-Obeidi, G. Toth, C. Gehrig and P. S. Hill

Department of Chemistry, University of Arizona
Tucson, Arizona 85721 U.S.A.

INTRODUCTION

Peptide hormones and neurotransmitters are carriers of biological information that initiate their biological effects by binding to receptors found on the surface of cells. A basic assumption of modern molecular biology is that the information content of a peptide hormone or neurotransmitter is related to the stereostructural and conformational properties of the peptide. However, the properties of most small (<20 residues) polypeptide hormones and neurotransmitters makes the rational design of such ligands difficult. These properties include: 1) a high degree of conformational flexibility with many conformations accessible at physiological temperatures; 2) a lack of high receptor selectivity; 3) ready biodegration; and 4) the complexity of peptide structures compared with most natural products that have been developed as potential drugs. Despite these difficulties, we and others have developed approaches that show great potential for the systematic rational design of peptide ligands. In this paper we outline some of the approaches that can lead to peptide ligands with high potency, high receptor selectivity, agonist or antagonist activity, high stability against enzymatic breakdown, and conformationally constrained structural properties that provide templates for topographical design and receptor mapping.

GENERAL CONSIDERATIONS

The design of peptide hormone and neurotransmitter ligands requires a highly interdisciplinary approach. Three primary research areas must be addressed simultaneously: 1. Chemistry - this includes synthetic methodology including asymmetric synthesis of unusual amino acids and peptidomimetics, and macrocyclic synthesis; state of the art purification and analysis methods; and an appreciation and understanding of the chemical and physical properties of polypeptides. 2. Biology - It is essential that excellent binding assays, and in vitro and in vivo bioassays be available and utilized from the beginning. Since specificity is often a major concern, and many hormones and neurotransmitters interact with multiple receptor subtypes, it often is necessary to have multiple assay systems that can distinguished or provide insight into those elements of structure, conformation and dynamics that are important to bioactivity. In our opinion, it is in this area where chemists and others interested in peptide and protein design most commonly fail. Often, though excellent synthetic work is done on unique structures, limited value comes of the work because of the poor or missing biological or biochemical studies. Surprisingly, this occurs in industry as well as in academia. A need for much closer collaboration between biologists and chemists is obvious, but neither group seems to have gotten the message to the extent that is required. 3. Structure, Biophysics and Molecular Modeling - an important goal in peptide and protein design is the development of a conformational structure for the molecule of interest that is important to its bioactivity. This requires state of the art spectroscopic and biophysical equipment including especially very high field 2D and 3D nuclear magnetic resonance spectrometers, circular dichroism and FT infrared spectrophotometers, and x-ray crystallography where possible. This equipment along with fast computers to collect, manipulate and help evaluate the data are needed. In addition, computer based programs are needed to help search for the structural, conformational, and dynamic properties that are needed to develop models for conformational structure-biological activity relationships. This information is then combined with computer assisted molecular design (CAMD) of new analogues that will test specific models of structure-bioactivity relationships.

With regard to the above, it is important to recognize that most peptide hormones and neurotransmitters are small, conformationally flexible molecules with multiple conformational states available under biological conditions (1-3). These intrinsic properties of peptides provide a special challenge in design, especially since, in addition to the multiple conformation problem, most peptide hormones are readily degraded in biological system, making their systematic long term study in biological assays difficult. Furthermore, we do not understand the general problem of peptide folding and the related problems of peptide - receptor/acceptor interaction (the docking problem), and the mechanism(s) of receptor transduction. Thus structural algorithms and the parameters utilized to predict a priori the conformations of peptides are flawed and must be utilized with the proper mixture of skepticism and chemical intuition.

Nonetheless, a useful and often successful approach has been developed utilizing the principle of conformational constraint (for overviews see 4-6). In this approach, a linear or already cyclic peptide constrained either locally or more globally to a relatively small family of conformations such that the peptide analogue or mimetic can be used as a template for further design. Though considerable success in designing peptide ligands has occurred using this approach (vide infra, 7,8), many approaches to peptide hormone and neurotransmitter analogue design have ignored the side chain moieties. Yet it has been well established in many peptide hormones and neurotransmitters that these moieties are critical to their biological activities. We believe that much more careful consideration of these groups can provide important new insights into peptide ligand design. For this purpose, we wish to distinguish between the conformational and topographical elements in peptide design (7,8). As generally utilized, the conformation of a peptide refers to its backbone conformation. Thus, conformation often refers to secondary structural features such as an α-helix, a β-sheet, a reverse turn structure, an extended structure, etc. On the other hand, topography refers to the relative, cooperative three dimensional arrangements of side chain groups in a peptide that determines its surface architecture. The significance of such structural considerations can be appreciated by recognizing that most amino acids have three distinct conformations that result from rotation about the C_α-C_β bond (χ_1) (Figure 1). Topographical design involves either biasing or fixing a side chain group to one of these side chain conformations. It is obvious that such constraints, if applied to even a few side chains in an otherwise conformationally identical peptide, would lead to peptide analogues with significantly different topographical properties. It would be expected that these analogues might have quite different biological properties.

In the remainder of this paper will briefly illustrate principles of peptide design that led to conformationally and topographically constrained peptides with specific biological activities.

DESIGN OF α-MELANOTROPINS WITH SUPERPOTENCY AND SUPER PROLONGED BIOACTIVITIES

α-Melanotropin (α-MSH, α-melanocyte stimulating hormone) is a linear 13-amino acid peptide, Ac-Ser-Tyr-Ser-Met-Glu-His-Phe-Arg-Trp-Gly-Lys-Pro-Val-NH_2. It has important peripheral biological activities, primarily involving pigmentation, and in addition, has a number of central effects related to memory, attention and thermal regulation. Utilizing the concept of conformation

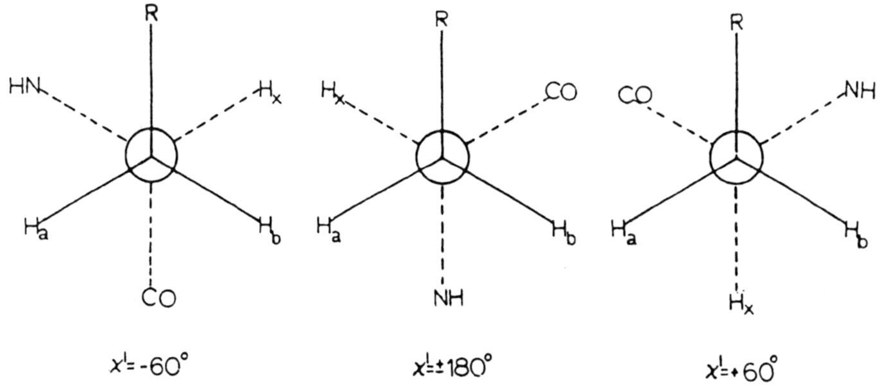

Figure 1. Low energy conformations about the C_α-C_β bond (χ^1 torsion angle) in amino acid residues corresponding to gauche(-) (-60°), trans (±180°) and gauche(+) (+60).

constraint by stabilization of a reverse turn structure (quasicyclization), we designed the analogue [Nle^4,D-Phe^7]α-MSH (Ac-Ser-Tyr-Ser-Nle-Glu-His-D-Phe-Arg-Trp-Gly-Cys-Pro-Val-NH$_2$) and demonstrated its superpotency and highly prolonged (days in vitro to weeks in vivo) activity in a number of bioassays (9,10). Furthermore, this analogue is highly stable to proteolytic enzyme degradation (11, for a review).

The results with the highly potent linear analogue led us to further test our hypothesis regarding its biological active conformation (a reverse turn structure about the D-Phe^7 residue) utilizing the concept of pseudoisosteric cyclization. In this concept, one considers a conformation from the standpoint of which distant (in a linear sequence sense) side chain moieties are brought in close proximity as a result of folding. One then designs in a covalent bond between two side chain moieties such that after formation of the covalent bond the same general stereostructural, lipophilic, and other chemical features are retained. This led us to design the pseudoisosteric analogue Ac-[Cys^4,Cys^{10}]α-MSH (Figure 2, Ac-Ser-Tyr-Ser-Cys-Glu-His-Phe-Arg-Trp-Cys-Lys-Pro-Val-NH$_2$). Note that in this cyclization, the Met^4 side chain group (-CH$_2$-CH$_2$-S-CH$_3$) and the Gly^{10} side chain group (-H) is replaced by the group (-CH$_2$-S-S-CH$_2$-) which is pseudisosteric (12). Interestingly, this analogue was superpotent in the classical frog skin assay, the bioactivity on which the design was based, but subsequent studies showed that it only had similar potency to α-MSH in mammalian systems, and a variety of subsequent structure-function studies did not lead to analogues that were markedly superior to Ac-[Cys^4,Cys^{10}]α-MSH (see 13 for a review).

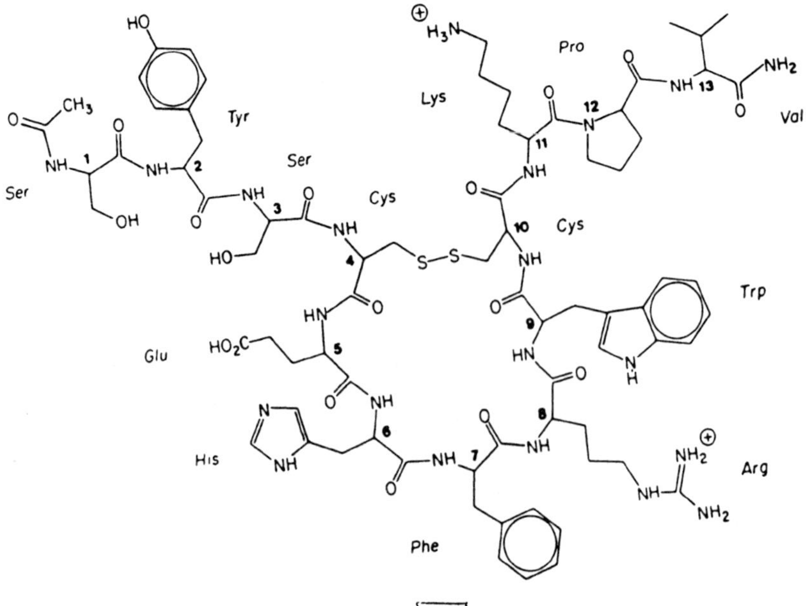

Figure 2. Structure of [Cys^4,Cys^{10}]α-melanotropin.

In the course of investigating the conformation properties of the linear and cyclic highly potent melanotropin analogues, we utilized quenched molecular dynamics to explore the conformational transitions and properties available to these peptides. In this approach, the molecule(s) of interest are "heated" to elevated temperatures (300°K to 1000°K) and molecular dynamics simulations are run either with or without constraints. Then, at specific intervals along the trajectory (e.g. every picosecond), energy minimization algorithms are utilized (quenched) to bring the molecule to a low energy conformation closely related to its structure at the elevated temperature. Examination of a series of such structures utilizing both linear and cyclic analogues led to the recognition that if the pro-S hydrogen on the Gly^{10} residue were replaced by a Lys side chain group ($-CH_2CH_2CH_2CH_2-NH_2$), and the Lys^{11} side chain group by a hydrogen, an analogue would result (Ac-Ser-Tyr-Ser-Nle-Glu-His-Phe-Arg-Trp-Lys-Gly-Pro-Val-NH_2) in which the Glu^5 and Lys^{10} side chain groups were in close proximity. This immediately led us to ask the question whether these two side chain groups might be covalently attached via a lactam bridge. Molecular mechanics calculations indicated that this should be possible within the context of a low energy, stable conformation. Both the linear (14,15) and the cyclic analogues were made (14,16). In addition, the length of the side chain groups in both the Glu^5 position and the Lys^{10} position were modified by addition or subtraction of methylene groups in the side chains of these residues. Thus at position 5, either an Asp or Glu residue was used, while at position 10, α,β-diaminoproponic acid, α,γ-diaminobutyric acid, ornithine, and lysine were substituted. Furthermore, since earlier studies had demonstrated that the N-terminal tripeptide -Ser-Tyr-Ser-, and the C-terminal tripeptide -Lys-Pro-Val- (or -Gly-Pro-Val- here) were not substantial contributors to the potency of melanotropin analogues, we also have prepared the truncated 4-10 linear and cyclic analogues of melanotropin. The results of these studies have been interesting. Almost all of the linear and cyclic analogues are as potent as or even more potent than α-MSH in the classical frog skin (*Rana pipiens*) or lizard skin (*Anolis carolinensis*) bioassays. Interestingly, we have shown in previous studies (reviewed in 17) that structure-activity relationship in the *Anolis* assay closely resembled bioactivities in mammalian assays, whereas the *Rana* assay did not. In this new class of compounds, the highest potency was found in the lizard skin assay. Furthermore, the cyclization generally increased potency by a factor of 10 or more in the assay, and several compounds were superagonists. Furthermore, cyclization also led to analogues with ultraprolonged biological activity (days *in vitro* to weeks *in vivo*). As an example, the cyclic analogue Ac-Nle-Asp-His-D-Phe-Arg-Trp-Lys-NH_2 Ac-[Nle^4,Asp^5,D-Phe^7,Lys^{10}]α-MSH is about 90 times more potent than α-MSH in the *Anolis* assay and has ultraprolonged activity in both the lizard and frog skin assays.

These studies demonstrate the power of utilizing both conformational and topographical consideration in the design of peptide hormone and neurotransmitter analogues. It is interesting to note that the cyclic analogues from above are about one half the size of the original peptide hormone and yet are superpotent, superprolonged acting, and likely highly stable to enzymatic degradation by proteases. The use of these compound in medicine and as templates for peptide mimetic design are of great current interest in our laboratories.

DESIGN OF ENKEPHALIN ANALOGUES WITH HIGH POTENCY AND SELECTIVITY FOR DELTA OPIOID RECEPTORS

The endogenous opioid peptides [Met^5]enkephalin (H-Tyr-Gly-Gly-Phe-Met-OH) and [Leu^5]enkephalin (H-Tyr-Gly-Gly-Phe-Leu-OH) can interact with a variety of different classes of opioid receptors that may elicit different physiological responses. Currently, at least three receptor subtypes, μ, δ, κ are generally accepted to exist, with two others, ε and σ, being proposed to be either receptor subtypes or separate receptors entirely.

However, isolation and careful study of these receptors has proven difficult in part because of the lack of adequately selective (10^3 fold greater binding of one receptor over another) analogues. Such selectivity is much preferred so that one can use biophysical and computer assisted methods to study the conformational and topographical requirements of a ligand for a specific membrane receptor with same assurance that information obtained is relevant to the ligand-receptor interaction.

The theme for our design of δ-selective ligands has been via the use of conformational and topographical constraints. The conformational constraint of the linear enkephalin involved side chain to side chain cyclization as in the case of the melanotropins. In addition, we used a second theme in our design, "topographical bias." In this latter method, we design analogues that will allow side chain moieties to assume a limited number of rotamer populations. As discussed below, this design approach was used only after conducting extensive biophysical (NMR) and computer-modeling studies to develop leads into what type of "biases" might be most appropriate.

Our original design for a conformationally constrained enkephalin analogue was based upon the concept of pseudoisosteric cyclization similar to that used for design of the superpotent melanotropin analogues (12). In the enkephalin molecule, the Gly^2 and the Met^5 could be substituted with a Cys residue to yield a pseudoisosteric cyclic analogue. It was previously shown (18) that [D-Cys^2,L-Cys^5]enkephalin amide possessed analgesic effects <u>in vivo</u>, and Schiller and co-workers (19) found that [D-Cys^2,L-Cys^5]enkephalinamide was potent, but not highly δ opioid receptor selective, in <u>in vitro</u> assays.

We thus decided to introduce further constraints in such cyclic analogues. It was previously found that penicillamine (β,β-dimethyl cysteine)-containing compounds reduced conformational flexibility when replaced in the tocin ring of oxytocin (1,20,21). Thus, we reasoned that the β,β-dimethyl substitution in the 2 and/or 5 positions of [Met^5]enkephalin could impart both local constraints (i.e. constrain the disulfide bridge to prefer the right-handed or left-handed helicity) as well as providing further constraints on the 14-membered disulfide-containing ring.

Synthesis of a series of analogues containing permutations of D- or L-cysteine or D- or L-penicillamine in the 2 and/or 5 positions, led in a short time to the synthesis of [D-Pen^2, D-Pen^5]-

enkephalin (DPDPE) which exhibited greater than a 3000 fold selectivity for the δ opioid receptor over the mu receptor (22).

Once this favorable conformational constraint was established and very high receptor selectivity achieved, we felt that compared to the acyclic enkephalins the conformational flexibility of the 14-membered ring of DPDPE would be sufficiently reduced so that physical studies could lead us to some reasonable conclusions regarding the conformational and topographical requirements for the δ-receptor.

We, therefore, undertook NMR studies of DPDPE in aqueous solvent (23). An important observation was that the diastereotopic alpha protons of Gly^3 were very non-equivalent ($\Delta\delta = 0.8$ ppm). This was most encouraging since it supported the notion that the system was indeed fairly constrained. Subsequently, it was found that the non-equivalence primarily was due to shielding and deshielding effects imparted by the peptide bond of residues 3 and 4 in the rigid backbone. In addition, nuclear Overhauser enhanced spectroscopy (NOESY) provided evidence for a transannular effect between the aromatic side chain moieties of Tyr^1 and Phe^4 and the β,β- dimethyl groups of D-Pen^2. Through extensive model building and energy minimization, two pairs of similar, energy minimized conformers were found. Despite dithering in the helicity of the disulfide bridge, it was found that the structures were amphiphilic in nature with the backbone forming a hydrophilic region (N-terminal carboxylate, peptide carbonyls) on one surface, and the side chains of Tyr^1, Phe^4, and D-Pen^2 forming a hydrophobic region on the other surface. The backbone best fit a Type IV β-turn (see Figure 3). Recently, these NMR experiments were repeated in the polar aprotic solvent DMSO-d^6 with results very similar to those found in water. This suggests that the conformational constraints used are sufficient to restrict the overall topography of DPDPE into a class of similar groups of conformations whatever the solvent environment (Prakash, Hruby, et al., unpublished results).

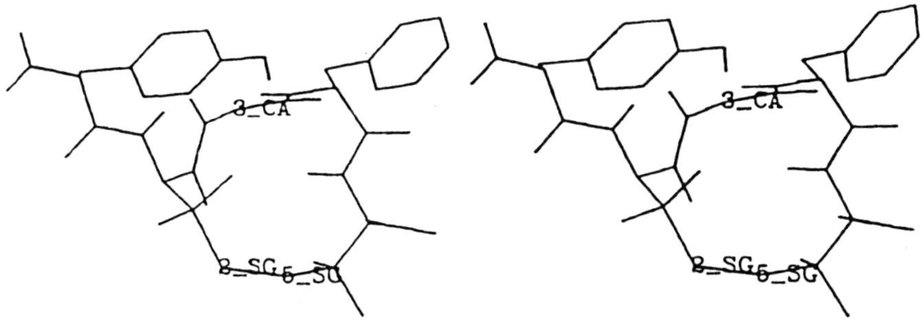

Figure 3. A proposed conformation of [D-Pen^2,D-Pen^5]enkephalin in solution based on NMR data and molecular mechanics calculations.

Recently, we have undertaken a dynamical search for possible low energy conformers that could satisfy the NMR criteria (24). Using a 200 picosecond dynamics trajectory followed by extensive minimization (quenching) (25) at one picosecond intervals, 200 different structures of DPDPE were generated. Although the distribution of conformers did not yield a skewed number of low energy preferred "crystal" states (rather, a Gaussian distribution was observed), the lowest energy conformers were qualitatively compared to determine if a "family" of conformers resided in this group. As was found in the previous NMR and modeling studies, amphilic structures once again predominated the low energy conformers. Again, the most common features were the distribution of hydrophobic (aromatic sidechains and penicillamine methyls) to one surface, and the backbone moieties to the other surface.

Our continuing efforts to design improved δ-opioid receptor selective peptides has directed us to analogues focused on the Phe^4 position. Although the aromatic side chain groups appear to prefer being on one surface relative to the disulfide ring, we hypothesized that the position of the aromatic rings relative to the backbone could be crucial in optimizing receptor selectivity. In other words, the χ_1 bond angle could play a major role as a receptor determinant. Thus, we have focused on the design and synthesis of Phe^4 analogue of DPDPE that could impart a "topographical biasing" to the side chain rotamer populations. To do this, we have synthesized the four diastereoisomers of [β-$MePhe^4$]DPDPE (i.e. the L-threo, L-erythro, D-threo, and D-erythro isomers). NMR evidence to date has been consistent with previous conclusions, that L-Phe^4 containing analogues maintain the large non-equivalence of the Gly^3 CH_2 protons, whereas D-amino acid analogues have CH_2 protons much more magnetically equivalent. Binding data has confirmed that the L-Phe^4 diastereoisomers, and more specifically [L-threo-Phe^4]DPDPE is the most potent and δ-selective compound. We are currently conducting NMR experiments to determine the predominant rotamer populations as well as to seek a correlation between the topography of the ring with the biological receptor selectivity.

In summary these studies indicate the design of peptide to include conformational constraints and topographical biasing in conjunction with extensive biophysical experiments and molecular modelling, can lead to highly potent and very selective analogues. The potential advantage of analogues with high receptor selectivity is that they will lead to a functional and physiological selectivity that will provide high therapeutic efficacy without the side effects and toxicity often produce by nonselective drugs. It also is hoped that future work will provide us with a detailed mapping of the δ-receptor as well as the rest of the opioid receptors.

DESIGN OF OXYTOCIN ANTAGONIST LIGANDS

Oxytocin has a long and storied history in the design of peptide hormones and neurotransmitters being the first peptide hormone to be isolated, its structure determined and proven by total synthesis. Oxytocin H-Cys-Tyr-Ile-Gln-Asn-Cys-Pro-Leu-Gly-NH_2 is a cyclic disulfide-containing peptide, that also contains an acyclic moiety. The acyclic moiety is not necessary for its agonist activity (the major bioactivities of oxytocin at peripheral receptors are its uterotonic-uterine contracting, and its milk ejecting activities), but the potency is greatly reduced (for reviews see 26-28). It also has a number of CNS activities (26). Oxytocin has served as a key peptide in the development of most strategies of peptide design including the following: 1. examination of the function groups that make up the molecule (-S-S, N-terminal NH_2, C-terminal $CONH_2$, Asn and Gln carboxamides, Tyr hydroxyl, etc.); 2. size of the ring - larger or smaller; 3. stereochemistry at individual amino acid residues; 4. linear vs. cyclic structure; 5. QSAR for peptides; 6. conformation-biological activity relationships; and 7. development of receptor antagonists. The latter activity will be the focus of our discussions here.

From the earliest studies, the search for oxytocin antagonist analogues was a central issue in the laboratories examining structure-activity relationships, and some of the chief strategies to search for antagonists have been developed from these early studies. Among these findings was the observation of partial agonist activity, and this was recognized to provide a lead for the design of antagonist analogues. In order to discover such leads, it became obvious that full dose-response curves are necessary to uncover partial agonist activity. Among the first good antagonist analogues was the 1-penicillamine-substituted analogue of oxytocin [Pen^1]oxytocin, H-Pen-Tyr-Ile-Gln-Asn-Cys-Pro-Leu-Gly-NH_2 (29). The only difference in structure between this antagonist and the native hormone is the replacement of the two β-protons of Cys^1 by β-methyl groups (Pen = β,β-dimethylcystine). This raised the question of the structural, conformation, and dynamic properties responsible for coverting the agonist hormone into an antagonist analogue. Conformational studies using NMR and other biophysical methods led to the conclusion that oxytocin and [Pen^1]oxytocin had different conformational and dynamic properties, and it was proposed that these properties were specifically responsible for antagonist activity (20). These ideas subsequently were tested in several ways, and ultimately led to the conclusion the agonist and antagonist analogues possess different modes of interaction with the oxytocin receptor, and hence will show different structure-biological activity relationships (for discussions, see 30,31). These results demonstrate two important ideas that must be considered in the design of peptide hormone or neurotransmitter antagonists. First, to obtain antagonist analogues, you need a lead. To obtain a lead you must have appropriate bioassays that will be capable of differentiating antagonist activity as well as partial agonist/antagonist activity. Second, agonists and antagonists (including competitive antagonists) have different modes of interaction with their receptors. Thus design of antagonists generally will involve different considerations than those used to design agonists. We would emphasize this latter point especially. Though in some cases, there is overlap in structure-activity relationships, the correspondence of agonist and antagonist potency is generally not one to one even in the most favorable cases. Time and again we note in the literature, investigators utilize the "lessons" learned from agonist design to try and improve antagonists. It is not surprising that often these approaches do not work. Thus, it is important when antagonist leads are discovered to examine new structural, conformational, and dynamic properties in designing better antagonist analogues. We will now demonstrate each of these features with selected examples.

Replacement of Tyr^2 by Leu^2 in oxytocin led to a huge decrease in biological potency (>1000 fold at the uterotonic receptor) (32). However, when a similar substitution was made in [Pen^1]oxytocin to give the analogue [Pen^1,Leu^2]oxytocin a more potent antagonist analogue was obtained (33). Even more dramatic is the recent finding of Manning and co-workers that certain linear analogues of oxytocin and vasopressin are potent receptor antagonists (34).

Conformational considerations also are very important. In the case of oxytocin agonists, it appears that a β-turn in the ring portion of oxytocin, and conformational flexibility at the disulfide bridge, the Tyr^2 side chain, and the Asn^5 side chain groups, are important to agonist activity. We have examined further the topographical requirements of the 2 position by substitution of L(D)-tetrahydroisoquinoline carboxylate [L(D)-Tic), a constrained analogue of phenylalanine (see Figure

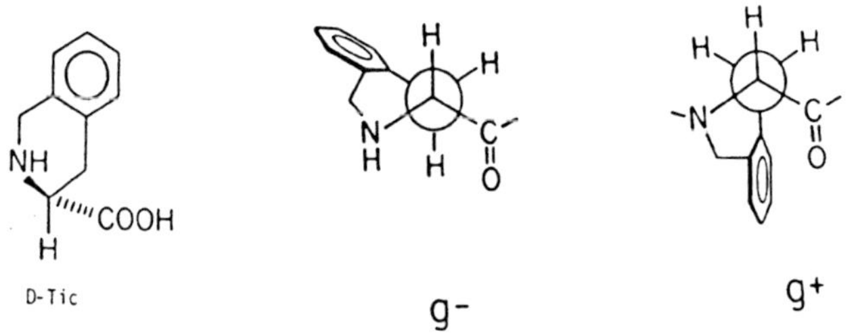

Figure 4. The structure of D-tetrahydroisoquinoline carboxylic acid (D-Tic) and its gauche(-) and gauche(+) conformations when part of a peptide.

4) that can exist only in gauche(+) and gauche(-) conformations at the χ_1 torsional angle. When the L-amino acid residue was substituted into the 2 position of oxytocin to give [Tic^2]oxytocin, essentially a complete loss in agonist activity was observed and instead the compound became a very weak antagonist. Based on NMR studies, the Tic^2 aromatic side chain group existed exclusively in the gauche(+) conformation. In the antagonist series, the [D-Tic^2]oxytocin analogue was found to have antagonist potency comparable to [Pen^1]oxytocin (35). Based on NMR and CD studies, the Tic^2 side chain group was found to be exclusively in the gauche(+) conformation, and molecular modeling indicated that a low energy conformation virtually identical to one found for [D-Phe^2]oxytocin, an antagonist, was observed (35). Further examination of new ways to examine the topographical features important to interaction with the oxytocin receptor antagonists is indicated, and should provide important new insights to oxytocin structure-activity relationship in the antagonist series.

Finally, we would emphasize the importance of dynamic considerations in the design of peptide antagonists. Peptide hormone and neurotransmitter biological activities require not only receptor binding, but also transduction, and we have suggested that the transduction can require structural features different from those important to receptor recognition. Thus, conformational and dynamic changes in the hormone generally must accompany hormone-receptor transduction to attain the biologically active conformation of the hormone-receptor complex. Recently, the X-ray structure of deamino-oxytocin was determined (36). An interesting and remarkable feature of the X-ray structure was that it actually consisted of two conformations, in which the major difference was the presence of a right handed disulfide bridge for one conformation and a left handed disulfide bridge for the other. The availability of the X-ray coordinates, the interesting dynamic properties of the peptide even in the crystalline state, and the intriguing suggestions regarding structure-activity relationships that came out of these studies, led us to computer assisted molecular modeling studies in which we examined further possibilities for side chain to side chain cyclizations that would be pseudoisosteric with specific side chain groups. In the course of these investigations, we noted that when a Lys^8 residue was placed in oxytocin, interactions with the Gln^4 side chain group were possible. We, therefore, decided to synthesize the monocyclic analogue of oxytocin [β-Mpa^1,Glu^4,Lys^4]oxytocin and then convert it to the bicyclic analogue β-Mpa^1,Glu^4,Cys^6,Lys^8]-oxytocin, by formation of a lactam bridge between the Glu^4 and Lys^8 side chain moieties (37). The results obtained were very interesting. The monocyclic analogue was found to have very weak agonist activity (about 1/1400 the potency of oxytocin), but was a full agonist in the rat uterus (oxytocic) assay. Based on the assumption that bioactivity is directly proportional to binding activity, the EC_{50} for interaction with the rat uterus oxytocin receptor is about $4x10^6$ molar. Bicyclization produced a dramatic change in both biological activity and receptor recognition. The bicyclic analogue had no agonist activity and instead was found to be a very potent antagonist with a pA_2 value of 8.2, (i.e. an EC_{50} of about $6x10^9$ molar). Thus, it binds almost 3 orders of magnitude better than the monocyclic agonist to the rat uterus receptor. Preliminary NMR studies indicate that the bicyclic oxytocin analogue is quite rigid and possesses a new and unique conformation among oxytocin analogues. This strongly suggest that the topographical features of the analogue are such that the binding elements are disposed favorably for interaction with the uterine receptor, but at the same time, they cannot rearrange themselves in three dimensional space for transduction of the biological response. Thus, this analogue and further modified analogues should provide unique insights into the differential requirements for agonist and antagonist activity. Finally, these results are consistent with the dynamic model of hormone action that we proposed some time ago (20).

CONCLUSIONS

In this paper we have briefly outlined structural, conformational, topographical, and dynamic approaches that can be utilized in the design of peptide ligands that interact with membrane bound receptors. Though considerable progress has been made and some general principles have emerged (some of which are illustrated in this paper), there is still much that needs to be learned to render these principles applicable in the general case. Especially relevant to the discussions of this paper are the need for more careful considerations of conformations and topographical features in peptide ligand design. There is an urgent need for the development of methods of asymmetric synthesis that will readily provide the unusual amino acids needed for the control of proper conformational and topographical features in highly constrained peptide systems. While we plan to pursue these methodologies ourselves, we are hopeful that our discussion here will encourage others. In our opinion, the scientific payoff will be substantial, providing not only new design principles for peptide ligands, but also fundamental new insights into the protein folding problem in its broadest sense.

ACKNOWLEDGEMENTS

This work was supported by grants of the U. S. Public Health Service DK17420, NS 19972 and DA 06284 and by a grant from the National Science Foundation. We thank Cheryl McKinley for careful typing and editing of this manuscript.

REFERENCES

1. V.J. Hruby, Structure and conformation related to the activity of peptide hormones, in: "Perspectives in Peptide Chemistry," A. Eberle, R. Geiger and T. Wieland, eds., S. Karger, Basel, pp. 207-220 (1981).
2. V.J. Hruby and H.I. Mosberg, Conformational and dynamic considerations in peptide structure-function studies, Peptides **3**: 329 (1982).
3. V.J. Hruby, Design of peptide superagonists and antagonists: conformational and dynamic considerations, in: "Conformationally Directed Drug Design," J.A. Vida and M. Gordon, eds., ACS Symposium Series No. 251, Washington, D.C., pp. 9-27 (1984).
4. V.J. Hruby, Conformational restrictions of biologically active peptides via amino acid side chain groups, Life Sciences **31**: 189 (1982).
5. C. Toniolo, Conformationally restricted peptides through short-range cyclization, Int. J. Peptide Protein Res. **35**: 287 (1990).
6. V.J. Hruby, F. Al-Obeidi and W. Kazmierski, Emerging approaches in the molecular design of receptor selective peptide ligands: Conformational, topographical and dynamic considerations, Biochemical J., in press (1990).
7. W. Kazmierski and V.J. Hruby, A new approach to receptor ligand design: Synthesis and conformation of a new class of potent and highly selective μ opioid antagonists utilizing tetrahydroisoquinoline carboxylic acid, Tetrahedron **44**: 697 (1988).
8. V.J. Hruby, Designing molecules: Specific peptides for specific receptors, Epilepsia **30** (Suppl. 1): S42 (1989).

9. T.K. Sawyer, P.J. Sanfilippo, V.J. Hruby, M.H. Engel, C.B. Heward, J.B. Burnett and M.E. Hadley, [Nle^4,D-Phe^7]α-Melanocyte stimulating hormone: A highly potent α-melanotropin with ultralong biological activity, Proc. Natl. Acad. Sci. U.S.A. **77**: 5754 (1980).

10. M.E. Hadley, B. Anderson, C.B. Heward, T.K. Sawyer and V.J. Hruby, Calcium-dependent prolonged effects on melanophores of [4-Norleucine, 7-D-Phenylalanine]-α-Melanotropin, Science **213**: 1025 (1981).

11. A.M. de L. Castrucci, M.E. Hadley and V.J. Hruby, Melanotropin enzymology, in: "The Melanotropic Peptides. Vol. I: Source, Synthesis Chemistry, Secretion, Circulation, and Metabolism," M.E. Hadley, ed., CRC Press, Boca Raton, pp. 171-182 (1988).

12. T.K. Sawyer, V.J. Hruby, P.S. Darman and M.E. Hadley, [4-Half-cystine, 10-half-cystine]α-Melanocyte stimulating hormone: A cyclic α-melanotropin exhibiting superagonist biological activity, Proc. Natl. Acad. Sci. U.S.A. **79**: 1751 (1982).

13. W.L. Cody, M.E. Hadley and V.J. Hruby, "The Melanotropic Peptides. Vol. III, Mechanisms of Action and Biomedical Applications," M.E. Hadley, ed., CRC Press, Boca Raton, pp. 75-92 (1988).

14. F. Al-Obeidi, M.E. Hadley, B.M. Pettitt and V.J. Hruby, Design of a new class of superpotent cyclic α-melanotropins based on quenched dynamic simulations, J. Am. Chem. Soc. **111**, 3413 (1989).

15. F. Al-Obeidi, V.J. Hruby, A.M. de L. Castrucci and M.E. Hadley, Design of potent linear α-melanotropin analogues modified in positions 5 and 10, J. Med. Chem. **32**: 174 (1989).

16. F. Al-Obeidi, A.M. de L. Castrucci, M.E. Hadley and V.J. Hruby, Potent and prolonged acting cyclic lactam analogues of α-melanotropin: Design based on molecular dynamics, J. Med. Chem. **32**: 2555 (1989).

17. V.J. Hruby, B.C. Wilkes, W.L. Cody, T.K. Sawyer and M.E. Hadley, Melanotropins: Structural, conformational and biological considerations in the development of superpotent and superprolonged analogs, Peptide Protein Res. **3**: 1 (1984).

18. D. Sarantakis, Peptides with morphine-like activity, U. S. Patent 4098781 (1979).

19. P.W. Schiller, B. Eggimann, J. DiMaio, C. Lemieux, and T.M.-D. Nguyen, Cyclic enkephalin analogs containing a cystine bridge, Biochem. Biophys. Res. Commun. **101**: 373 (1981).

20. J.-P. Meraldi, V.J. Hruby and A.I.R. Brewster, Relative conformational rigidity in oxytocin and [1-penicillamine]oxytocin: A proposal for the relationship of conformational flexibility to peptide hormone agonism and antagonism, Proc. Natl. Acad. Sci. U.S.A. **74**: 1373 (1977).

21. H.I. Mosberg, V. J. Hruby and J.P. Meraldi, Conformational study of the potent peptide hormone antagonist [1-penicillamine, 2-leucine]-oxytocin in aqueous solution, Biochemistry, **20**, 2822 (1981).

22. H.I. Mosberg, R. Hurst, V.J. Hruby, K. Gee, H.I. Yamamura, J.J. Galligan and T.F. Burks, Bis-penicillamine enkephalins possess highly improved specificity toward delta opioid receptors, Proc. Natl. Acad. Sci. U.S.A., **80**, 5871 (1983).

23. V.J. Hruby, L.-F. Kao, B.M. Pettitt and M. Karplus, The conformational properties of the delta opioid peptide [D-Pen^2, D-Pen^5]enkephalin in aqueous solution determined by NMR and energy minimization calculations, J. Am. Chem. Soc., **110**, 3351 (1988).
24. B.M. Pettitt, T. Matsunaga, F. Al-Obeidi, C. Gehrig, V.J. Hruby and M. Karplus, Dynamical search for bis-penicillamine enkephalin conformations: Quenched molecular dynamics, Biopolymers, submitted
25. R. Elber and M. Karplus, Multiple conformational states of proteins: A molecular dynamics analysis of myoglobin, Science **235**: 318 (1987).
26. V.J. Hruby, M.S. Chow and D.D. Smith, Conformational and structural considerations in oxytocin receptor binding and biological activity, Ann. Rev. Pharmacol. Toxicol. **30**: 501 (1990).
27. C.W. Smith, ed., "The Peptides: Analysis, Synthesis, Biology. Vol. 8. Chemistry, Biology, and Medicine of Neurohypophyseal Hormones and Their Analogs," Academic Press, New York (1987).
28. K. Jost, M. Lebl and F. Brtnik, eds., "Handbook of Neurophypophyseal Hormone Analogs," Vol. I, Parts 1 and 2, Vol. II, Parts 1 and 2, CRC Press, Boca Raton (1987).
29. H. Schulz and V. du Vigneaud, Synthesis of 1-L-penicillamine-oxytocin, 1-D-penicillamine-oxytocin and 1-deaminopenicillamine-oxytocin, potent inhibitors of the response of oxytocin, J. Med. Chem. **9**: 647 (1966).
30. V.J. Hruby and H.I. Mosberg, Structural, conformational and dynamic considerations in the development of peptide hormone antagonists, in: "Hormone Antagonists," M.K. Agarwal, ed., W. de Gruyter, Berlin, pp. 433-474 (1982).
31. V.J. Hruby, Implications of the x-ray structure of deamino-oxytocin to agonist/antagonist-receptor interactions, Trends in Pharmacol. Sci. **8**: 336 (1987).
32. V.J. Hruby and V. du Vigneaud, Synthesis and some pharmacological activities of [2-L-valine]-oxytocin and [2-L-leucine]-oxytocin, J. Med. Chem. **12**: 731 (1969).
33. V.J. Hruby, K.K. Deb, D.M. Yamamoto, M.E. Hadley and W.Y. Chan, [1-Penicillamine, 2-leucine]oxytocin: Synthesis, pharmacological and conformational studies of a potent peptide hormone inhibitor, J. Med. Chem. **22**: 7 (1979).
34. M. Manning, J. Przybylski, A. Olma, W.A. Klis, M. Kruszynski, N.C. Wo, G.H. Pelton and W.H. Sawyer, No requirement of cyclic conformation of antagonists in binding to vaspressin receptors, Nature **329**: 839 (1987).
35. M. Lebl, P. Hill, W. Kazmierski, W.Y. Chan, L. Karasova, J. Slaninova, I. Fric and V.J. Hruby, Conformationally restricted analogues of oxytocin: Stabilization of inhibitor conformation, Int. J. Peptide Protein Res., in press (1990).
36. S.P. Wood, I.J. Tickle, A.M. Treharne, J.E. Pitt, Y. Mascarenkas, J. Y. Li, J. Husain, S. Cooper, T.L. Blundell, V.J. Hruby, A. Buku, A.J. Fishman and H.R. Wyssbrod, Crystal structure analysis of deamino-oxytocin: Conformational flexibility and receptor binding, Science **232**: 633 (1986).

37. P.S. Hill, D.D. Smith, J. Slaninova and V.J. Hruby, Bicyclization of a weak oxytocin agonist produces a highly potent oxytocin antagonist, J. Am. Chem. Soc. **112**: 3110 (1990).

STEREOSELECTIVE SYNTHESIS OF BIOLOGICALLY AND PHARMACOLOGICALLY IMPORTANT CHEMICALS WITH MICROBIAL ENZYMES

Sakayu Shimizu and Hideaki Yamada

Department of Agricultural Chemistry
Kyoto University
Kitashirakawa, Sakyo-ku, Kyoto 606
Japan

INTRODUCTION

In recent years, the most significant development in the field of synthetic chemistry has been the application of biological systems to chemical reactions. Reactions catalyzed by enzymes or enzyme systems display far greater specificities than more conventional forms of organic reactions, and of all the reactions available, some of which have been shown to be useful for synthetic or biotechnological applications.

We have rcently been carring out studies on the synthesis of various biologically, pharmacologically or chemically useful coenzymes, amino acids and other chemicals, using microbial enzymes as catalysts (1-6). Here, we summarize the results of our recent works on the production of D-pantothenate and L-carnitine, which included the evaluation of useful enzymatic reactions for the synthesis, screening of potent microorganisms as catalysts, characterization of the responsible enzymes for the conversions and determination of reaction conditions for the practical preparation.

SYNTHESIS OF D-(+)-PANTOTHENIC ACID AND RELATED COMPOUNDS

At present, commercial production of pantothenate depends exclusively on chemical synthesis. The conventional chemical process involves reactions yielding racemic pantoyl lactone from isobutyraldehyde, formaldehyde and cyanide, optical resolution of the racemic pantoyl lactone to D-(-)-pantoyl lactone with quinine, quinidine, cinchonidine, brucine and so on and condensation of D-(-)-pantoyl lactone with β-alanine. This is follwed by isolation as the calcium salt and drying to obtain the final product. A problem of this chemical process apart from the use of poisonous cyanide is the troublesome resolution of the racemic pantoyl lactone and the reracemization of the remaining L-(+)-isomer. Therefore, most of recent studies in this area have been concentrated to development of efficient method to obtain D-(-)-pantoyl lactone.

To skip this resolution-reracemization step, several microbial or enzymatic methods have been proposed. They are roughly fall into two types based on the starting substrate used (1,7).

Use of Prochiral Ketones

Recently, we developed an efficient one-pot synthesis method of ketopantoyl lactone, in which it is synthesized from isobutyraldehyde, sodium methoxide, diethyl oxalate and formalin (Fig. 1). The reaction is performed in one step at room temperature with a yield of 81.0% (8). Ketopantoyl lactone is a very promising starting material for the synthesis of D-(-)-pantoyl lactone, because it might permit several microbiological approaches leading to D-(-)-pantoyl lactone or D-(+)-pantothenate, as shown in Fig. 2. Thus, we assayed variety of microorganisms as to their reducing ability using several prochiral carbonyl compounds, such as ketopantoyl lactone, ketopantoic acid, ethyl 2'-ketopantothenate and 2'-ketopantothenonitrile.

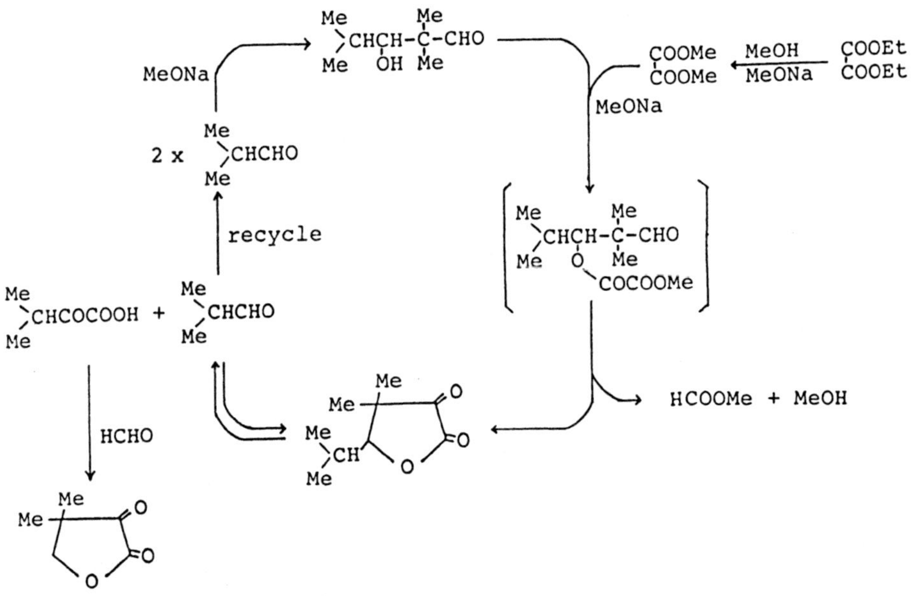

Fig. 1. The proposed reaction pathway for the one-pot synthesis of ketopantoyl lactone from isobutyraldehyde, diethyl oxalate and formalin.

Conversion of ketopantoyl lactone to D-(-)-pantoyl lactone. This conversion was assayed at pH 4-6 by incubating ketopantoyl lactone (10 mg/ml) in the culture broth, which had been grown with each test microorganism, for 2 days at 28°C. Many microorganisms were found to convert the added ketopantoyl lactone to pantoyl lactone (Fig. 3a). However, the ratios of D- and L-isomers of formed pantoyl lactone were randomly distributed among the strains tested and the stereospecificity shown by the tested strains almost completely showed no relation to the genera or sources. For example, *Mucor racemosus* produced almost specifically the L-isomer with more than 85% molar yield. On the other hand, *Mucor javanicus* yielded a racemic mixture with 59% yield. Among 9 strains of *Rhodotorula glutinis*, which produced pantoyl lactone with greater than 90% molar yields, 5 gave racemic mixtures, 2 gave the L-isomer predominantly and the remaining 2 gave the D-isomer with more than 70% ee (9).

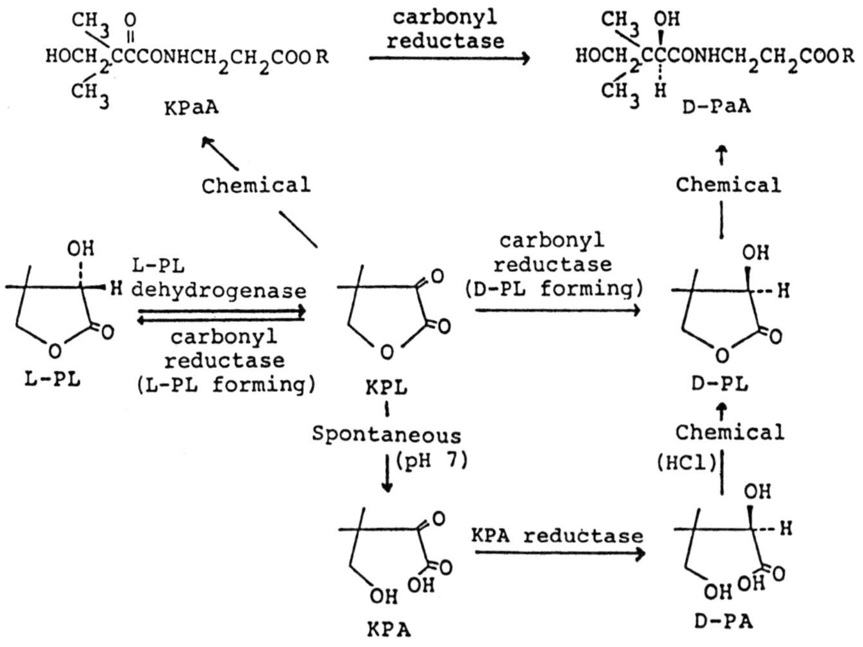

Fig. 2. Reactions involved in the enzymatic transformation to D-(-)-pantoyl lactone or D-(+)-pantothenate. L-PL, L-(+)-pantoyl lactone; D-PL, D-(-)-pantoyl lactone; KPL, ketopantoyl lactone; KPA, ketopantoic acid; D-PA, D-(-)-pantoic acid; KPaA, 2'-ketopantothenate; D-PaA, D-(+)-pantothenate.

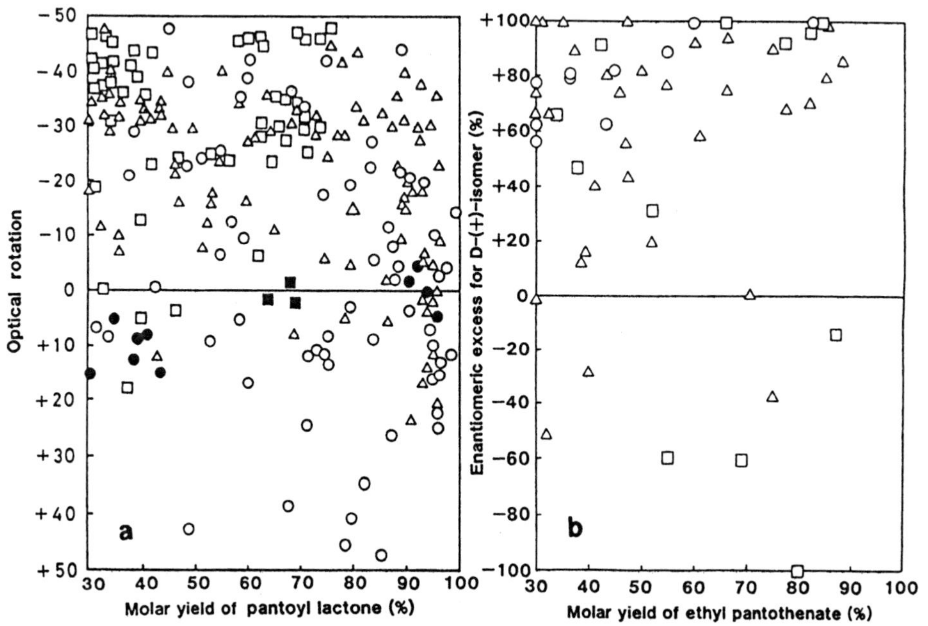

Fig. 3. Diversity of microbial reduction of ketopantoyl lactone (a) and ethyl 2'-ketopantothenate (b). △ , yeasts; ○ , molds; □ , bacteria; ■ , actinomycetes; ● , basidiomycetes.

Practical stereospecific reduction of ketopantoyl lactone to D-(-)-pantoyl lactone was carried out with washed cells of *Rhodotorula minuta* or *Candida parapsilosis* as a catalyst and glucose as energy for the reduction. About 50 or 90 g/liter of D-(-)-pantoyl lactone (94 or 98% ee, respectively) was produced with a molar yield of nearly 100% by *Rhodotorula minuta* and *Candida parapsilosis*, respectively (Fig 4) (8,10).

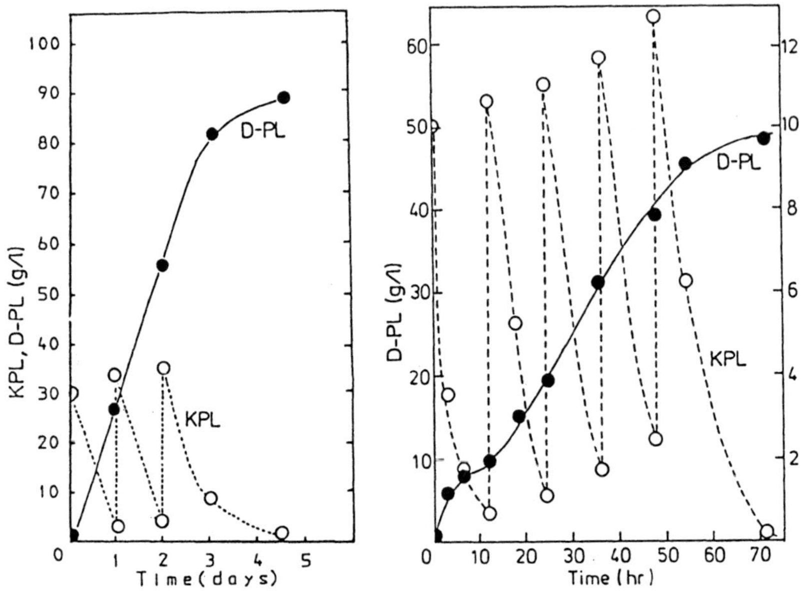

Fig. 4. Stereoselective conversion of ketopantoyl lactone (KPL) to D-(-)-pantoyl lactone (D-PL) by *Candida parapsilosis* (left) and *Rhodotorula minuta* (right).

The enzyme catalyzing the asymmetric reduction of ketopantoyl lactone was isolated in a crystalline form from the cells of *Candida parapsilosis* and characterized in some detail (11, also see Table 1 and 2). It is a novel NADPH-dependent carbonyl reducatse with a molecular mass of about 40,000. In addition to ketopantoyl lactone, the enzyme catalyzes a variety of cyclic diketones, including derivatives of ketopantoyl lactone, isatin, camphorquinone and so on and gave the corresponding *R*-alcohols (11,12). We named the enzyme "conjugated polyketone reducatse", since the enzyme catalyzes only the reduction of conjugated polyketones as follows,

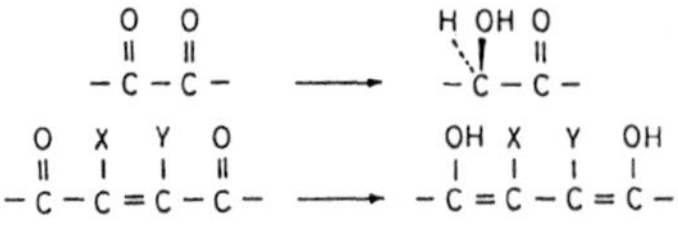

X, Y = H or alkyl

The enzyme yielding the antipode, L-(+)-pantoyl lactone, was also isolated from *Mucor ambiguus* cells (13). It is also a kind of "conjugated polyketone reductase" and consists of two polypeptide chains with an identical molecular mass of about 27,500 (see Table 1 and 2). The occurrence of two kinds of enzymes that show similar substrate specificity but are different each other in their stereospecificity may be one of the possible reasons why the reduction of unnatural ketopantoyl lactone resulted in the formation of the D- and L-isomers in varying ratios as shown in Fig. 3a.

Table 1. Properties of the Carbonyl Reductases Purified from Various Microorganisms

	Polyketone reductase (*C.parapsilosis*)	Polyketone reductase (*M.ambiguus*)	KPA reductase (*P.maltophilia*)	K-PaOEt reductase (*C.macedoniensis*)	Aldehyde reductase (*S.salmonicolor*)
Native Mr	37,000	56,000	116,000	45,000	37,000
Subunit Mr	41,600	27,000	30,500	42,000	37,000
$s_{20,w}$ (S)	4.8	-	7.75	-	2.85
pI	6.3	6.4	3.5	5.5	4.7
Absorption maximum (nm)	278	276	276	278	278
E (1%)	8.3	-	20.0	-	12.2
COOH amino acid	-	-	Phe	Thr	Lys
Km (mM)	0.33 (KPL)	0.71 (KPL)	0.40 (KPA)	2.50 (K-PaOEt)	0.36 (CAAE)
Vmax (μmol/min/mg)	481 (KPL)	541 (KPL)	1,310 (KPA)	120 (K-PaOEt)	144 (CAAE)
Cofactor	NADPH	NADPH	NADPH	NADPH	NADPH
Optimum pH	7.0	6.0	6.0	6.5	7.0
Optimum temp.(°C)	40	40	37	40	60
pH stability	6.0-7.5	5.5-7.0	6.0-10	4.5-10.5	6.5-8.5
Thermal stability	42%(40°C,10min)	75%(45°C,10min)	90%(60°C,10min)	100%(55°C,10min)	100%(45°C,10min)
Inhibitor	quercetin	quercetin	-	-	dicoumarol
Reaction mechanism	ping-pong	ping-pong	ordered Bi-Bi	-	ordered Bi-Bi
Enzyme formation	constitutive	constitutive	constitutive	constitutive	constitutive
Reference	(11)	(13)	(14)	-	(26)

KPA, ketopantoic acid; KPL, ketopantoyl lactone; K-PaOEt, ethyl 2'-ketopantothenate; CAAE, etheyl 4-chloro-acetoacetic acid.

Table 2. Comparison of Substrate Specificities of the Carbonyl Reductases Purified from Various Microorganisms[a]

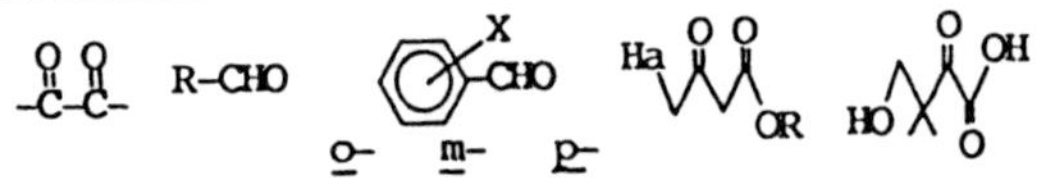

		R-CHO	o-	m-	p-		
K-PaOEt reductase (*C. macedoniensis*)	○		○			○	
Polyketone reductase (*C. parapsilosis*)	○						
(*M. ambiguus*)	○						
Aldehyde reductase (*S. salmonicolor*)		○	○	○	○	○	
KPA reductase (*P. maltophilia*)							○

[a] Open circles indicate that enzymes can reduce the substrates indicated in the table. Ha = Cl, Br, F or N_3.

Since the above-mentioned enzymes require NADPH for the reduction of ketopantoyl lactone, there must be regenerating reaction(s) for NADPH coupled with the reduction reaction in the cells. Experiments using cell-free extracts of *Candida parapsilosis* demonstrated that hexokinase, glucose 6-phosphate dehydrogenase and 6-phosphogluconate dehydrogenase were involved in the regeneration of NADPH. Cellular levels of these enzymes in *Candida parapsilosis* were almost the same as those in other common yeasts such as baker's yeast and brewer's yeast, while considerable higher activity of "conjugated polyketone reductase" was detected in this yeast. Figure 5 outlines the mechanism for this reduction.

Conversion of ketopantoic acid to D-(-)-pantoic acid. The stereospecific reduction of ketopantoic acid to D-(-)-pantoic acid by ketopantoic acid reductase (EC 1.1.1.169), which is involved in the pantothenate biosynthesis pathway, is also a promising reaction for the

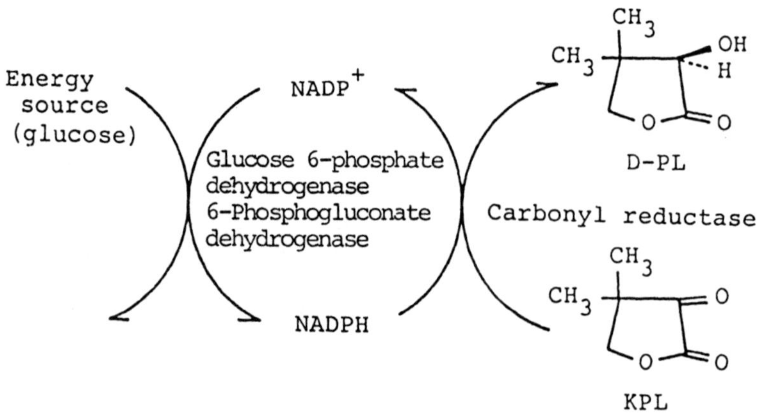

Fig. 5. The proposed mechanism for the reduction of ketopantoyl lactone (KPL) to D-(-)-pantoyl lactone (D-PL) coupled with the regeneration of NADPH.

same purpose, because the ring-opening of ketopantoyl lactone to yield ketopantoic acid is easy and ketopantoic acid reductase shows absolute stereospecificity for D-(-)-pantoic acid (14). When the same screeining as described above was performed at pH 7 to 8, under which the added ketopantoyl lactone underwent rapid and spontaneous hydrolysis to ketopantoic acid, we observed a quite different distribution profile of the reducing activity. As expected, most of the microorganisms which showed high reducing activity were found to produce only the D-isomer. Through this screening, we found that many bacteria belonging to the genus Agrobacterium almost specifically produce the D-(-)-isomer (>96% ee) in high yields. On incubation with a soil isolate, Agrobacterium sp. S-246, the yield of D-(-)-pantoic acid reached 119 g/liter (molar yield, 90%; optical purity, >98% ee) (15).

Ketopantoic acid reducatse was isolated in a crystalline form from one of the potent D-(-)-pantoic acid producers, Pseudomonas maltophilia and characterized in some detail (14, see also Table 1 and 2). It is an NADPH-dependent enzyme and is strictly specific to ketopantoic acid. The observation that mutants lacking this enzyme require either D-(-)-pantoic acid or pantothenic acid for growth and the revertants regain this activity indicates that it is involved in the pantothenate biosynthesis.

Reduction of 2'-ketopantothenate derivatives. We recently found that the rate of condensation of ketopantoyl lactone or D-(-)-pantoyl lactone with ethyl β-alanine, that yields ethyl 2'-ketopantothenate or ethyl D-(+)-pantothenate, respectively, is quite fast in comparison with that of the condensation of ketopantoyl lactone or D-(-)-pantoyl lactone with β-alanine, and that it proceeds more stoichiometrically (16). Since the enzymatic hydrolysis of ethyl D-(+)-pantothenate has already been established (17), if the stereoselective reduction of ethyl 2'-ketopantothenate to ethyl D-(+)-pantothenate is possible, both the troublesome resolution and the incomplete condensation might be avoided at the same time. Thus, we assayed the reducing ability toward ethyl 2'-ketopantothenate of a variety of microorganisms (see Fig. 3b). Several yeast strains, such as Pichia aganobii, Hansenula miso and Candida macedoniensis, exhibited high ability to reduce the 2'-ketopantothenate ester. Under the optimal conditions, Candida macedoniensis converted ethyl 2'-ketopantothenate (80 g/liter) almost specifically to ethyl D-(+)-pantothenate (>98% ee), with a molar yield of 97.2% (18). In a similar manner, 2'-ketopantothenonitrile (50 g/liter) was converted to D-(+)-pantothenonitrile (93.6% ee), with a molar yield of 95.6%, on incubation with Sporidiobolus salmonicolor cells as a catalyst (19).

The enzyme catalyzing these conversions was isolated from Candida macedoniensis and characterized in some detail (see Table 1 and 2). The enzyme shows broad substrate specificity; not only conjugated polyketones, but also aromatic aldehydes and 4-haloacetoacetic acid esters are reduced. The enzyme gives only the D-isomer on reduction of ethyl 2'-ketopantothenate, whereas it gives a mixture of the D- and L-isomers in a ratio of 4:1 on reduction of ketopantoyl lactone. The reduction product from ethyl 4-chloroacetoacetate is ethyl (S)-4-chloro-3-hydroxybutanoate (95% ee).

These enzymatic methods are simple and require no reracemization step, which is necessary for the conventional chemical resolution.

Use of Racemic Pantoyl Lactone

Racemic pantoyl lactone can also be used as a starting substrate (20,21). We found that Nocardia asteroides cells specifically oxidizes the L-(+)-isomer in a racemic mixture of pantoyl lactone to ketopantoyl lactone, which is then converted to D-(-)-pantoyl lactone by the reduction with Candida parapsilosis cells as described above. In these two enzymatic steps, the coexisting D-(-)-isomer remains without any modification (see Fig. 2). Under suitable conditions, 72 g/liter of D-(-)-pantoyl lactone was obtained from 80 g/liter of DL-pantoyl lactone (20). Agrobacterium sp. 246 also can used in place of Candida parapsilosis. Similar specific oxidation and reduction reactions can also be carried out with a single microorganism as catalyst (21). On incubation with washed cells of Rhodococcus erythropolis, D-(-)-pantoyl lactone in the reaction mixture reached 18.2 g/liter with a molar yield of 90.5% (optical purity, 94.4% ee). This unique conversion proceeds through the successive reactions as follows: (1) the enzymatic oxidation of L-(+)-pantoyl lactone to ketopantoyl lactone, the same enzyme as that in N. asteroides has been suggested to be the responsible enzyme for this oxidation, (2) the rapid and spontaneous hydrolysis of ketopantoyl lactone to ketopantoic acid, and (3) the enzymatic reduction of the ketopantoic acid to D-(-)-pantoic acid. The enzyme catalyzing this reduction seemed to be ketopantoic acid reductase, because Rhodococcus erythropolis cells could not utilize ketopantoyl lactone as substrate, different from Candida parapsilosis.

The enzyme catalyzing specific oxidation of L-(+)-pantoyl lactone to ketopantoyl lactone was purified from Nocardia asteroides cells. It is a flavoprotein with a molecular mass of about 600,000. Non covalently bound FMN was identified as the responsible cofactor for the oxidation.

Stereospecific hydrolysis of pantoyl lactone. Optical resolution of unmodified DL-pantoyl lactone can be carried out by a specific fungal hydrolase, as shown in Fig. 6. We found that many mold strains belonging to the genera Fusarium, Giberella and Cylindrocarpon specifically hydrolyze D-(-)-pantoyl lactone. When Fusarium oxysporum cells were incubated in 70% (w/v) aqueous solution of DL-pantoyl lactone for 24 h at 30°C with automatic pH control (pH 6.8-7.2), about 90% of the D-isomer was hydrolyzed. The resultant D-(-)-pantoic acid in the reaction mixture showed a high optical purity (96% ee) and the coexisting L-isomer remained without any modification. Further studies are now in progress in our laboratory.

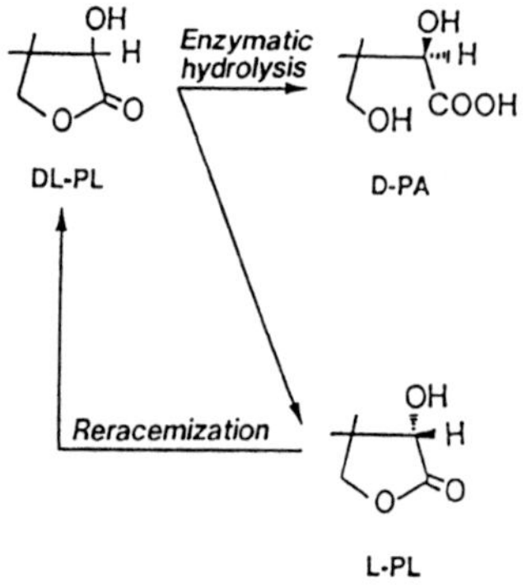

Fig. 6. Enzymatic resolution of racemic pantoyl lactone. DL-PL, DL-pantoyl lactone; L-PL, L-(+)-pantoyl lactone; D-PA, D-(-)-pantoic acid.

Many chemical, enzymatic and microbiological approaches have been made in the past few years to develop practically useful processes for the production of L-(-)-carnitine. Several stereospecific biological reactions, such as hydroxylation of γ-butyrobetaine, reduction of dehydrocarnitine (22), hydration of crotonobetaine (23) and reduction of 4-haloacetoacetic acid esters (24,25), have been suggested to have high practical potential for this purpose.

In the evaluation studies of biochemical reactions useful for L-(-)-carnitine biotransformations, we found that a novel aldehyde reductase produced by a yeast, Sporobolomyces salmonicolor, is a suitable catalyst for the reduction of 4-haloacetoacetic acid esters to the corresponding (R)-4-halo-3-hydroxybutanoic acid esters (26), promising chiral building blocks in carnitine synthesis (Fig. 7). Based on this finding, we

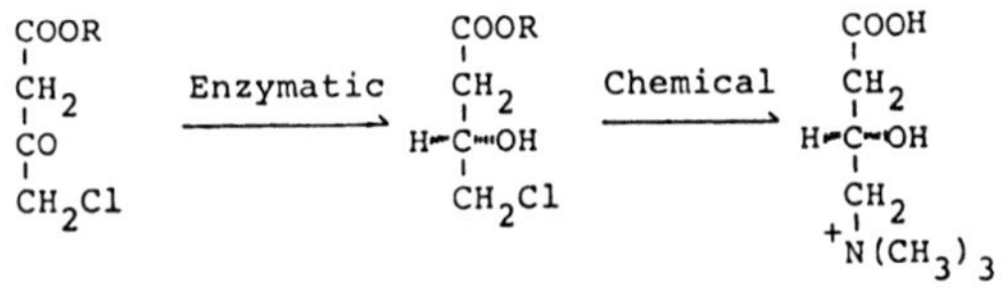

Fig. 7. Chemicoenzymatic route for the synthesis of L-(-)-carnitine.

developed a practical procedure, in which the reduction reaction with the aldehyde reductase is carried out in a two phasic system involving water and an organic solvent. This system yields optically pure (R)-4-halo-3-hydroxybutanoic acid esters in extremely high yields (27,28).

Distribution of Enantioselectivity of 4-Haloacetoacetic Acid Ester Reduction in Microorganisms

The screening to obtain suitable microorganisms that produce enzyme for the reduction was performed by incubating washed cells, which had been grown in an appropriate medium, in 20 mM potassium phosphate buffer (pH 6.0) containing 1% methyl 4-chloroacetoacetate and 5% glucose for 2 days at 28°C. Among 600 stock cultures including bacteria, actinomycetes, yeasts, molds and basidiomycetes tested, only three strains, i.e. Sporobolomyces salmonicolor, Micrococcus luteus and Cellulomonas sp. were found to produce the (R)-enantiomer predominantly (~30% ee) in high molar conversions (73, 49 and 49%, respectively). Other 28 strains showing more than 30% conversion almost specifically yielded the undesired (S)-enantiomer (80-95% ee).

Through the screening study, we selected Sporobolomyces salmonicolor as a potent catalyst for the reduction. To improve optical purity of the product, various conditions for cultivation of the yeast and preparation of the catalyst were tested. Through these tests, we found that heating of the cells and addition of an organic solvent, such as ethyl acetate, n-butyl acetate or acetone, to the reaction mixture efficiently increase optical purity of the product (50-70% ee). This improvement was found to be due to specific inactivation of undesired enzymes yielding the (S)-enantiomer by these treatments.

Enzymatic Synthesis of (R)-4-Chloro-3-hydroxybutanoic Acid Ester in a Water-Organic Solvent Two Phasic System

When the extracted crude *Sporobolomyces* enzyme, i.e. aldehyde reductase (26), was heated, precipitated with acetone, and used as a catalyst together with glucose dehydrogenase (*Bacillus* sp.) as a cofactor regenerator, 48.1 g/liter of (*R*)-4-chloro-3-hydroxybutanoic acid ethyl ester was produced from ethyl 4-chloroacetoacetate (molar yield, 52.1%; optical purity, 86% ee) (see Table 3).

Problem with this synthesis was that both the substrate and product inhibited the enzymatic reactions. To remove this inhibition, the reaction was carried out in a two phasic system composed of water and an organic solvent as outlined in Fig 8. We selected *n*-butyl acetate as the most suitable organic solvent, because both the substrate and product were easily extracted to the organic layer and the enzymes (i.e. aldehyde reductase and glucose dehydrogenase) were very stable against the organic solvent. All other organic solvents tested were disadvantagous in that they showed poorer partition efficiencies for the substrate and/or product or inactivated the enzymes. For example, ethyl acetate, which showed essentially the same partition efficiencies to those of *n*-butyl acetate, caused complete inactivation of the both enzymes on incubation for 48 h at 38°C. On the other hand, losses of the enzyme activities in the case of *n*-butyl acetate were 57.9 and 26.7% for aldehyde reductase and glucose dehydrogenase, respectively. Table 3 compares reduction efficiency in *n*-butyl acetate-water two phasic system to those in ethyl acetate-water two phasic system and an aqueous system. The results clearly demonstrate that the *n*-butyl acetate system is superior to others in showing high coenzyme turnover, high product yield and high catalytic efficiencies of the enzymes. Furthermore, both aldehyde reductase and glucose dehydrogenase in this system were still active after the reaction. Based on the above results, reaction in a bench-scale reactor containing *n*-butyl acetate (1600 ml), water (1600 ml), aldehyde reductase (5000 units), glucose dehydrogenase (15000 units), $NADP^+$ (142 µmol), glucose (1900 mmol), NaCl (320 mmol) and ethyl 4-chloroacetoacetate (each 270 mmol was added at 0, 24 and 48 h) was carried out at 25-27°C and pH 6.5-6.8. After 120 h, the amount of (*R*)-4-chloro-3-hydroxybutanoic acid ester in the reaction mixture reached 770 mmol (138 g) (molar yield, 95%; optical purity, 86% ee).

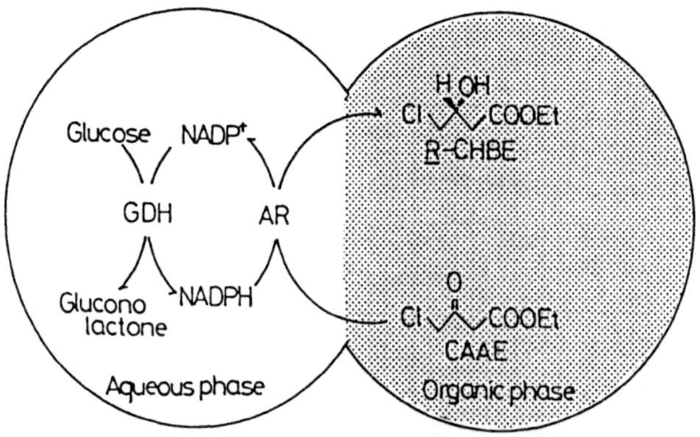

Fig. 8. Outline of the stereospecific reductio of ethyl 4-chloroacetoacetate (CAAE) by *Sporobolomyces* aldehyde reductase (AR) with glucose dehydrogenase (GDH) as a cofactor regenerator in a water-organic solvent two phasic system. CHBE, ethyl 4-chloro-3-hydroxybutanoate.

Table 3. Comparison of Reduction Efficiencies in Water-Organic Solvent Two Phasic System to Those in an Aqueous System[a]

Conditions and results		Monophase water	Biphase ethyl acetate	Biphase n-butyl acetate
AR	(units)	800	800	156
GDH	(units)	1070	1070	536
$NADP^+$	(mmol)	0.0084	0.0084	0.0048
CAAE	(mmol)	38.8	38.8	38.8
Glucose	(mmol)	40	40	40
NaCl	(mmol)	14	14	12
H_2O	(ml)	70	70	60
Organic solvent	(ml)	0	100	50
Temperature	(°C)	25	25	26
Time	(h)	48	48	99
pH		7.0	7.0	6.5-7.0
pH controller		NaOH	NaOH	Na_2CO_3
CAAE remained	(mmol)	3.76	0.04	0.12
CHBE formed	(mmol)	20.2	37.0	36.7
	(mol%)	52.1	95.3	94.5
	(g/l H_2O)	48.1	88.0	102
Optical purity	(% ee)	86	86	86
Efficiency				
$CHBE/NADP^+$	(mol/mol)	2406	4402	7640
CHBE/AR	(mg/unit)	4.2	7.7	39.2
CHBE/GDH	(mg/unit)	3.2	5.8	11.4
Remaining activity				
AR	(%)	0	0	80
GDH	(%)	0	0	38

[a] See (27,28) for details. AR, aldehyde reductase; GDH, glucose dehydrogenase; CAAE, ethyl 4-chloroacetoacetate; CHBE, ethyl 4-chloro-3-hydroxybutanoate.

Characterization of Aldehyde Reductase from *Sporobolomyces salmonicolor*

The enzyme catalyzing this stereospecific reduction was isolated in a crystalline form from the cells of *Sporobolomyces salmonicolor* and characterized in some detail (26, see also Table 1 and 2). It is a kind of aldehyde reductase with a molecular mass of about 37,000. The enzyme absolutely requires NADPH as a cofactor. Besides 4-haloacetoacetic acid esters, *p*-nitrobenzaldehyde, and a variety of other aromatic and aliphatic aldehydes were found to be reduced well by the enzyme. In *Sporobolomyces salmonicolor*, the enzyme comprised 2-6% of the total extractable proteins. Such a high content of the enzyme was suggested to be one of the reasons why the yeast produces the (*R*)-enantiomer predominantly.

REFERENCES

1. H. Yamada and S. Shimizu, Microbial and enzymatic processes for the production of biologically and chemically useful compounds, *Angew. Chem. Int. Ed. Eng.* 27:622 (1988).
2. H. Yamada and S. Shimizu, Biotechnology - microbial conversion, *in*: "Ullman's Encyclopedia of Industrial Chemistry," Vol. A4, VCH Verlagsgesellschaft, Weinheim, pp. 150-170 (1985).

3. S. Shimizu and H. Yamada, Microbial and enzymatic processes for the production of pharmacologically important nucleosides, Trends Biotechnol. 2:137 (1984).
4. H. Yamada and S. Shimizu, Microbial enzymes as catalysts for synthesis of biologically useful compounds in: "Biocatalysts in Organic Synthesis," J. Tramper, H. C. van der Pals and P. Linko, eds., Elsevier, Amsterdam, pp. 19-37 (1985).
5. S. Shimizu and H. Yamada, Coenzymes, in: "Biotechnology," Vol. 4, H.-J. Rehm and G. Reed, eds., VCH Verlagsgesellschaft, Weinheim, pp. 159-184 (1986).
6. T. Nagasawa and H. Yamada, Microbial transformations of nitriles, Trends Biotechnol. 7:153 (1989).
7. S. Shimizu and H. Yamada, Pantothenic acid (vitamin B_5), coenzyme A and related compounds, in: "Biotechnology of vitamins, pigments and growth factors," E. J. Vandamme, ed., Elsevier Applied Science, London, pp. 199-219 (1989).
8. S. Shimizu, H. Yamada, H. Hata, T. Morishita, S. Akutsu and M. Kawamura, Novel chemoenzymatic production of D-(-)-pantoyl lactone, Agric. Biol. Chem. 51:289 (1987).
9. S. Shimizu, H. Hata and H. Yamada, Reduction of ketopantoyl lactone to D-(-)-pantoyl lactone by microorganisms, Agric. Biol. Chem. 48:2285 (1984).
10. H. Hata, S. Shimizu and H. Yamada, Enzymatic production of D-(-)-pantoyl lactone from ketopantoyl lactone, Agric. Biol. Chem. 51:3011 (1987).
11. H. Hata, S. Shimizu, S. Hattori and H. Yamada, Ketopantoyl-lactone reductase from Candida parapsilosis: purification and characterization as a conjugated polyketone reductase, Biochim. Biophys. Acta 990:175 (1989).
12. H. Hata, S. Shimizu, S. Hattori and H. Yamada, Stereoselective reduction of diketones by a novel carbonyl reductase from Candida parapsilosis, J. Org. Chem. in press.
13. S. Shimizu, S. Hattori, H. Hata and H. Yamada, A novel fungal enzyme, NADPH-dependent carbonyl reductase, showing high specificity to conjugated polyketones, purification and characterization, Eur. J. Biochem. 174:37 (1988).
14. S. Shimizu, M. Kataoka, M. C. M. Chung and H. Yamada, Ketopantoic acid reductase of Pseudomonas maltophilia 845, purification, characterization and role in pantothenate biosynthesis, J. Biol. Chem. 263:12077 (1988).
15. M. Kataoka, S. Shimizu and H. Yamada, Novel enzymatic production of D-(-)-pantoyl lactone through the stereoselective reduction of ketopantoic acid, Agric. Biol. Chem. 54:177 (1990).
16. K. Sakamoto, S. Kita, T. Morikawa, S. Shimizu, and H. Yamada, Preparation of ethyl 2'-ketopantothenate, Japanese patent application H1-45407 (1989).
17. S. Shimizu, K. Sakamoto and H. Yamada, Studies on the enzymatic hydrolysis of pantothenate esters (in Japanese) Nippon Nogeikagaku Kaishi 62:283 (1987).
18. M. Kataoka, S. Shimizu, Y. Doi and H. Yamada, Stereoselective reduction of ethyl 2'-ketopantothenate to ethyl D-(+)-pantothenate with microbial cells as a catalyst, Appl. Environ. Microbiol. submitted.
19. M. Kataoka, S. Shimizu, Y. Doi, K. Sakamoto and H. Yamada, Microbial production of chiral pantothenonitrile through stereospecific reduction of 2'-ketopantothenonitrile, Biotechnol. Lett. submitted.
20. S. Shimizu, S. Hattori, H. Hata and H. Yamada, Stereoselective enzymatic oxidation and reduction system for the production of D-(-)-pantoyl lactone from a racemic mixture of pantoyl lactone, Enzyme Microb. Technol. 9:411 (1987).

21. S. Shimizu, S. Hattori, H. Hata and H. Yamada, One-step microbial conversion of a racemic mixture of pantoyl lactone to optically active D-(-)-pantoyl lactone, Appl. Environ. Microbiol. 53:519 (1987).
22. J-P. Vandecasteele, Enzymatic synthesis of L-carnitine by reduction of an achiral precursor, the problem of reduced nicotinamide adenine dinucleotide recycling, Appl. Environ. Microbiol. 39:327 (1980).
23. M. Seim and H-P. Kleber, Synthesis L-(-)-carnitine by hydration of crotonobetaine by entrobacteria, Appl. Microbiol. Biotechnol. 27:538 (1988).
24. B. Zhou, A. S. Gopalan, F. VanMiddlesworth, W-R. Shieh and C. J. Sih, Stereochemical control of yeast reductions. 1, asymmetric synthesis of L-carnitine, J. Am. Chem. Soc. 105:5925 (1983).
25. C-H. Wong, D. G. Drueckhammer and H. M. Sweers, Enzyme vs. fermentative synthesis: thermostable glucose dehydroganase catalyzed regeneration of NAD(P)H for use in enzymatic synthesis, J. Am. Chem. Soc. 107:4028 (1985).
26. H. Yamada, S. Shimizu, M. Kataoka and T. Miyoshi, A novel NADPH-dependent aldehyde reductase, catalyzing asymmetric reduction of β-keto acid esters, from Sporobolomyces salmonicolor, purification and characterization, FEMS Microbiol. Lett. in press.
27. S. Shimizu and H. Yamada, Stereospecific reduction of 3-keto acid esters by a novel aldehyde reductase of Sporobolomyces salmonicolor in a water-organic solvent two phasic system, Ann. N. Y. Acad. Sci. in press.
28. S. Shimizu, M. Kataoka, M. Katoh, T. Morikawa, T. Miyoshi and H. Yamada, Stereospecific reduction of ethyl 4-chloro-3-oxobutanoate by a microbial aldehyde reductase in an organic solvent-water two phasic system, Appl. Environ. Microbiol. submitted.

DESIGN AND DEVELOPMENT OF ENZYMATIC ORGANIC SYNTHESIS

C.-H. Wong

Department of Chemistry
Research Institute of Scripps Clinic
La Jolla, CA 92037

INTRODUCTION

The rate acceleration and selectivity of various enzymatic reactions operated under mild conditions are the major attractive features of enzymes for use as catalysts in organic synthesis. Many natural and unnatural enzymatic reactions have been demonstrated for multigram scale synthesis of chiral organic molecules[1]. The number of enzymes isolated (about 2,500), however, only represents approximately 2% of the total number of enzymes which may exist in nature. So far, there have been only about 50 enzymes exploited for use in organic synthesis. With an increasing number of enzymes available, the synthetic methods based on enzyme catalysis are being extended from the preparation of small chiral molecules to the synthesis of more complex molecules such as oligosaccharides, polypeptides, nucleotides and their conjugates. This review describes the most recent developments in my laboratory in this area with emphasis on the synthesis of carbohydrates and polypeptides.

Recombinant Fructose-1,6-diphosphate (FDP) Aldolase from *E. coli*

There have been more than 20 aldolases reported, each of which catalyzes a distinct aldol reaction. The type-I Schiff-base forming FDP aldolase from rabbit muscle is the one most often used in organic synthesis[2]. To explore the synthetic utility of the type-II zinc-containing FDP aldolase (Figure 2) from microorganisms, we have recently reported the cloning and overexpression of *E. coli* FDP aldolase in *E. coli*[3]. A typical 2-L fermentation of the recombinant *E. coli* would produce 10g of crude extracts, which contain approximately 40,000 units of FDP aldolase (1 unit of enzyme would catalyze the aldol cleavage of 1 umol FDP per minute). The enzyme is a dimer with a molecular weight of ~80,000. The sequence has recently been determined based on the cDNA sequence[4]. It is relatively more stable than the rabbit muscle aldolase. In the presence of 0.3 mM Zn^{++}, the half life at 25°C is about 60 days compared to 2.5 days for the rabbit enzyme. The active-site zinc ion can be replaced with Co^{++}. The Co^{++} -enzyme, however, is slightly less active and less stable than the Zn^{++} -enzyme. Figure 3 summarizes the stability and activity of both the Co^{++} - and Zn^{++} - enzymes[5]. Preliminary studies indicate that the Zn^{++} -enzyme has similar substrate specificity to the rabbit muscle enzyme. Both enzymes, however, have little sequence homology. Work is in progress to determine the X-ray crystal structure of the zinc enzyme for the next stage of research to engineer the substrate specificity.

Chemical Aspects of Enzyme Biotechnology
Edited by T.O. Baldwin *et al.*, Plenum Press, New York, 1990

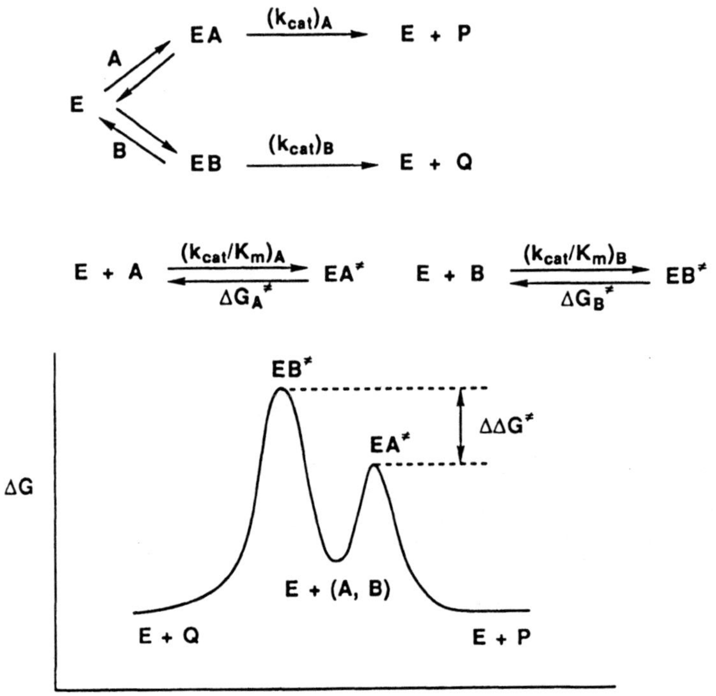

$$\frac{v_A}{v_B} = \frac{(k_{cat}/K_m)_A\,[A]}{(k_{cat}/K_m)_B\,[B]}\ , \quad \frac{(k_{cat}/K_m)_A}{(k_{cat}/K_m)_B} = \exp(-\Delta\Delta G^{\neq}/RT) = E$$

$$\frac{\ln\,[(1-c)(1-ee_B)]}{\ln\,[(1-c)(1+ee_B)]} = \frac{\ln\,[1-c(1+ee_P)]}{\ln\,[1-c(1-ee_P)]} = E$$

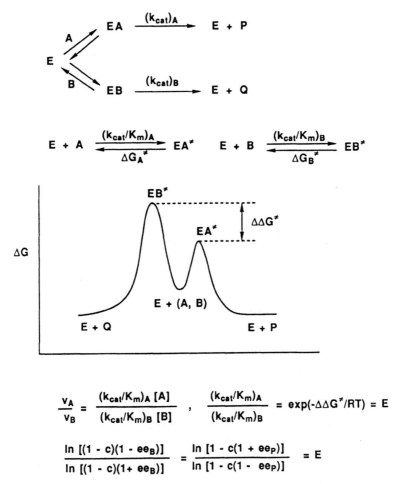

Fig. 1. Enzyme-catalyzed stereoselective transformation. A and B could be an enantiomeric pair, or two enantiotopic or diastereotopic groups or faces of a single molecule. E: enzyme; P: product from A; Q: product from B; E': enantioselectivity value; C: percent of conversion; ee_P: product ee; ee_B: ee of recovered B.

Synthesis of Azasugars

To illustrate the synthetic utility of the Zn^{++} -FDP aldolase, we have carried out the synthesis of several azàsugars, including deoxynojirimycin, deoxymannojirimycin, fagomine, and 1,4-dideoxy-1,4-imino-D-arabinitol (Figure 4). The syntheses involve an enzymatic aldol condensation followed by a catalytic reductive amination[3]. The overall yield was in the range of 50-75%. The chiral aldehydes used in the aldol reactions were prepared via a lipase-catalyzed enantioselective hydrolysis of racemic 3-azido-2-acetoxypropanal diethyl acetal. This strategy is being extended to the synthesis of azasugars corresponding to N-acetylglucosamine and N-acetylmannosamine (Figure 5) and, with the use of other aldolases, to the azasugars corresponding to other monosaccharides commonly found in glycoproteins. Azasugars are inhibitors of glycosidases involved in di- and polysaccharide degradation and glycoprotein processing. They hold potential value for the treatment of symptoms or diseases associated with these biochemical processes.

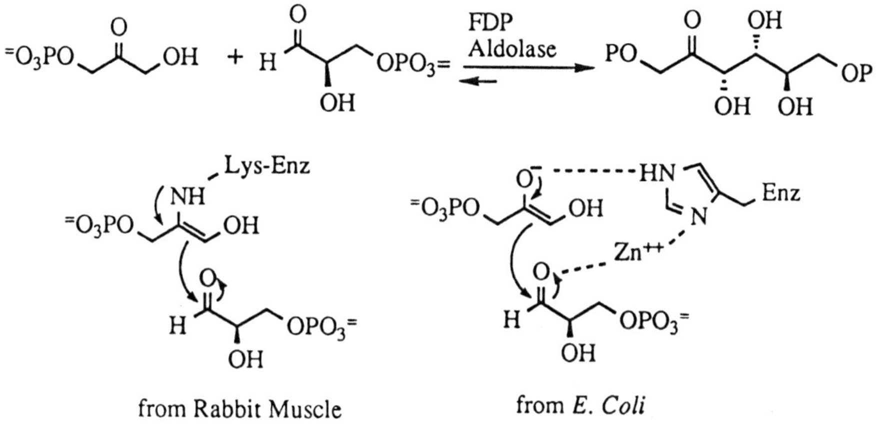

Fig. 2. FDP aldolase from rabbit muscle and *E. coli*.

$$EZn_2 \rightleftarrows EZn + Zn\,(II) \quad K_1 = 10^{-8.9}M \qquad ECo_2 \rightleftarrows ECo + Co\,(II) \quad K_1 = 10^{-7.6}M$$

$$EZn \rightleftarrows E + Zn\,(II) \quad K_2 = 10^{-11.8}M \qquad ECo \rightleftarrows E + Co\,(II) \quad K_2 = 10^{-10.9}M$$

$$t_{1/2}{}^{60°C} = 1.5\text{ h} \quad V_{max}{}^{20°C} = 56\text{ U/mg} \qquad t_{1/2}{}^{60°C} = 0.5\text{ h}, \quad V_{max}{}^{20°C} = 35\text{ U/mg}$$

Fig. 3. Activity, stability and dissociation constants of Zn^{++}- and Co^{++}-FDP aldolase from *E. coli* at 20°C, pH 8.

In the synthesis of N-acetylglucosamine and N-acetylmannosamine types of azasugars, the aldehyde substrates were prepared from the aziridine precursor via a Lewis acid assisted azide opening of the aziridine.

The FDP aldolase and two other aldolases (i.e. fuculose 1-phosphate aldolase and rhamnulose 1-phosphate aldolase) require dihydroxyacetone phosphate as a substrate. There are three methods commonly used for the preparation of dihydroxyacetone phosphate: from dihydroxyacetone by chemical or enzymatic synthesis, and from FDP by *in situ* enzymatic generation[2]. Dihydroxyacetone phosphate can also be replaced with dihydroxyacetone and catalytic amounts of inorganic arsenate[1]. We have recently developed an improved procedure for the chemical synthesis which is outlined in Figure 6. The major improvement is the use of a trivalent phosphorylating reagent, dibenzyl-N,N-diethyl phosphoramidite, in the presence of triazole or tetrazole. The phosphorylating reagent was easily prepared from PCl_3, benzyl alcohol and diethylamine. It is stable and can be kept for several months without decomposition or disproportionation. The overall yield of dihydroxyacetone phosphate from dihydroxyacetone is about 55%. The process does not require chromatography and the intermediates are crystalline and easy to handle.

Recombinant 2-Deoxyribose-5-phosphate Aldolase from E. coli

This enzyme catalyzes the aldol condensation between acetaldehyde and D-glyceraldehyde 3-phosphate to 2-deoxyribose 5-phosphate. It is unique among the aldolases that it accepts two aldehydes as substrates. It has been cloned and overexpressed in *E. coli*[6]. A 6-L fermentation produced approximately 124,000 units of the aldolase. The purified enzyme showed optimal activity at pH 7.5 with V_{max} = 55 U/mg. Examination of the substrate specificity of the enzyme indicates that acetone, fluoroacetone, and propionaldehyde are also accepted by the enzyme as the donor substrate. The rates, however, are about 1% that of the natural reaction. Chloroacetaldehyde and glycoaldehyde are not donor substrates. It is interesting that the enzyme does not catalyze the self

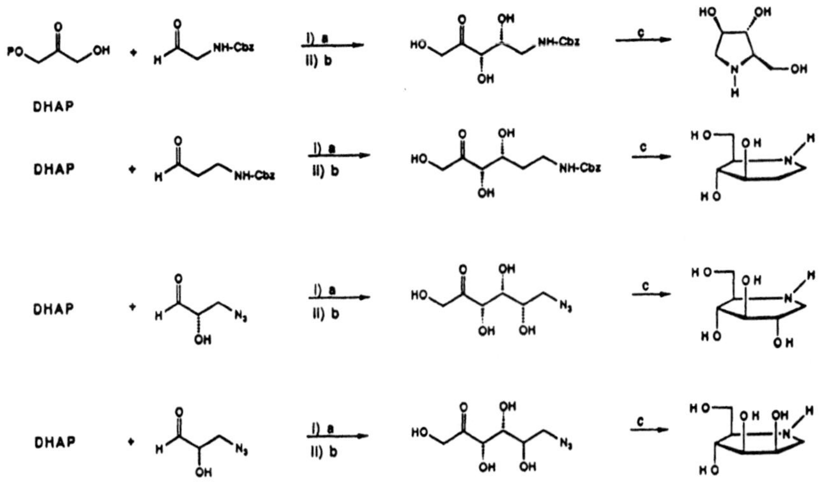

Fig. 4. Chemo-enzymatic synthesis of azasugars. a) FDP aldolase; b) acid phosphatase; c) H_2/Pd-C.

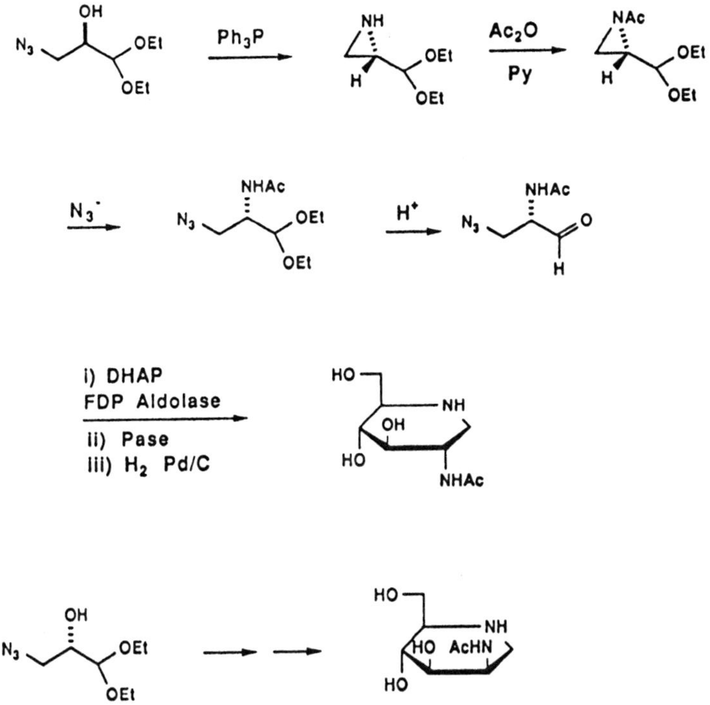

Fig. 5. Chemo-enzymatic synthesis of azasugars corresponding to N-acetylglucosamine and N-acetylgalactosamine.

condensation of acetaldehyde but does for propionaldehyde. With fluoroacetone as donor, the reaction occurs regioselectively at the nonfluorinated carbon. Figure 7 illustrates a representative synthesis of (S)-4-hydroxy-5-methylhexan-2-one, (S)-1-fluoro-4-hydroxy-5-methylhexan-2-one, and 2-deoxyribose-5-phosphate. In the synthesis of 2-deoxyribose-5-phosphate, D-glyceraldehyde 3-phosphate was generated *in situ* from dihydroxyacetone phosphate catalyzed by triosephosphate isomerase or prepared from (R)-glycidaldehyde diethyl acetal. Work is in progress to extend the use of this enzyme to the synthesis of other β-hydroxy compounds and to further investigate the substrate specificity of the enzyme.

Peptide Synthesis

Synthetic peptide chemistry has been well developed for the laboratory synthesis of various sizes of peptides. Synthesis of polypeptides greater than 100 amino acids in length based on the solid phase method, or on the strategy of segment condensation, however, often suffers from problems such as racemization, low yield, and poor solubility. The biological method based on recombinant DNA technology has been the choice for the synthesis of large protein molecules and homologs with altered characteristics for structure-function study. Enzymatic methods based on proteases combined with chemical synthetic methods are attractive alternatives because enzymatic reactions are catalytic, regio- and stereoselective, and racemization-free, and require minimal side-chain protection. Further, the reactions can be carried out in a mixture of organic solvent and water to improve the solubility of peptide segments. With these advantages, however, come the disadvantages that the amidase activity of proteases causes secondary hydrolysis of the growing peptide chain and that the substrates of proteases are generally limited to natural L-amino acids.

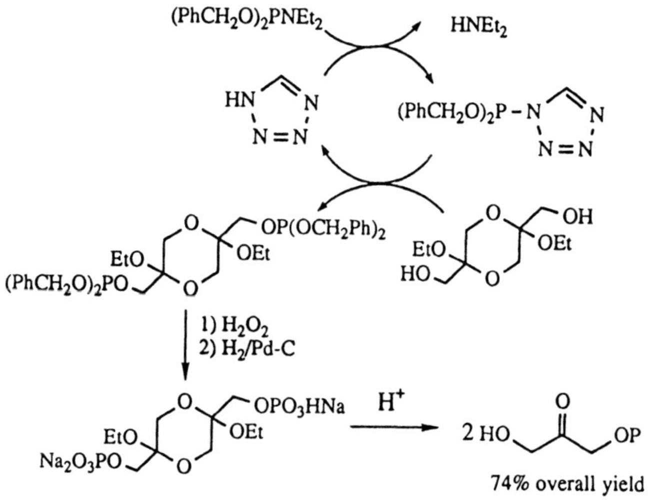

Fig. 6. An improved synthesis of dihydroxyacetone phosphate involving a catalytic cycle of tetrazole for phosphorylation with a trivalent phosphorylating reagent.

Strategies of Protease-Catalyzed Peptide Coupling

Two strategies are often used in protease-catalyzed peptide synthesis: one is the direct reversal of the catalytic hydrolysis of peptides (*i.e.* thermodynamic approach) and the other is the aminolysis of N-protected amino acid or peptide esters (*i.e.* kinetic approach). The thermodynamic approach is an endergonic process, and manipulation of reaction conditions is required to increase the product yield. The addition of water-miscible organic solvents to increase the pK value of the carboxyl component (thus increasing the substrate concentration), the use of a biphasic system, reverse micelles, or anhydrous media containing a minimal amount of water or water mimics, or the selection of appropriate N- or C-protecting groups to reduce the solubility of products are often employed. The kinetic approach is faster, and the product yield can be improved by manipulating the reaction conditions as that used in the thermodynamic approach. Aminolysis, however, requires the use of esters as substrates and is limited to those enzymes (*e.g.* chymotrypsin, trypsin, papain, and subtilisin) which form an acyl intermediate. The synthesis of ester substrates for enzymatic peptide coupling can be accomplished by chemical (by solid phase or solution phase) or enzymatic methods as illustrated in a recent report on subtilisin catalyzed peptide segment coupling[7].

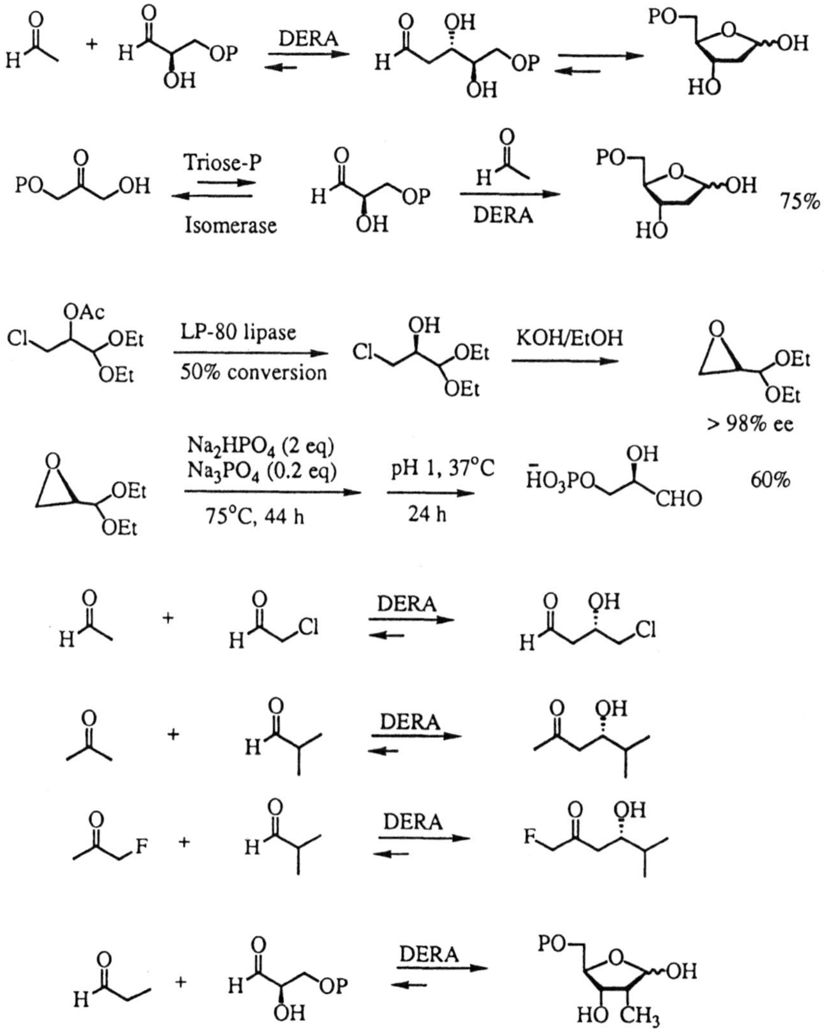

Fig. 7. Synthesis with 2-deoxyribose 5-phosphate aldolase (DERA).

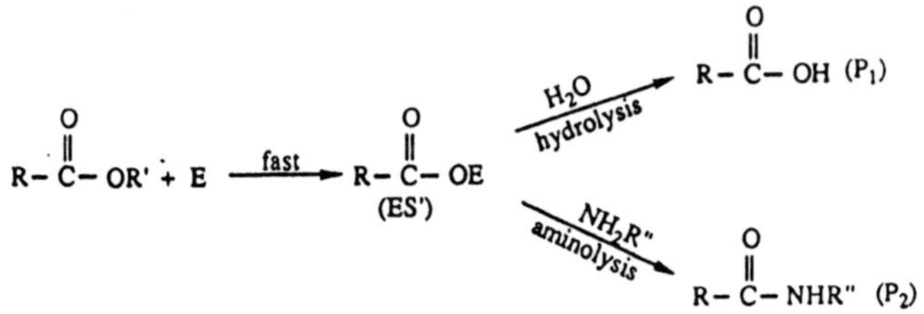

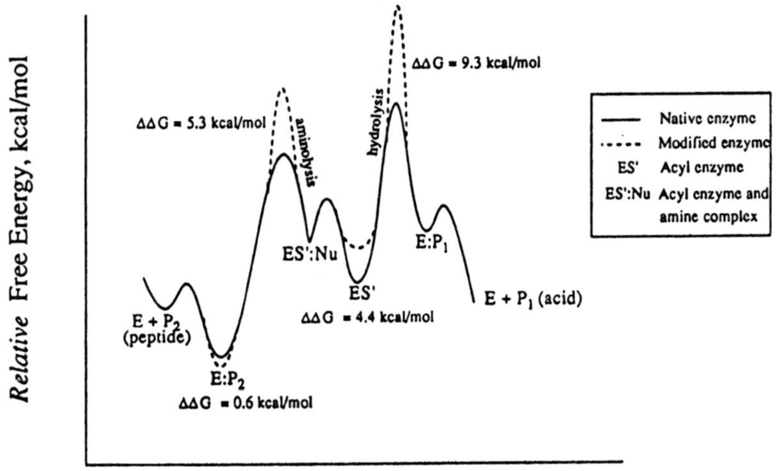

Fig. 8. Free energy diagrams for serine protease-catalyzed aminolysis in water (solid line) compared with that in the presence of 50% DMF. The energy diagram for aminolysis catalyzed by the methylated enzyme in water is similar to the aminolysis catalyzed by the co-solvent modified native enzyme.

Control of Amide Cleavage Activity in Serine Protease Catalyzed Aminolysis

The hydrolyses of esters and amides catalyzed by serine or thiol proteases have similar mechanisms but different rate-determining steps. The formation of an acyl intermediate in amide hydrolysis is rate-determining and pH independent, while deacylation of the acyl intermediate in ester hydrolysis is the rate-determining step of general base catalysis and thus is pH dependent. The significant enhancement of esterase vs amidase activity of such proteases at higher pH has been utilized in the stepwise synthesis and fragment coupling of peptides via aminolysis, where hydrolysis of the growing polypeptide chain was inhibited[8]. Alternatively, esterases without amidase activities (such as lipases) can be used as catalyst for aminolysis[9-10]. The rate, however, is very slow. Serine and thiol proteases have been found to behave similarly as nonproteolytic esterases when water-miscible organic solvents such as dimethylformamide (DMF), dioxane, or acetonitrile are added to the aqueous enzyme solution[11]. The different effect of water-miscible organic solvents on the esterase and amidase activities of trypsin has been known for some time. The application of such a solvent effect to the kinetically controlled peptide synthesis, however, has only recently been exploited[11]. Further investigation of this phenomenon by reducing the water content in a subtilisin catalyzed aminolysis revealed that the enzymatic activity decreased with decrease in water content and became catalytically inactive in anhydrous DMF. Although the enzyme is reasonably stable in anhydrous DMF, the aminolysis reaction is too slow to be useful for practical synthesis. In the presence of 50-70% DMF, the enzyme is quite stable and active for aminolysis, while the amidase activity is insignificant. To further investigate the effect of organic solvents on catalysis, the kinetic parameters (k_{cat} and K_m) for chymotrypsin and subtilisin catalyzed hydrolysis of ester and amide substrates were determined[7,12]. It was found that organic solvents affect both catalysis and binding. Both k_{cat} and k_{cat}/K_m for the amide as well as the ester hydrolysis decrease as the content of organic solvent increases. The rate of decrease for the amide hydrolysis, however, is faster than that for the ester hydrolysis. Amides are generally more tightly bound to the enzyme as the organic solvent is added to the solution except in the case of hydrophobic amides. The binding of ester substrates, however, is affected less by organic solvents. A typical free energy diagram for chymotrypsin catalysis in an aqueous solution and in a 50% DMF solution is shown in Figure 8. The transition-state energy for both the ester and amide hydrolysis increases when the organic solvent is added to the solution. The acyl-intermediate, however, is less stable in the presence of DMF compared to that in the aqueous solution. The energy barrier for the formation of an acyl-enzyme-nucleophile complex in the presence of DMF is lower, and aminolysis is thereby more favorable than hydrolysis. In the direction of amide cleavage, the rate-determining energy barrier for the formation of an acyl intermediate is substantially high, accounting for the irreversible nature of the aminolysis process. Introduction of a methyl group to the ε-2 N of the active-site His of chymotrypsin resulted in a significant change in the enzymatic catalysis. The methylated enzyme (MeCT) favors aminolysis over hydrolysis[13]. The effects of methylation on the enzyme kinetics are very similar to that of organic solvents as indicated in the free-energy diagram (Figure 8). Previous studies on the acylation and deacylation reactions catalyzed by MeCT indicated that a functional group with $pK_a = 7$ was involved as a general base[14]. Our recent C-NMR study indicated the presence of an O-acyl intermediate in the MeCT-catalyzed ester hydrolysis[13]. All these results suggest that the unmethylated N (δ-1) of the active-site His acts as a general base for aminolysis. This process may require ring-flipping of the methylated imidazole, a process first suggested by Henderson[14] and later supported by studies on solvent isotope effects[15], proton inventories[16], and model systems[17]. The slight change in the orientation of the active-site groups Asp, His, Ser after methylation and ring-flipping may account for the favorable aminolysis vs hydrolysis for MeCT (Figure 9). Other serine proteases may be modified similarly to have new enzymatic activities useful for peptide segment coupling. It is worth noting that subtilisin can also be converted to an acyl transferase via modification of the active-site serine to cysteine[18] or selenocysteine[19]. An alternative approach to avoid the problem associated with the endopeptidase activity of proteases is to use exopeptidases as catalysts for peptide coupling. Carboxypeptidase Y, for example, has been successfully

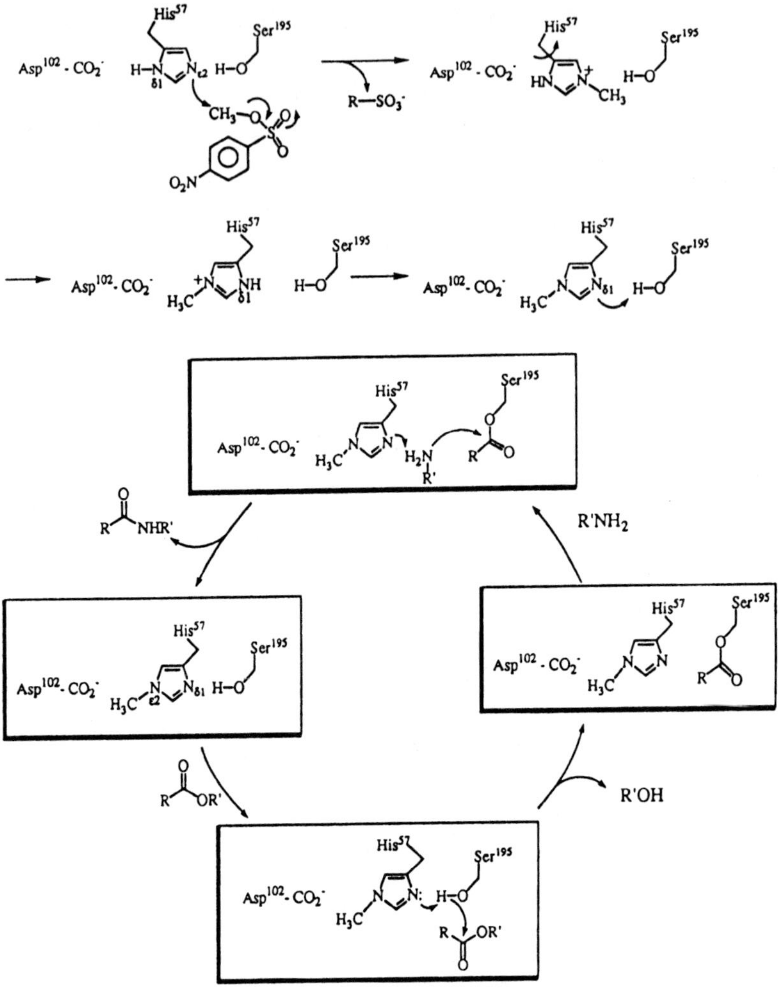

Fig. 9. Mechanisms for active-site methylation of chymotrypsin and aminolysis catalyzed by the methylated enzyme.

used in the stepwise synthesis of peptides from the N to C terminus[20]. Nonproteases involved in protein synthesis also hold potential[21]. These enzymes require ATP or GTP to activate the carboxyl group of an amino acid and seem to accept various amino acid nucleophiles for amide bond formation.

Issues of Stability

One limitation to the usefulness of most enzymes in synthesis is their intrinsic instability in many unnatural environments required for organic reactions. Many techniques such as immobilization have been developed to improve the stability of enzymes for large-scale reactions. A solution to this problem is to create a stable protein catalyst via amino acid substitution. More stable enzymes can be obtained via selection from thermophilic species or via site-directed mutagenesis. A subtilisin mutant derived from subtilisin BPN' via six site-specific mutations was 100 times more stable than the wild-type enzyme in aqueous solution and 50 times more stable than the wild-type in anhydrous DMF, and was applied to the synthesis of peptides, chiral amino acids and sugar derivatives[7]. Figure 10 illustrates the use of this mutant for regioselective acetylation of nucleosides.

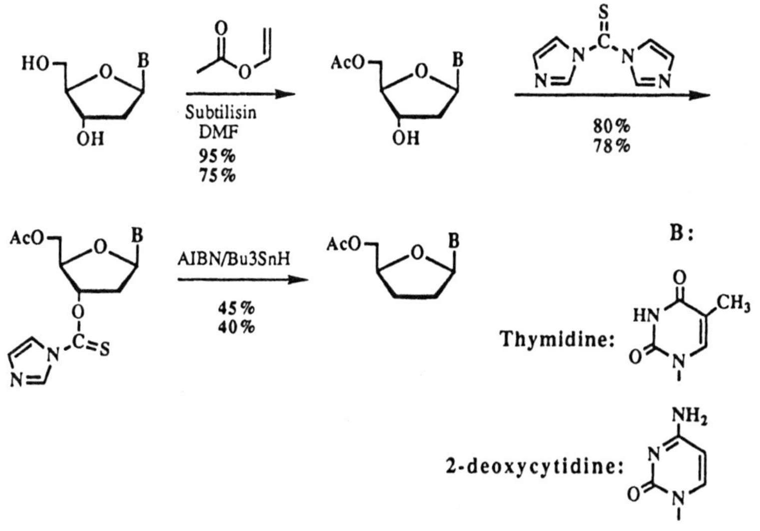

Fig. 10. Regioselective acetylation of nucleosides catalyzed by subtilisin 8350 for the synthesis of 2,3-dideoxynucleosides.

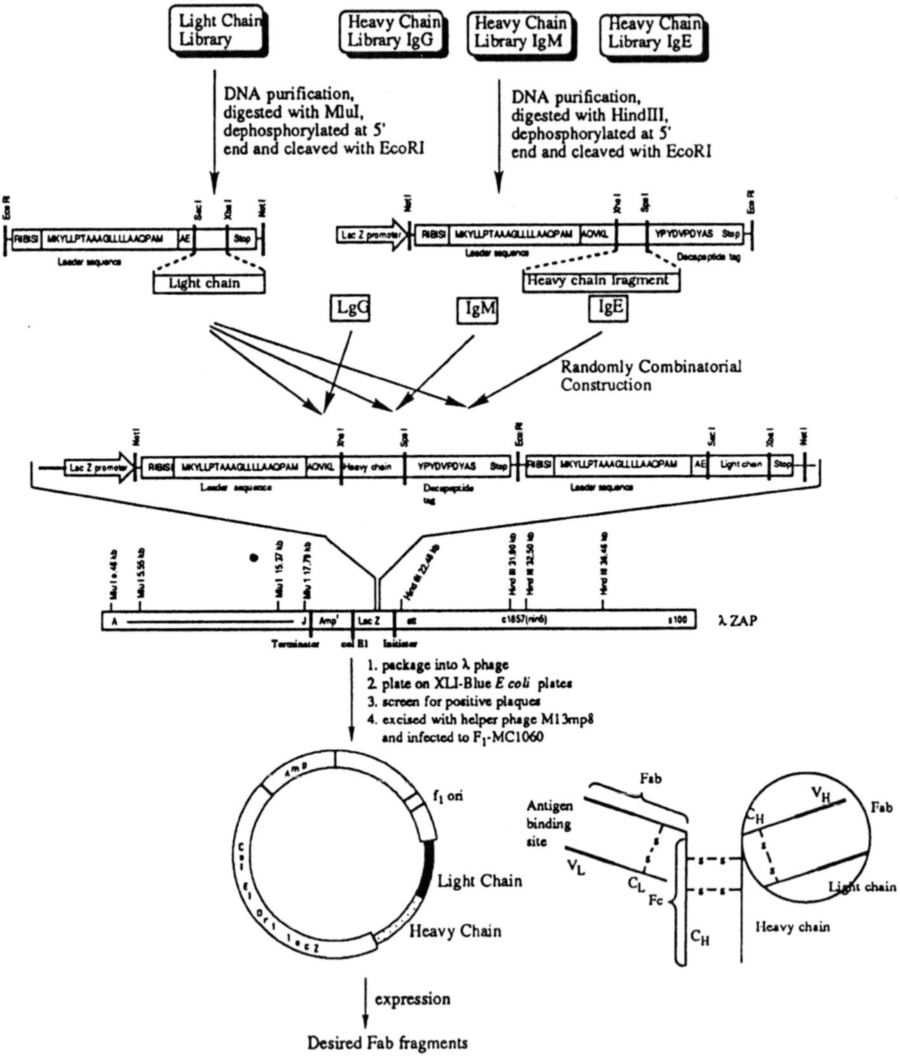

Fig. 11. Construction of recombinatorial Fab fragments for catalysis.

PROSPECTS

It seems clear that many biological catalysts including those from nature and from rational design will continuously be explored for organic synthesis. The technology of recombinant DNA and site-directed mutagenesis together with the advanced instrumentation available for structure determination, molecular modeling, and design will play very important roles in the rational development of catalysts for organic reactions. The recent breakthrough in the field of catalytic antibody based on a phage vector system has made possible the production of tailor-made Fab fragments without going through the hybridoma technology (Figure 11)[22]. As more and more new enzymatic catalysts and new enzymatic reactions are developed, organic synthesis based on biocatalysis is expected to combine with various chemical methods to tackle the new generation of synthetic problems appearing at the interface between chemistry and biology.

ACKNOWLEDGEMENTS

This research was supported by NSF and NIH. I thank the many coworkers, who are listed in the references, for their contribution in the area.

REFERENCES

1. C.-H. Wong, Enzymatic catalysts in organic synthesis, *Science* 244:1145 (1989).
2. M.D. Bednarski, E.S. Simon, N. Bischofberger, W-D. Fessner, M.J. Kim, W. Lees, T. Saito, H. Waldmann, G.M. Whitesides, Rabbit muscle aldolase as a catalyst in organic synthesis, *J. Am. Chem. Soc.* 111:627 (1989).
3. C.H. von der Osten, A.J. Sinskey, C.F. Barbas, III, R.L. Pederson, Y.-F. Wang, C.-H. Wong, Use of a recombinant bacterial fructose-1,6-diphosphate aldolase in aldol reactions, *J. Am. Chem. Soc.* 111:3924 (1989).
4. P.R. Alefounder, S.A. Baldwin, R.N. Perham, N.J. Short, Cloning, sequence analysis and over-expression of the gene for the class II fructose-1,6-biphosphate aldolase of *E. coli*, *Biochem. J.* 257:529 (1989).
5. Y.L. Chen, Metalloenzyme chemistry, Ph.D. Thesis, Texas A&M University (1989).
6. C.F. Barbas, III, Y.-F. Wang, C.-H. Wong, Deoxyribose 5-phosphate aldolase as a synthetic catalyst, *J. Am. Chem. Soc.* 112:2013 (1990).
7. C.-H. Wong, S.-T. Chen, W.J. Hennen, J.A. Bibbs, Y.-F. Wang, J.L.-C. Liu, M.W. Pantoliano, M. Whitlow, P.N. Bryan, Enzymes in organic synthesis: use of subtilisin and a highly stable mutant derived from multiple site-specific mutations, *J. Am. Chem. Soc.* 112:945 (1990).
8. C.F. Barbas, III, C.-H. Wong, Papain catalyzed peptide synthesis: control of amidase activity and the introduction of unusual amino acids, *J. Chem. Soc. Chem. Comm.* 532-534 (1987).
9. J.B. West, C.-H. Wong, Use of nonproteases in peptide synthesis, *Tetrahedron Lett.* 28:1629 (1987).
10. A.L. Margolin, A.M. Klibanov, Peptide synthesis catalyzed by lipases in anhydrous organic solvents, *J. Am. Chem. Soc.* 109:3802 (1987).
11. C.F. Barbas, III, J.R. Matos, J.B. West, C.-H. Wong, A search for peptide ligase: cosolvent-mediated conversion of proteases to esterases for irreversible synthesis of peptides, *J. Am. Chem. Soc.* 110:5162 (1988).
12. J.B. West, W.J. Hennen, J.L. Lalonde, J.A. Bibbs, Z. Zhong, E.F. Meyer, C.-H. Wong, Enzymes as synthetic catalysts: mechanistic and active-site considerations of natural and modified chymotrypsin, *J. Am. Chem. Soc.*, in press.
13. J.B. West, J. Scholten, N.J. Stolowich, J.L. Hogg, A.I. Scott, C.-H. Wong, Modification of proteases to esterases for peptide synthesis: methylchymotrypsin, *J. Am. Chem. Soc.* 110:3709 (1988).
14. R. Henderson, Catalytic activity of α-chymotrypsin in which histidine-57 has been methylated, *Biochem. J.* 124:13 (1971).
15. L.D. Byers, D.E. Koshland, Jr., On the mechanism of action of methyl chymotrypsin, *Bioorganic Chemistry* 7:15 (1978).

16. J.D. Scholten, J.L. Hogg, F.M. Raushel, Methyl chymotrypsin catalyzed hydrolyses of specific substrate esters indicate multiple proton catalysis is possible with a modified charge relay triad, *J. Am. Chem. Soc.* 110:8246 (1988).
17. J. Rebek, Jr., Recognition and catalysis using molecular clefts, *Chemtracts - Organic chemistry* 2:337 (1989).
18. T. Nakatsuka, T. Sasaki, E.T. Kaiser, Peptide segment coupling catalyzed by the semisynthetic enzyme subtilisin, *J. Am. Chem. Soc.* 109:3808 (1987).
19. Z.-P. Wu, D. Hilvert, Conversion of a protease into an acyl transferase: selenosubtilisin, *J. Am. Chem. Soc.* 111:4513 (1989).
20. F. Widmer, K. Breddam, J.T. Johansen, Carboxypeptidase Y catalyzed peptide synthesis using amino acid alkyl esters as amine components, *Carlsberg Res. Comm.* 45:453 (1980).
21. H. Nakajima, S. Kitabatake, R. Tsurutani, K. Yamamoto, I. Tomioka, K. Imahori, Peptide synthesis catalyzed by aminoacyl-tRNA synthetases from Bacillus stearothermophilus, *Int. J. Peptide Protein Res.* 28:179 (1986).
22. W.D. Huse, L. Sastry, S.A. Iverson, A.S. Kang, M. Alting-Mees, D.R. Burton, S.J. Benkovic, R.A. Lerner, Generation of a large combinatorial library of the immunoglobulin repertoire in phage lambda, *Science* 246:1275 (1989).

ENZYMES AS CATALYSTS IN CARBOHYDRATE SYNTHESIS

Eric J. Toone,† Yoshihiro Kobori,† David C. Myles,† Akio Ozaki,† Walther Schmid,† Claus von der Osten,# and Anthony J. Sinskey,# George M. Whitesides†*

†Department of Chemistry
Harvard University
Cambridge, MA 02138

#Department of Biology
Massachusetts Institute of Technology
Cambridge, MA 02139

INTRODUCTION

The field of carbohydrate chemistry is perhaps one of the most thoroughly investigated of all disciplines of chemistry. Thousands of investigators, from the time of Emil Fischer, have probed the structure, function and reactivity of mono-, oligo-, and polysaccharides. Nevertheless, carbohydrate chemistry and biology are currently undergoing a major renaissance. In addition to their well known roles as energy storage vehicles and structural components of cells, carbohydrates are now recognized as vital components in biological recognition phenomena.[1-3] The involvement of carbohydrates has been implicated in many biological processes, including cell-cell recognition in growth and differentiation, clearance of aging cells from circulation, and cell sorting and targeting. Cell-surface carbohydrates are specific attachment sites for certain pathogenic viruses, bacteria, and parasites, and for several soluble bacterial toxins.[4-7] With the realization of these new roles has come a renewed demand for improved synthetic routes to carbohydrates.

Synthetic methodologies for the preparation of monosaccharides, oligosaccharides and polysaccharides have been pursued by legions of chemists for over a century.[8-11] Although impressive progress has been made, carbohydrates continue to be an extremely challenging group of compounds to prepare. Recently, enzymes have been applied to the field of carbohydrate synthesis.[12] Enzymes have been utilized as catalysts in other areas of organic synthesis for many years.[13-15] Many of the advantages of enzyme-based synthesis are especially relevant to the preparation of carbohydrates:

i) Compatibility with aqueous media. Carbohydrates are soluble primarily in aqueous solution. Many of the reactions available to the organic chemist, however, are incompatible with water. As a result, carbohydrates must be modified to solublize them in organic media. Enzymes operate in aqueous solution, at or near room temperature, and at neutral pH. Complex protection/deprotection schemes can therefore be avoided.

ii) Specificity. The single most attractive feature of enzymes as catalysts in synthesis is their specificity. This specificity is manifested in three major ways. First, enzymes display absolute chemo-, or reaction, specificity. Carbohydrates frequently possess a variety of sensitive functionalities, which may be susceptible to undesirable modification during manipulation with conventional reagents. Second, enzymes in many cases display high regioselectivity. Since carbohydrates are composed of a number of functionalities of similar reactivity, the ability to operate selectively on a single functional group is highly desirable. This characteristic is an especially important issue in oligosaccharide syntheses, where, for example, a branched-chain carbohydrate may contain more than one identical monosaccharide unit. Finally, enzymes display great stereoselectivity. Carbohydrates contain much stereochemical information in the form of multiple stereogenic centers. This chirality must be correctly installed.

Probably no group of compounds is better suited to enzyme-based synthesis than the carbohydrates. We have developed synthetic methodologies for the synthesis of mono-, oligo- and polysaccharides. This review presents some of the more recent work on the use of aldolases and transketolase for monosaccharide synthesis, and of the enzymes of the Leloir pathway for oligosaccharide synthesis.

MONOSACCHARIDE SYNTHESIS

1. Aldolases

A. FDP Aldolases. The "Inversion Strategy"

The aldolases (E.C. 4.1.2.x) are an ubiquitous group of enzymes that catalyze the interconversion of higher aldoses and ketoses and their respective components. Two groups of aldolases are found in Nature.[16] Type 1 aldolases, present in mammalian and higher plant systems, utilize a lysine amino functionality to catalyze aldol condensation *via* a Schiff base intermediate. Type 2 aldolases, isolated primarily from prokaryotic sources, are Zn^{2+}-dependent enzymes.

Since the early 1950's, aldolases have been utilized as catalysts for the synthesis of monosaccharides.[12] The best studied of the aldolases is a fructose-1,6-diphosphate (FDP) aldolase from rabbit muscle (E.C. 4.1.2.13, frequently referred to as RAMA).[17] *In vivo*, this enzyme catalyzes the interconversion of FDP and dihydroxyacetone phosphate (DHAP) and D-glyceraldehyde-3-phosphate (Scheme 1).

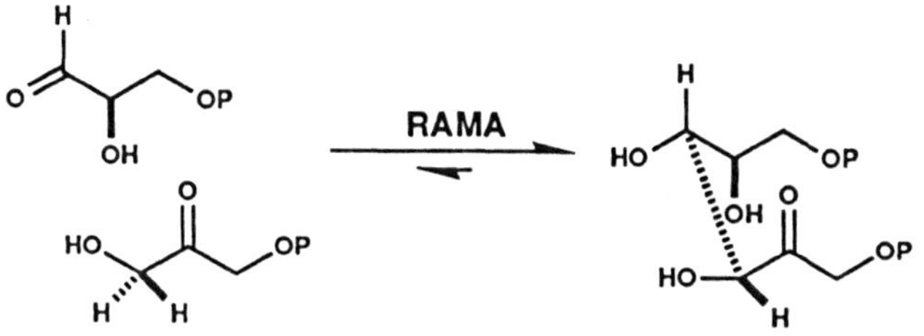

Scheme 1. The RAMA-Catalyzed Aldol Condensation

The enzyme displays absolute stereospecificity for the two new stereogenic centers formed during the reaction: the vicinal diol formed during the condensation is always D-*threo*. RAMA will accept a wide range of aldehydes as the electrophilic component of the reaction, although the enzyme appears to be absolute in its requirement for DHAP as the nucleophile.[12,17] A number of useful monosaccharides have been prepared using RAMA. More recently, a bacterial FDP-aldolase has been cloned and overproduced. In initial studies, the bacterial enzyme is reported to have similar substrate specificity to RAMA.[18]

Although clearly useful as a catalyst in monosaccharide synthesis, RAMA suffers a number of serious drawbacks. Most notably, RAMA produces only ketoses, while a number of important naturally occurring structures are aldose sugars. Furthermore, only a single diastereomeric product is available. Much of the recent work on monosaccharide synthesis in our group has focussed on strategies to overcome these limitations.

One potential route to aldose sugars is the so-called "inversion strategy" (Scheme 2).[19] The idea behind this methodology is to utilize a monoprotected dialdehyde as the electrophile in an enzyme-catalyzed aldol condensation. Monoprotected dialdehydes of this nature are conveniently prepared using a vinyl group as a masked aldehyde functionality. Ozonolysis is then used to release the desired carbonyl.

Following aldolase-catalyzed condensation, the ketone functionality is reduced stereospecifically, and the protected aldehyde is exposed to yield a new aldose sugar. This methodology accomplishes two goals: the generation of aldose sugars, and the placement of the vicinal diol formed during the enzyme-catalyzed aldol condensation in a position other than C3/C4. We have demonstrated the utility of this methodology through the synthesis of two monosaccharides, L-xylose (**2**, Scheme 3) and 2-deoxy-D-*arabino*-hexose (**5**, Scheme 4).[19]

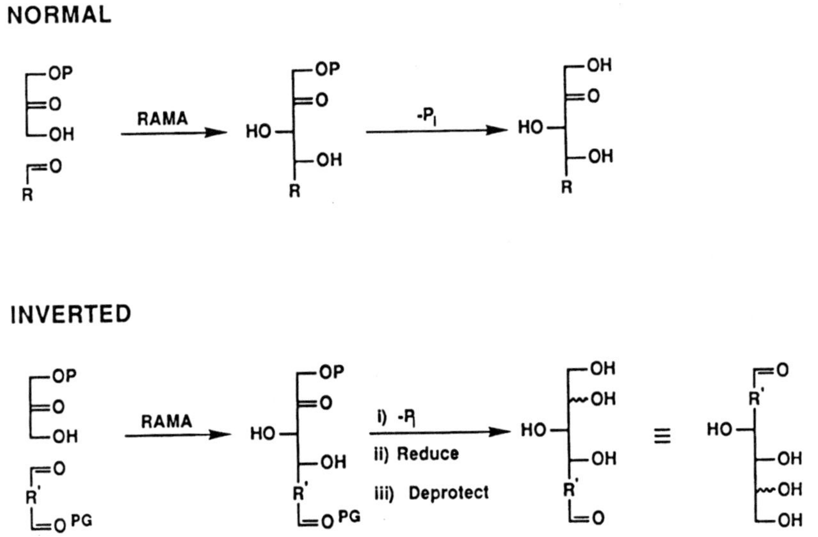

Scheme 2. The Inversion Strategy.

The stereospecific reduction of the ketone functionality generated in the RAMA adduct is another vital component of the inversion strategy. We have found iditol dehydrogenase (E.C. 1.1.1.14), from *Candida utilis*, also known as sorbitol or polyol dehydrogenase) to be a useful catalyst for this reduction. Together, Schemes 3 and 4 demonstrate the accessibility of ***either*** diastereomer using a single dehydrogenase. Iditol dehydrogenase-catalyzed reduction of adduct **1** produces exclusively the 2R polyol **1** (Scheme 3). The 2S configuration is available using a two-step procedure: the ketone is first reduced *non-specifically* with a hydride reducing agent, and the undesired 2R enantiomer is oxidized enzymatically to the original ketone (Scheme 4). The ketone can then be recycled, to produce a new diastereomeric mixture of polyols.

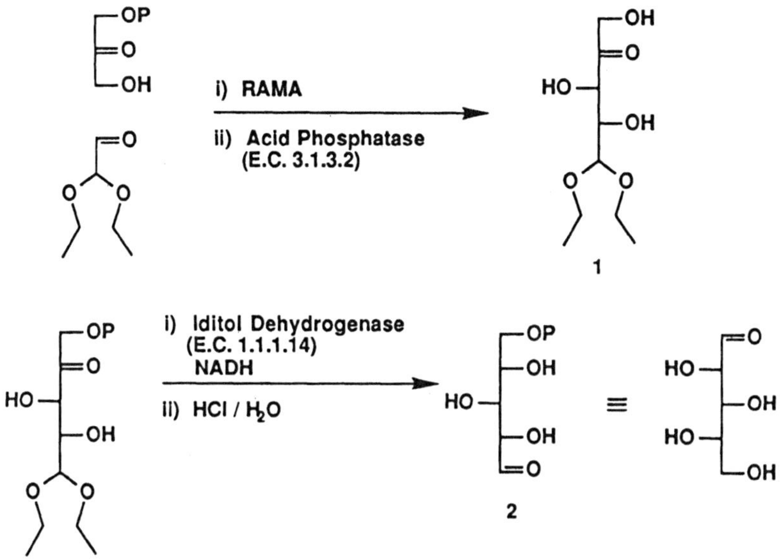

Scheme 3. Synthesis of L-xylose

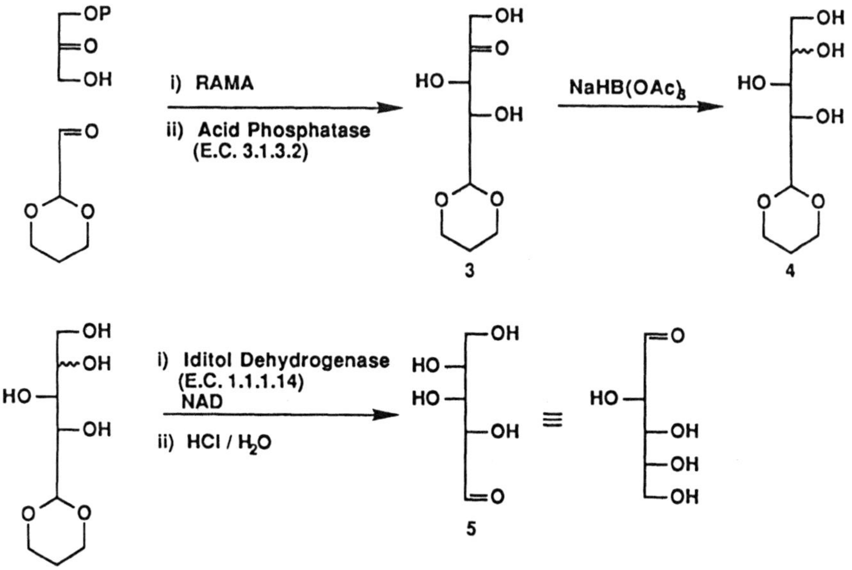

Scheme 4. Synthesis of 2-deoxy-D-*arabino*-hexose.

B. Fuculose-1-Phosphate Aldolase

Another potential route to other stereochemistries at C3 and C4 relies on the availability of other aldolases. Approximately 25 aldolases have been identified to date from various sources.[12] Of the four possible diastereomeric products resulting from an aldolase-catalyzed condensation of DHAP with an aldehyde, enzymes have been identified which generate three stereochemically distinct products (Scheme 5). We have begun investigations on the utility of bacterial aldolases that generate products analogous to but stereochemically distinct from those generated by FDP-aldolases. We have now begun an evaluation of a bacterial fuculose-1-phosphate (fuc-1-P) aldolase as a catalyst for monosaccharide synthesis.[20]

Unlike FDP aldolases, which are synthesized by most organisms under all growth conditions, fuc-1-P aldolase is an inducible enzyme; it is expressed only when cells are grown on fucose as a carbon source.

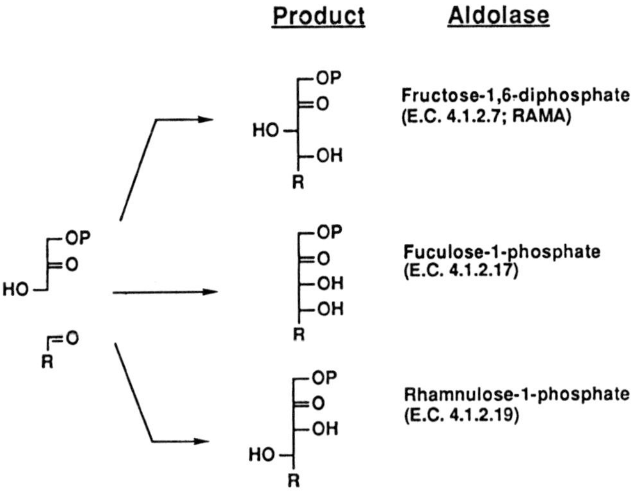

Scheme 5. Identified DHAP Aldolases.

Even under these conditions, the amount of enzyme produced is small, on the order of <1 unit of enzyme per gram of wet cells (a unit of enzyme catalyzes the conversion of 1 μmol of the substrate to products per minute, under optimal conditions of temperature and pH).[21] In order to produce the thousands of units of enzyme we required for carbohydrate synthesis, overproduction of the enzyme through recombinant technology seemed the only feasible route. We have overexpressed fuc-1-P aldolase from *E. coli*, using a *tac* promoter. Approximately 300 units of fuc-1-P aldolase are obtained per gram of wet cells. The enzyme has been purified to crystallinity (Table 1), although crude preparations can also be used synthetically. Fuculose-1-phosphate aldolase is a Type 2 aldolase, and is zinc-dependent. In initial substrate specificity studies, the enzyme shows broad substrate specificity with regard to the aldehyde component (Table 2).

Table 1. Cloned Fuc-1-P Aldolase.

Crude cell extracts:

E. coli JM105 (FDP-Aldolase)	**6.6 U/mg**
E. coli A03 (No FDP-Aldolase)	**1.7 U/mg**

Purification

	Protein Yield (mg)	Activity (U)	Specific Activity (U/mg)	(%)
Cell-Free Extracts	1063	5.8×10^3	5.5	100
$(NH_4)_2SO_4$ (40-60%)	755	5.4×10^3	7.1	93
DEAE-Sepharose	393	3.9×10^3	9.9	67

We have now extended the list of aldehydes examined for activity to approximately 40, and the enzyme appears to be a useful catalyst for the preparation of monosaccharides. Investigations on the specificity of the enzyme for the nucleophilic component remain to be completed.
The enzyme has also been used to prepare D-ribulose, on a 10 mmol scale (Scheme 6). The product pentose was identical to authentic commercial material.

Synthetic routes to monosaccharides utilizing all of the aldolases discussed here require DHAP. DHAP is commercially available, although at \$120/mmol DHAP clearly must be prepared for synthetic purposes. There are three methods generally used for the preparation of DHAP for use in aldolase-catalyzed syntheses: from fructose-1,6-diphosphate (FDP), by chemical synthesis from dihydroxyacetone, and by enzymatic synthesis from dihydroxyacetone using glycerol kinase (Schemes 7 and 8). The

Table 2. Substrate Specificity of Fuc-1-P Aldolase

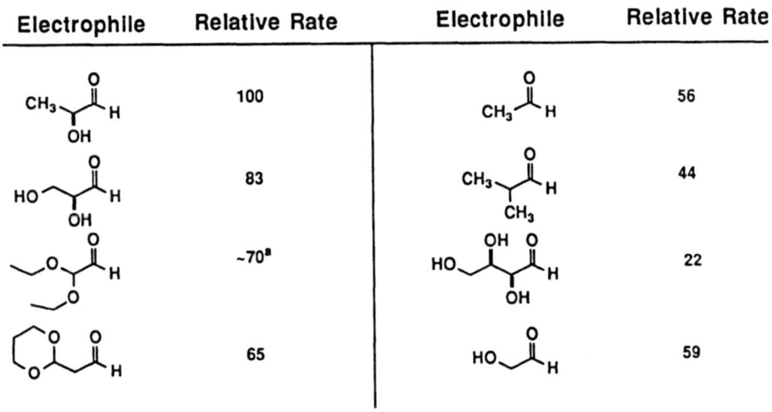

[a] Substrate insoluble; approximate rate.

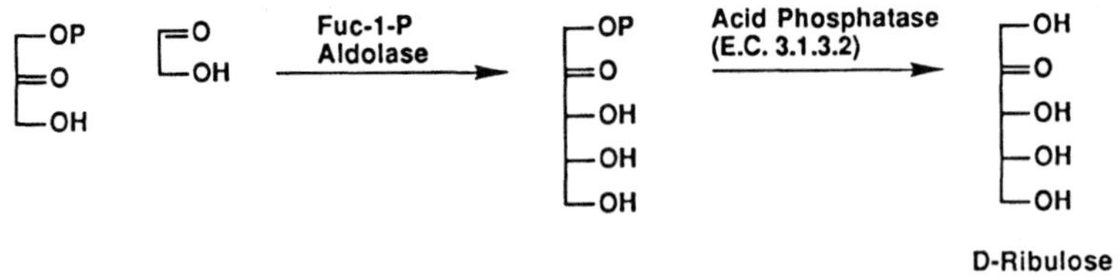

Scheme 6. Synthesis of D-Ribulose.

relative merits of these three methodologies have been discussed elsewhere.[12] Briefly, synthesis from fructose-1,6-diphosphate has the advantages of low cost and simplicity. The major disadvantage is the presence of large amounts of FDP during purification of products. Furthermore, this method is obviously only applicable to syntheses using FDP-aldolases. Chemical synthesis of DHAP provides high purity material, but the synthesis is technically awkward for amounts greater than 10 mmol. This method originally published[22] has recently been improved.[23] The major enhancements are the use of ion exchange resin as the acid in the formation of dihydroxyacetone cyclic dimer ethyl ketal, and the use of triethylorthoformate to drive this reaction (Scheme 7).

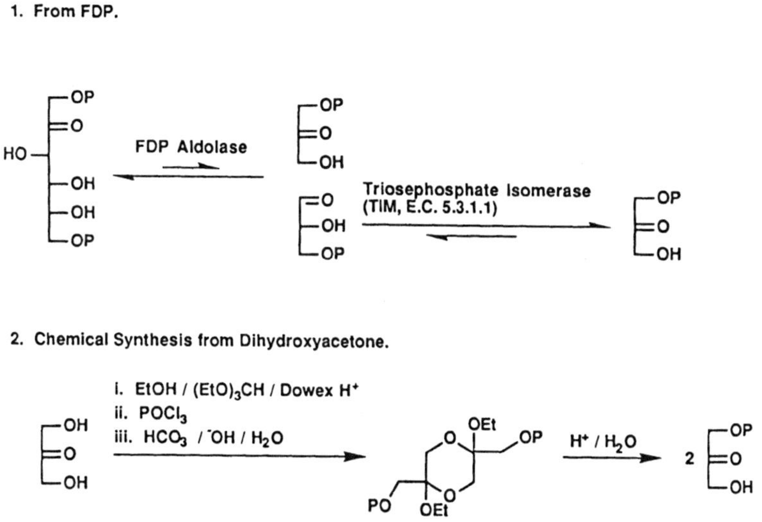

SCHEME 7. Synthesis of DHAP.

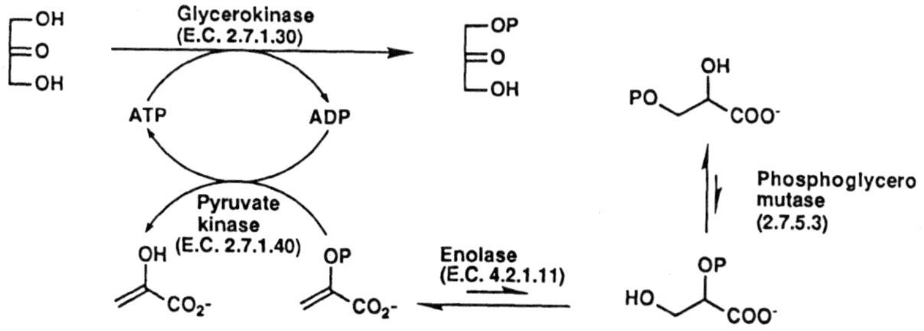

Scheme 8. Enzymatic Synthesis of DHAP.

The enzymatic synthesis of DHAP, on the other hand, is capable of generating product on a mole scale.[24] Glycerol kinase (E.C. 2.7.1.30) transfers a phosphate from ATP to commercially available dihydroxyacetone. ATP must be regenerated, and regeneration schemes for ATP in syntheses of DHAP have been reported using both acetyl phosphate with acetyl kinase, and phosphoenolpyruvate (PEP) with pyruvate kinase.[24] These schemes still require synthesis of the ultimate phosphate donors, acetyl phosphate or PEP. More recently, we have developed a method for the generation of PEP *in situ* from commercially available D-3-phosphoglycerate.[25] This synthesis of PEP can be used to regenerate ATP in syntheses of DHAP (Scheme 8). The net result is a one-pot synthesis of DHAP using only commercially available starting materials. The quantity of each of the four enzymes used represents approximately $10 per enzyme, while D-3-PGA is commercially available for $400/mole. This methodology represents an extremely attractive synthetic approach to large quantities of DHAP.

2. Transketolase

In addition to aldolases, the other major group of enzymes used *in vivo* to generate and degrade monosaccharides are the transaldolases and transketolases.[26] One of these enzymes, transketolase (E.C. 2.2.1.1), is a potentially useful catalyst for the synthesis of carbohydrates. The enzyme catalyzes the transfer of a two-carbon unit from hydroxypyruvate to a variety of aldehydes (Scheme 9).

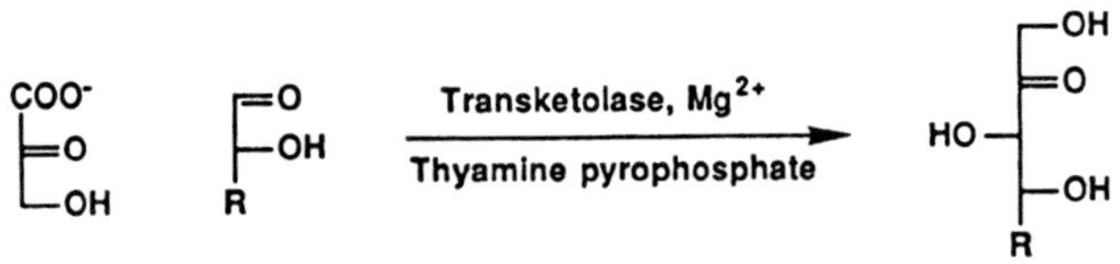

SCHEME 9. Transketolase.

Transketolase is a thyamine pyrophosphate dependant enzyme, and also requires divalent zinc as a cofactor. Transketolase from baker's yeast is commercially available for $1.5/unit. Transketolase can also be readily isolated from spinach; approximately 150 units per kilogram of spinach can be recovered.[28]

Transketolase has received scant attention in the past.[29,30] We are now investigating the enzyme in more detail. Although transketolase generates a vicinal diol with the same configuration as those produced by FDP-aldolases, there are two significant advantages to transketolase-catalyzed syntheses: the enzyme does not require DHAP, and the products are not phosphorylated. Substrate specificity studies to date have shown the enzyme will accept a wide range of aldehydes as substrates (Table 3).[31]

TABLE 3. Substrate Specificity of Transketolase.

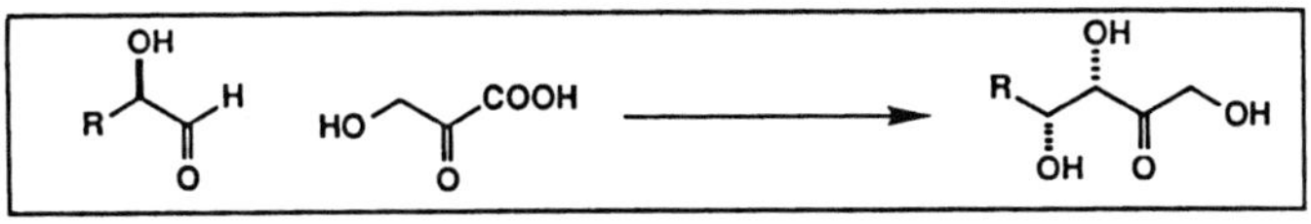

Aldehyde Substrates:

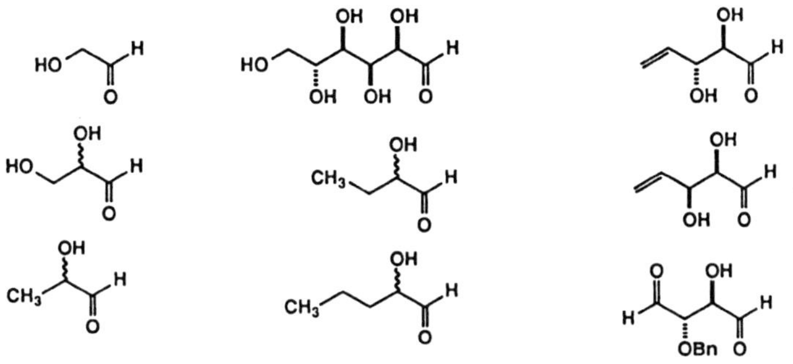

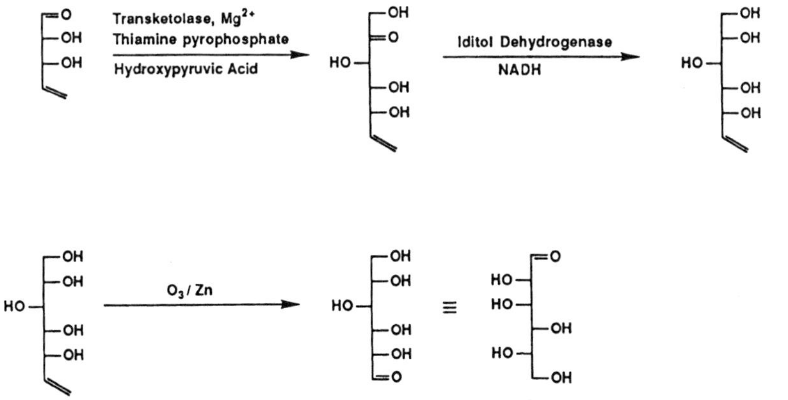

L-Gulose

SCHEME 10. Transketolase-Based Synthesis of L-Gulose.

The enzyme seems to be absolute in its requirement for a 2R hydroxyl group in the aldehyde; mixtures of aldehydes epimeric at C2 can be resolved using transketolase. The enzyme appears to have no preference for configuration beyond C2. To date, no keto acid has been found which is accepted by the enzyme in place of hydroxypyruvate. Transketolase has been used for a synthesis of L-gulose (Scheme 10).[32] It seems certain that transketolase will be a useful enzyme for the synthesis of monosaccharides.

GLYCOSIDIC BOND FORMATION. THE PREPARATION OF ACTIVATED MONOSACCHARIDES

1. Introduction

Another area of carbohydrate chemistry where enzymes are now finding extensive use is in the formation of glycosidic linkages. Although great progress has been made in the development of chemical methodologies for glycosidic bond formation, few general methods exist to generate high yields of a single anomeric product.[10,32] Furthermore, the exquisite specificity required in the synthesis of many important glycoconjugates makes chemical synthesis formidable.

Nature uses two basic motifs to activate monosaccharides for transfer to nascent chains.[1] In several systems, glycosyl phosphates are used as activated sugars to form glycosidic linkages. Sucrose phosphorylase and trehalose phosphorylase are examples of enzymes of this type: both have been used synthetically.[33,34] More commonly, monosaccharides are activated as nucleoside phosphate sugars. In mammalian biochemistry, eight activated monosaccharides are commonly encountered: uridine-5'-diphosphoglucose (UDPGlc), uridine-5'-diphosphoglucuronic acid (UDPGlcUA), uridine-5'-diphosphogalactose (UDPGal), uridine-5'-diphospho-N-acetylglucoamine (UDPGlcNAc), uridine-5'-diphospho-N-acetylgalactosamine (UDPGalNAc), guanidine-5'-diphosphomannose (GDPMan), guanidine-5'-diphosphofucose (GDPFuc), and cytidine-5'-monophospho-N-acetylneuraminic acid (CMPNeuAc). Monosaccharides are transferred from these activated sugars to nascent oligosaccharide chains by the glycosyl transferases. Collectively, the enzymes that form the nucleoside phosphate sugars and transfer them to growing carbohydrate chains are referred to as the enzymes of the Leloir pathway (named for the Argentinean biochemist who first elucidated this biosynthetic route). We have been active for several years in the study of glycosidic bond formation using Leloir pathway enzymes. Most recently, we have been engaged in elucidating practical routes to large quantities (>1g) of the important nucleoside phosphate sugars.

2. Synthesis of Nucleoside Triphosphates

The synthesis of most of the nucleoside diphosphate sugars is accomplished *in vivo* by coupling a sugar-1-phosphate with a nucleoside triphosphate (Scheme 11). An exception is CMPNeuAc, in which case N-acetylneuraminic acid is coupled directly with CTP (Scheme 12).

$$\text{Sugar-1-P} + \text{XTP} \rightarrow \text{XDP-Sugar}$$

Scheme 11. Biosynthesis of Nucleoside Diphosphate Sugars.

$$\text{NeuAc} + \text{CTP} \rightarrow \text{CMP-NeuAc}$$

Scheme 12. Biosynthesis of CMPNeuAc.

Clearly, a source of nucleoside triphosphates (XTP's) is vital for enzymatic syntheses of nucleoside phosphate sugars. We have compared chemical and enzymatic methodologies for the preparation of XTP's.[35] Although numerous chemical methodologies are available in the literature, we have

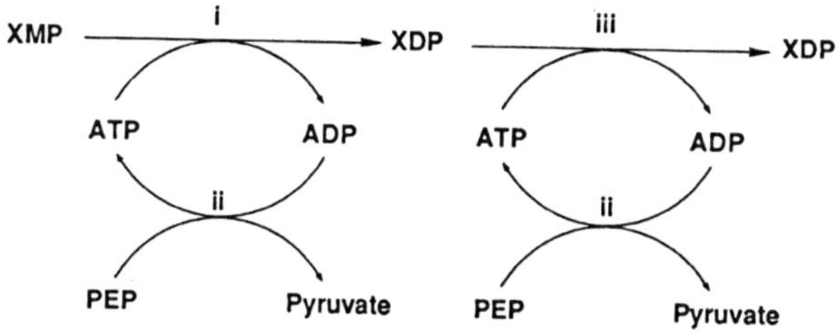

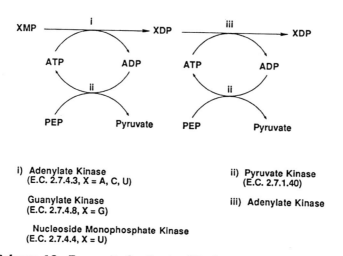

i) Adenylate Kinase
(E.C. 2.7.4.3, X = A, C, U)

Guanylate Kinase
(E.C. 2.7.4.8, X = G)

Nucleoside Monophosphate Kinase
(E.C. 2.7.4.4, X = U)

ii) Pyruvate Kinase
(E.C. 2.7.1.40)

iii) Adenylate Kinase

Scheme 13. Enzymatic Synthesis of Nucleoside Triphosphates.

found that enzymatic methods for the preparation of XTP's are operationally much simpler, especially for large scale production. Scheme 13 outlines the most effective enzymatic route to each of the nucleoside triphosphates from the corresponding monophosphate.

All of the nucleoside monophosphates are available from yeast RNA digests at low cost. Adenylate kinase (E.C. 2.7.4.3), which catalyzes the equilibrium between adenosine mono-, di- and triphosphate, has been used extensively for the production of ATP.[24,36] Although adenylate kinase will accept all of the nucleoside diphosphates as substrates, its substrate specificity for monophosphates is much more restrictive. Nonetheless, cytidine monophosphate can be phosphorylated at a synthetically useful rate using adenylate kinase.[37] The monophosphates of both uridine and guanosine are not accepted by adenylate kinase. GTP is best prepared using guanylate kinase, while UTP can be prepared by chemical deamination of CTP.[35] In all cases, the ultimate phosphate donor is PEP, which can be prepared from 3-PGA.

3. Synthesis of UDPGlcUA

Glucuronic acid is used *in vivo* in mammalian systems as a conjugate for xenobiotic removal.[38] The uronic acids (glucuronic and iduronic acid) are also components of the glycosamineglycans (referred to as mucopolysaccharides in older texts), including heparin, chondroitin, and hyaluronic acid. Iduronic acid (the C5 epimer of glucuronic acid) is generated in vivo from glucuronic acid *after* transfer by a glycosyl transferase.

UDPGlcUA is produced from UDPGlc by a nicotinamide-dependent dehydrogenase. UDPGlc dehydrogenase from bovine liver is commercially available. The enzyme is expensive ($10/U), and in our hands was extremely unstable: we found commercial enzyme preparations unsuitable for gram-scale production of UDPGlcUA. UDPGlcUA can also be isolated in straightforward fashion from bovine liver according to literature methods.[39] This preparation furnishes approximately 200U of enzyme per kilogram of liver. We have used crude enzyme (ammonium sulfate fractionation from crude liver homogenate) to prepare UDPGlcUA on a one gram scale (Scheme 14).[40] UDPGlcUA was purified from the crude product mixture by both gel filtration and ion exchange chromatography. The overall yield from UDPGlc was 76% (91% mass recovery, 84% pure by enzymatic analysis).

We have also prepared UDPGlcUA from UDPGlc *via* a platinum catalyzed air oxidation (Scheme 15). This technique suffers two major drawbacks: the catalyst is rapidly deactivated under the reaction conditions, and the rate of oxidation is very slow at pH values <8. This latter concern is significant since UDPGlc is unstable in basic media. It was recognized in previous studies on the oxidation of glucose-1-phosphate to glucuronic acid-1-phosphate that catalyst deactivation was also a serious problem.[41,42] Literature methods exist to prepare catalysts on carbon supports which are resistant to deactivation,[41,42] and this method might be useful under different conditions.

We were unable to generate synthetically useful amounts of UDPGlcUA *via* a platinum-catalyzed oxidation, and at this time oxidation of UDPGlc using crude UDPGlc dehydrogenase from bovine liver is the most useful route to gram-scale quantities of UDPGlcUA.

4. CMPNeuAc

One of the most important sugars in mammalian biochemistry is N-acetylneuraminic acid, which commonly terminates glycoconjugate oligosaccharides. N-Acetylneuraminic acid is synthesized in vivo by an

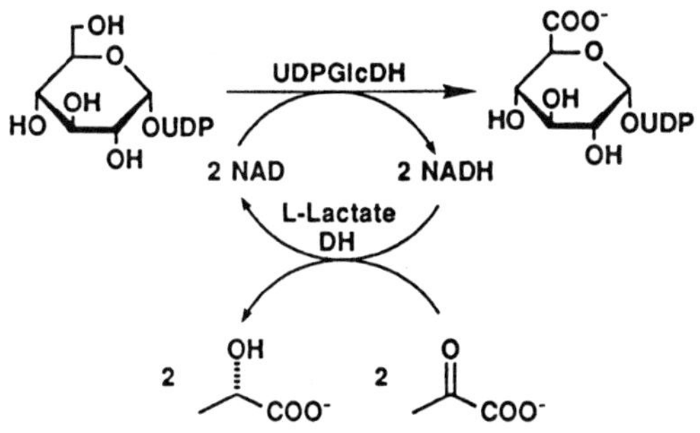

SCHEME 14. Enzymatic Synthesis of UDPGlcUA.

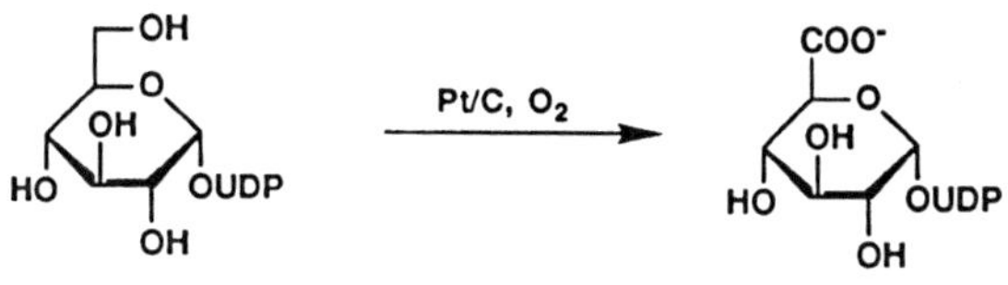

SCHEME 15. Platinum-Catalyzed Oxidation of UDPGlc.

aldolase-catalyzed condensation between N-acetylmannosamine and pyruvate (Scheme 16). CMPNeuAc is synthesized by CMPNeuAc synthetase (E.C. 2.7.7.43), from N-acetylneuraminic acid and CTP. Although preparations of CMPNeuAc have been reported, all are inconvenient for large scale (>250 mg) preparations. A total synthesis of CMPNeuAc from the inexpensive reagents N-acetylglucosamine, CMP, and pyruvate is outlined in Scheme 16.[43]

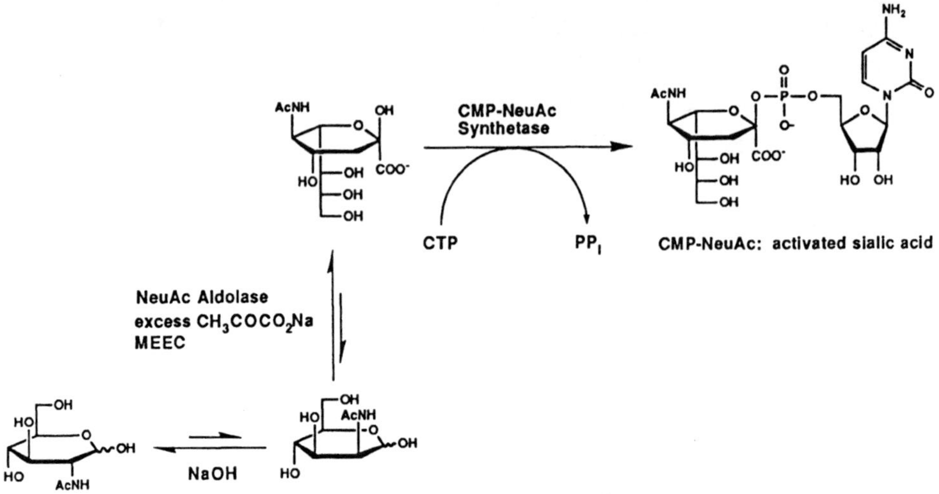

SCHEME 16. Enzymatic Synthesis of CMPNeuAc.

N-acetylneuraminic lyase (E.C. 4.1.3.3) is commercially available. The enzyme has been cloned and overexpressed. N-Acetylneuraminic lyase accepts several analogues of N-acetylmannosamine as the electrophile, and has been used in the synthesis of unnatural sialic acids. The final enzyme required for the preparation of CMPNeuAc, CMPNeuAc synthetase (E.C. 2.7.7.43), is isolable from calf brain.

CONCLUSIONS

Enzyme-based carbohydrate synthesis is now clearly established as a useful technology. Its importance will continue to grow, as the demand for unnatural synthetic carbohydrates grows. Although enzyme-based oligosaccharide synthesis is currently limited by the availability of the glyosyl transferases, modern molecular biology will soon provide the necessary enzymes.

ACKNOWLEDGMENTS

This work was carried out by several members of the Whitesides group, past and present. The work presented here was funded by NIH, grants GM30367 and GM39589.

REFERENCES

1. Sharon, N. *Complex Carbohydrates*; Addison-Wesley: Reading, MA, 1975.
2. Horowitz, M. I.; Pigman, W., Eds. *The Glycocongugates, Vol I - II*; Academic: New York, 1977-1978.
3. Horowitz, M. I., Ed. *The Glycoconjugates, Vol III - IV*; Academic: New York, 1982.

4. Schauer, R. *Adv. Carbohydr. Chem. Biochem.* **1982**, *40*, 131.
5. *The Lectins: Properties, Functions and Applications in Biology and Medicine*; Liener, I. E.; Sharon, N.; Goldstein, I. J., Eds.; Academic: New York, 1986.
6. Paulson, J. C. In *The Receptors*; Cohn, P. M., Ed.; Academic: New York, 1985; Vol. 2, p 131.
7. Sairam, M. R. In *The Receptors*; Cohn, P. M., Ed.; Academic: New York, 1985; Vol. 2, p 307
8. Schmidt, R. R. *Angew. Chem. Int. Ed. Engl.* **1986**, *25*, 212.
9. Kunz, H. *Angew. Chem. Int. Ed. Engl.* **1987**, *26*, 294.
10. Paulsen, H. *Angew. Chem. Int. Ed. Engl.* **1982**, *21*, 155.
11. Lemieux, R. U. *Chem. Soc. Rev.* **1978**, *7*, 423.
12. Toone, E. J.; Simon, E. S.; Bednarski, M. D.; Whitesides, G. M. *Tetrahedron* **1989**, *45*, 5365.
13. Jones, J. B. *Tetrahedron* **1986**, *42*, 3351.
14. *Enzymes in Organic Synthesis (Ciba Foundation Symposium 111)*; Porter, R.; Clark, S., Eds.; Pitman: London, 1985.
15. Akiyama, A.; Bednarski, M. D.; Kim, M.-J.; Simon, E. S.; Waldmann, H.; Whitesides, G. M. *CHEMTECH* **1988**, 627.
16. Horecker, L.; Tsolas, O.; Lai, C. Y. In *The Enzymes*, Boyer, P. D. Ed.; Academic: New York, 1972; Vol. VII, p 213.
17. Bednarski, M. D.; Simon, E. S.; Bischofberger, N.; Fessner, W.-D.; Kim, M.-J.; Lees, W.; Saito, T.; Waldmann, H.; Whitesides, G. M. *J. Am. Chem. Soc.* **1989**, *111*, 627.
18. von der Osten, C. H.; Sinskey, A. J.; Barbas, C. F.; Pederson, R. L.; Wang, Y.-F.; Wong, C.-H. *J. Am. Chem. Soc.*, **1989**, *111*, 3924.
19. Borysenko, C.; Spaltenstein, A.; Straub, J.; Whitesides, G. M. *J. Am. Chem. Soc.* **1989**, *111*, 9275.
20. Ozaki, A.; Toone, E. J.; von der Osten, C. H.; Sinskey, A. J.; Whitesides, G. M. *J. Am. Chem. Soc.* **1990**, submitted for publication.
21. Ghalamor, M. A.; Heath, E. C. *J. Biol. Chem.* **1962**, *237*, 2427.
22. Effenberger, F.; Straub, A. *Tetrahedron Lett.* **1987**, *28*, 1641.
23. Schmid, W.; Whitesides, G. M. unpublished results.
24. Crans, D. C.; Kazlauskas, R. J.; Hirschbein, B. L.; Wong, C. -H.; Abril, O.; Whitesides, G. M. *Methods Enzymol.* **1987**, *136*, 263.
25. Simon, E. S.; Grabowski, S.; Whtiesides, G. M. *J. Am. Chem. Soc.*, **1989**, *111*, 8920
26. Racker, E. In *The Enzymes*; Boyer, P. D.; Lardy, H.; Myrbäch, K., Eds.; Academic: New York; 1961; Vol. 5, p 397.
27. Villifranca, J. J.; Axelrod, B. *J. Biol. Chem.*, **1971**, *246*, 3126.
28. Bolte, J.; Demuynck, C.; Samaki, H. *Tetrahedron Lett.* **1987**, *28*, 5525.
29. Mocali, A.; Aldinucci, D.; Paoletti, F. *Carbohydr. Res.* **1985**, *143*, 288.
30. Kobori, Y.; Myles, D.; Whitesides, G. M. unpublished results.
31. Myles, D.; Whitesides, G. M. unpublished results.
32 Paulsen, H. *Chem. Soc. Rev.* **1984**, *13*, 15.
33. Haynie, S. L.; Whitesides, G. M. *Appl. Biotech. Biochem.* **1990**, *23*, 155.
34. Waldman, H.; Gygax, D.; Bednarski, M. D.; Shangraw, R.; Whitesides, G. M. *Carbohydr. Res.*, **1986**, *157*, c4.
35. Simon, E. S.; Grabowski, S.; Whitesides, G. M. *J. Org. Chem.* **1990**, submitted.
36. Hirschein, B. :.; Mazenod, F. P.; Whtiesides, G. M. *J. Org. Chem.*, **1982**, *47*, 3765.
37. Simon, E. S.; Bednarski, M. D.; Whitesides, G. M. *Tetrahedron Lett.*, **1988**, *29*, 1123.
38. *Enzymatic Bases of Detoxication*; Jakoby, W. B. Ed.; Academic: New York, 1980; Vol. II.
39. Zalitis, J.; Feingold, D. S.; *Arch. Biochem. Biophys.* **1969**, *12*, 457.
40. Toone, E. J.; Whitesides, G. M. unpublished results.
41. Van Dam, H. E.; Dieboom, A. P. G.; ban Bekkum, H. *Appl. Catal.*, **1987**, *33*, 367.
42. Van Dam, H. E.; Dieboom, A. P. G.; ban Bekkum, H. *Appl. Catal.*, **1987**, *33*, 373.
43. Simon. E. S.; Bednarski, M. D.; Whitesides, G. M. *J. Am. Chem. Soc.* **1988**, *110*, 7159.

EXPLOITING ENZYME SELECTIVITY FOR THE SYNTHESIS OF BIOLOGICALLY ACTIVE COMPOUNDS

Alexey L. Margolin

Merrell Dow Research Institute, Indianapolis Center, 9550 N. Zionsville Rd., Indianapolis, IN 46268

INTRODUCTION

Over the last several years we have been witnessing the proliferation of enzymatic synthetic methods in organic chemistry. More and more organic chemists without any special training in enzymology use enzymes as catalysts on a routine basis. Enzymes have a number of properties useful in organic synthesis, but their greatest advantage over standard chemical techniques is selectivity. Different levels of enzyme selectivity - substrate specificity, regio-, chemo- or stereoselectivity - may be successfully exploited in the synthesis of complex, polyfunctional molecules. The optimal situation would be to have a catalyst which can accomodate as many subtrates as possible (broad substrate specificity) and then carry out the catalytic transformation in a selective way. The recent discovery that substrate specificity and enantioselectivity of enzymes may be dramatically altered, and sometimes predictably controlled by changing the reaction medium creates new opportunities for organic chemistry.[1-5]

In this paper I will show how it is possible to exploit different levels of enzyme selectivity in the synthesis of biologically active compounds.

ENZYME-CATALYZED ACYLATION OF CASTANOSPERMINE

The alkaloid castanospermine is a potent inhibitor of the endoplasmic reticulum enzyme α-glucosidase I. It prevents removal of glucose residues during the normal processing of glycoproteins and therefore is highly bio-logically active.[6] The recent wave of interest in this compound is due to the fact that castanospermine may be of clinical value as a drug in the treatment of Acquired Immune Deficiency Syndrome (AIDS).[7] It has been reported that several O-acyl derivatives of castanospermine are as much as twenty times more active than castanospermine itself in inhibiting human immunodeficiency virus (HIV) replication.[8] These findings make the synthesis of castanospermine analogs an extremely important and urgent goal.

Castanospermine ((1S,6S,7R,8R,8aR)-1,6,7,8-tetrahydroxyoctahydroindolizine) contains four secondary hydroxyl groups of similar reactivity and therefore represents a challenging target for regioselective modification. The synthesis of individual esters of castanospermine as well as other aminosugars and carbohydrates usually requires 5-6 steps involving protection and deprotection of neighboring hydroxyl groups.

Recently, lipases and subtilisin have successfully been used for the regioselective acylation and deacylation of a number of carbohydrates and their derivatives such as mono- and disaccharides, conjugates of sugars with aglycons, etc.[9-13] In all reported examples, enzymes exhibit a predominant preference toward a primary hydroxyl group in both acylation and deacylation reactions. When the primary hydroxyl group is protected, several lipases regioselectively catalyze the acylation of secondary hydroxyl groups in monosaccharides.[14,15] It should be noted, however, that despite the obvious progress in selective modification of monosaccharides, this methodology has so far only been applied to the synthesis of model compounds. We further developed this methodology and applied it to the synthesis of castanospermine analogs[16]. Subtilisin Carlsberg (EC 3.4.21.4) has been used as a catalyst for enzymatic transesterification of castanospermine (Fig 1).

A typical experimental procedure is illustrated below by using the synthesis of 1-O-butyryl-castanospermine as an example. We dissolved 170 mg (0.90 mmol) of castanospermine in 15 mL of warm anhydrous pyridine followed by addition of 396 mg (1.80 mmol) of 2,2,2-trichloroethylbutyrate. Then 75 mg of subtilisin was added. The suspension was shaken at 45°C and 260 rpm for four days; periodically aliquots were withdrawn and analyzed by TLC. Then the enzyme was removed by filtration, pyridine was evaporated under vacuum and the product was isolated and purified by radial silica-gel chromatography. This procedure resulted in 192 mg (0.74 mmol; 82% isolated yield) of a white crystalline compound. The data of ^{1}H NMR, ^{13}C NMR, low and high resolution MS, and elemental analysis were consistent with the structure of 1-O-butyryl-castanospermine. Encouraged by this result we decided to test the scope of this approach with other acylating agents (Table 1).

Subtilisin catalyzes the synthesis of a wide variety of castanospermine analogs. It is possible to regulate the hydrophobicity of the acylating group (acetyl, butyryl, octanoyl), or to incorporate an aromatic moiety (phenylacetyl benzoyl), or L- or even D-amino acids (phenylalanyl, L- and D-alanyl). These results clearly indicate that in organic solvent, subtilisin possesses a broad substrate specificity

TABLE 1. PREPARATIVE REGIOSELECTIVE SYNTHESIS OF 1-O-ACYLCASTANOSPERMINE CATALYZED BY SUBTILISIN IN ANHYDROUS PYRIDINE

Acylating agent	reaction time (hr)	isolated yield (%)
$H_3C(CH_2)_2CO_2CH_2CCl_3$	92	82
$H_3C(CH_2)_6CO_2CH_2CF_3$	96	23
$H_3CCO_2CH{=}CH_2$	84	91
$C_6H_5CO_2CH{=}CH_2$	96	65
$C_6H_5CH_2CO_2CH{=}CH_2$	96	30
N-Ac-L-Phe-OCH_2CH_2Cl	96	49
N-Ac-L-Ala-OCH_2CH_2Cl	96	44
N-CBZ-L-Ala-OCH_2CH_2Cl	120	64
N-Ac-D-Ala-OCH_2CH_2Cl	120	31

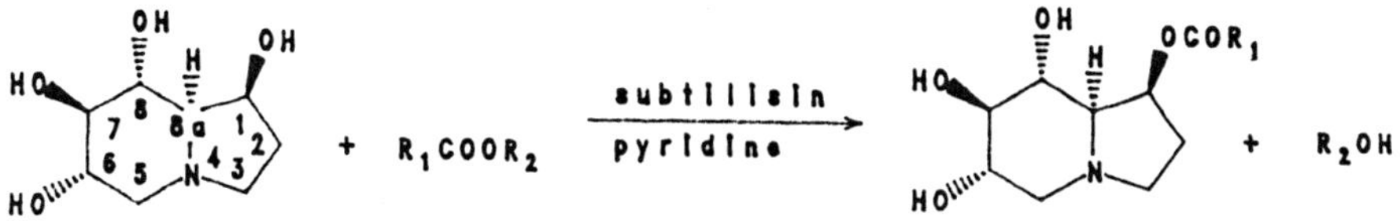

R_1 is alkyl, aryl or α-aminoalkyl

R_2 is chloroethyl, 2,2,2-trichloroethyl,

2,2,2-trifluoroethyl or vinyl

Fig. 1 Subtilisin-Catalyzed Acylation of Castanospermine

and at the same time a high regioselectivity. (In all cases no formation of di-, tri-, or mono-O-acyl derivatives other than 1-O-acylcastanospermine was observed). This combination of broad substrate specificity and high regioselectivity in organic solvents makes subtilisin a powerful tool for organic synthesis.

It should be noted that subtilisin is remarkably stable under reaction conditions. After 5 days' reaction in pyridine at 45°C, the enzyme does not lose its catalytic activity and can be reused. This fact may significantly decrease the cost of the process on a large scale.

The high regioselectivity of subtilisin allows for preparation of a variety of 1-O-acyl derivatives without the formation of byproducts; on the other hand the same high regioselectivity precludes the synthesis of other castanospermine analogs. Unfortunately, few other enzymes are active in pyridine or dimethylformamide (the only organic solvents where castanospermine has reasonable solubility). Out of many enzymes tested only porcine pancreatic lipase was catalytically active in pyridine but to a much smaller extent than subtilisin. When the same experimental procedure was employed for the reaction of castanospermine with 2,2,2-trichloroethyl butyrate catalyzed by porcine pancreatic lipase it resulted in the synthesis of 6-O-butyryl and 7-O-butyryl castanospermine with an isolated yield of 11% and 6% respectively. These data indicate that although lipase has different regioselectivity, it cannot be used as an efficient catalyst in the preparative acylation of castanospermine. Here we describe another approach which may solve the aforementioned problem.

1-O-Acyl derivatives of castanospermine are well soluble in a variety of organic solvents. The use of tetrahydrofuran (THF) as a reaction medium, for example, allows us to exploit the regioselectivity of several lipases as well as subtilisin. Porcine pancreatic lipase (PPL), lipase from *Chromobacterium viscosum* (CV), and subtilisin Carlsberg have been used as catalysts for further enzymatic acylation of 1-O-acylcastanospermine (Fig 2).

Several 1-O-acyl derivatives of castanospermine have been used as substrates for enzymatic modification. (Table 2) It is clear that the regioselectivity of subtilisin is different from that of lipases tested. Subtilisin shows a preference for acylation of the OH-group at the C-6 position, while in most cases lipases prefer the OH-group at C-7. This effect is especially striking when a bulky phenylacetyl group is attached to the hydroxyl group at the C-1 position (entry 1). In this case the regioselectivity of subtilisin for the C-6 position is at least 80 times more than that of PPL. It is important to mention that in no cases have triesters or 1,8 diesters of castanospermine been found. The yield of 1,6-isomer formation catalyzed by subtilisin is low and does not exceed 20%. On the other hand the reactions catalyzed by lipases are more efficient. Amoung lipases tested, lipase CV is especially active. The reactions catalyzed by this enzyme usually result in combined isolated yield of 80%.

In order to overcome the high regioselectivity of subtilisin

TABLE 2. ENZYME-CATALYZED REGIOSELECTIVE MODIFICATION OF 1-O-ACYL-CASTANOSPERMINE IN ANHYDROUS THF. Reprinted with permission from A.L. Margolin, D.L. Delinck and M.R. Whalon *J. Am. Chem. Soc.* 1990. Copyright 1990 of American Chemical Society.

Substrate	Acylating agent	Enzyme	Product (%) 1,6-isomer	Product (%) 1,7-isomer	Regioselectivity
1-O-Phenylacetyl-castanospermine	2,2,2-Trichloro-ethyl butyrate	Subtilisin	23	-	>20
		Lipase CV	30	58	0.52
		PPL	12	52	0.23
1-O-Butyryl-castanospermine	2,2,2-Trichloro-ethyl butyrate	Subtilisin	20	8	2.30
		Lipase CV	7	72	0.10
		PPL	10	57	0.18
1-O-Acetyl-castanospermine	2,2,2-Trichloro-ethyl butyrate	Subtilisin	10	3	2.63
		Lipase CV	26	54	0.48
		PPL	15	47	0.32
1-O-Butyryl-castanospermine	Vinyl phenyl-acetate	Lipase CV	39	36	1.10

in pyridine we tried to enzymatically hydrolyze these di-O-acyl derivatives of castanospermine to form mono-O-acyl compounds other than 1-O-acyl castanospermine.

Several lipases, proteases and esterases were tested for their ability to hydrolyze 1,7-di-O-butyrylcastanospermine in aqueous solution. Two enzymes - porcine liver esterase (PLE) and subtilisin - turn out to be active. The most interesting fact, however, is the reverse regioselectivity of these two enzymes. While PLE preferentially cleaves off the butyryl group from the C-7 position, with regioselectivity (C1:C7) better than 1:25, subtilisin hydrolyzes the ester bond at the C-1 position with regioselectivity more than 25:1.

For synthetic purposes this property of subtilisin is very important because it makes possible a three-step enzyme-catalyzed synthesis of 7-O-butyrylcastanospermine.(Fig 3)

ENZYME-CATALYZED ACYLATION OF 1-DEOXYNOJIRIMYCIN AND THE PROBLEM OF CHEMOSELECTIVITY

Unlike castanospermine another highly bioactive alkaloid 1-deoxynojirimycin has primary and secondary hydroxyl groups and also a potentially reactive amino function. Indeed, if an excess of an activated ester is used, the final product consists of both esters and amide derivatives. The same procedure developed for castanospermine modification has been used for the esterification of 1-deoxynojirimycin[17].(Fig 4) When a small excess (1.5 eq) of an acylating agent trichloroethyl butyrate (TCEB) is used (b) the reaction gives predominantly mono-ester. When a large excess (6 eq) of an activated ester is used (c) a nearly complete conversion of the alkaloid takes place with predominant formation of the diester. Similar behaviour can be expected in non-enzymatic reaction as well. What distinguishes enzyme-catalyzed process from its chemical counterpart is the chemoselectivity of enzymatic catalysis. While without enzyme an activated ester reacts with both the primary hydroxyl and a secondary amine, subtilisin directs the reaction to only hydroxyl groups In no case the formation of an amide has been observed.

Another area where enzyme chemoselectivity plays an important role is peptide synthesis. We studied a reaction of monochloroethyl butyrate with 6-amino-1-hexanol in *tert*-amyl alcohol and have shown that lipase from *Aspergillus niger* catalyzes the acylation of OH group 37 times faster than that of the NH_2 group. (Table 3)[4]

In contrast, when the same ester of *N*-acetyl-L-phenylalanine was used as the acylating agent, the reactivity of the hydroxyl group in the enzymatic coupling was less than one fifth of that of the amino group. This surprising reversal of chemoselectivity was also observed with other biocatalysts such as porcine pancreatic lipase and *Pseudomonas sp.* lipoprotein lipase.

The first entry in Table 3 depicts lipase-catalyzed synthesis of 6-aminohexyl butyrate on a mmol scale. The N-acylation of 6-amino-1-hexanol with *N*-Ac-L-Phe-OCH_2CH_2Cl

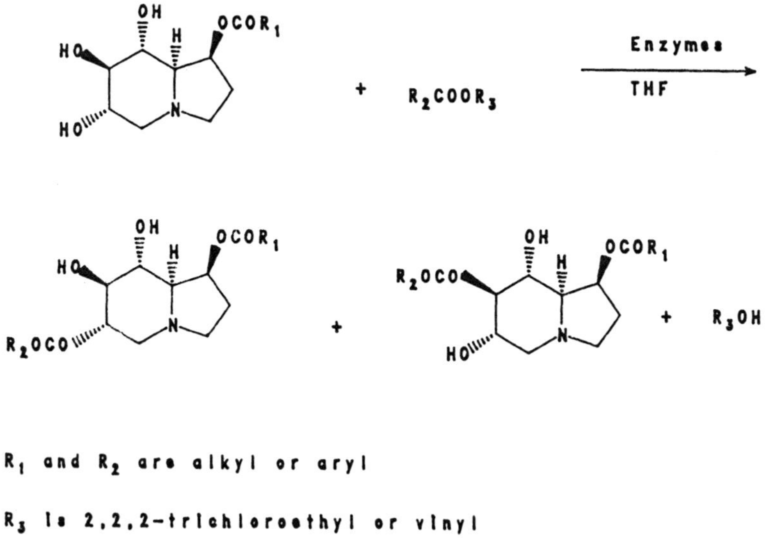

Fig 2 Enzyme-Catalyzed Synthesis of Castanospermine Diesters

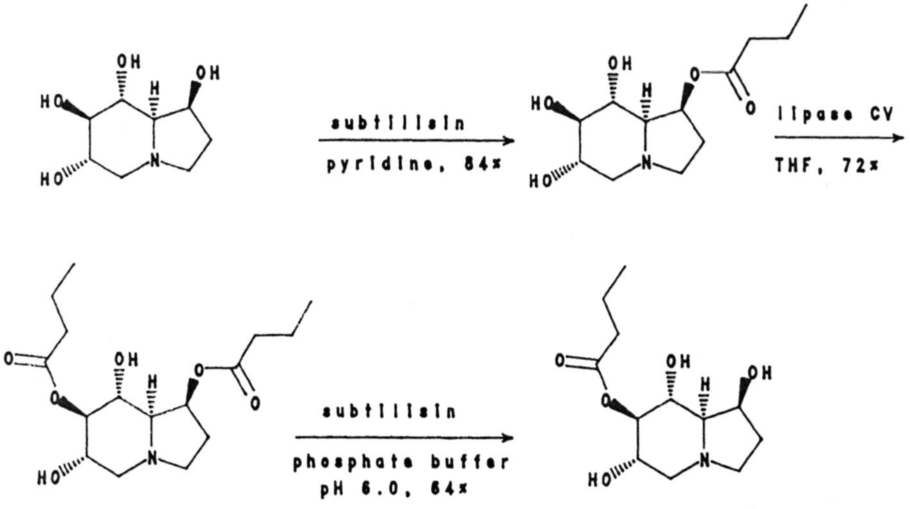

Fig 3 Three-step Enzymatic Synthesis of 7-O-Butyrylcastanospermine

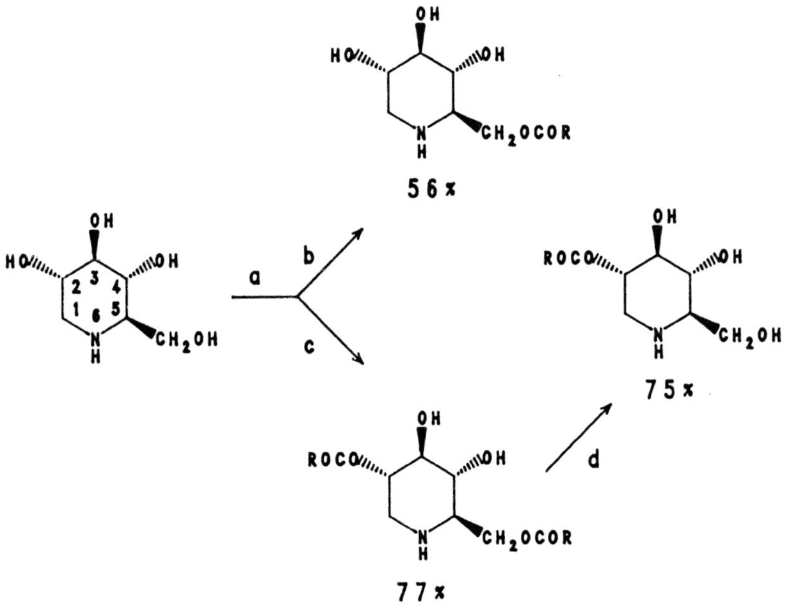

Fig 4 Enzyme-Catalyzed Regioselective Acylation and Deacylation of 1-Deoxynojirimycin

(a) subtilisin 5mg/ml, 45°, pyridine (b) 1.5 eq of TCEB (c) 6 eq of TCEB (d) subtilisin 2mg/ml, 0.1M phosphate (pH 6.0), 6 hr, r.t.

Table 3. Chemoselective Enzymatic Acylation of Aminoalcohols in *tert*-Amyl Alcohol. Reprinted with permission from N. Chinsky, A.L. Margolin and A.M. Klibanov, *J. Am. Chem. Soc.*, III:386 (1989). Copyright 1989 American Chemical Society.

acylating agent (mmol)	aminoalcohol (mmol)	enzyme	bond formed	product	isolated yield, %
$ButOCH_2CH_2Cl$ (18)	6-amino-1-hexanol (15)	lipase	C-O	$ButO-(CH_2)_6-NH_2$	47
N-Ac-L-Phe-OCH_2CH_2Cl (8.4)	6-amino-1-hexanol (7)	subtilisin Carlsberg	C-N	N-Ac-L-Phe-$NH(CH_2)_6-OH$	74
$ButOCH_2CH_2Cl$ (18)	*trans*-4-aminocyclo hexanol (15)	lipase	C-O	$ButO-C_6H_{10}-NH_2$	42
N-CBZ-L-Tyr-OCH_2CH_2Cl (20)	L-Thr-NH_2 (20)	subtilisin Carlsberg	C-N	N-CBZ-L-Tyr-L-Thr-NH_2	73
N-CBZ-L-Thr-OCH_2CH_2Cl (20)	L-Thr-NH-Ind (20)	subtilisin Carlsberg	C-N	N-CBZ-L-Thr-L-Thr-NH-Ind	81
N-Ac-D-Ala-OCH_2CH_2Cl (3.7)	L-Ser-NH-Nph (3.7)	subtilisin BPN'	C-N	N-Ac-D-Ala-L-Ser-NH-Nph	64

catalyzed by lipases was much slower than the O-acylation, thereby making the preparative conversion less attractive. Fortunately, we observed that another hydrolase, the protease subtilisin Carlsberg, was extremely reactive in this process, while still exhibiting a 50-fold preference toward the NH_2 over the OH group. Therefore, this enzyme was employed instead of lipase for the preparative N-acylation of 6-amino-1-hexanol (second entry in Table 3).

It is noteworthy that lipase-catalyzed acylation of 6-amino-1-hexanol occurs with chemoselectivity opposite to that of the chemical reaction: incubation of the aminoalcohol with butyryl chloride (200 mM each) in *tert*-amyl alcohol at 45°C for 3 h resulted, expectedly, in the nearly quantitative acylation of the NH_2 group exclusively. This chemoselectivity was not limited to 6-amino-1-hexanol as shown in Table 3; *trans*-4-aminocyclohexanol was O-butyrylated on a preparative scale with *Aspergillus niger* lipase as a catalyst. In addition, the nucleoside adenosine was butyrylated by subtilisin in dimethylformamide exclusively at the OH group, while the chemical acylation of such compounds occurs only at the NH_2 group.[10]

The aforedescribed NH_2-chemoselectivity of subtilisin when acylating with *N*-Ac-L-Phe-OCH_2CH_2Cl was found to be rather general with respect to both the amino acid in the acylating agent and the nucleophile molecule. Consequently, derivatives of hydroxy amino acids could be enzymatically acylated without the need for protection of OH groups. As an illustration, we utilized both subtilisin Carlsberg and BPN' to prepare dipeptide fragments of the actapeptide T which had been reported[18] to possess anti-HIV activity. The last three entries in Table 3 indicate that the subtilisins catalyzed formation of only the natural peptide bond.

ENANTIOSELECTIVITY OF PROTEASES IN ORGANIC SOLVENTS AND SYNTHESIS OF PEPTIDES CONTAINING D-AMINO ACIDS

A number of biologically active peptides, including important antibiotics, synthetic vaccines, and enkephalins and other hormones, contain D-amino acid residues. Proteases, on the other hand, exhibit a profound preference toward L-amino acids. This means that enzymatic methods are not generally applicable to peptides containing D-amino acids. The only exception has been the use of D-amino acid derivatives as nonspecific nucleophiles competing with water for the acyl enzyme formed in the reaction between a protease and a protected L-amino acid ester[19]. This approach, however, is inherently lmited to the enzymatic incorporation of a single D-amino acid into the C-terminal position of a peptide.

We have found that upon transition from aqueous solution to anhydrous organic solvents, the discriminating stereoselectivity of several proteases drastically relaxes[3]. Consequently these enzymes have been used to prepare numerous peptides containing D-amino acids in various positions[2,3]. We measured the ratio of specificity constants $(k_{cat}/K_m)_L/(k_{cat}/K_m)_D$ for subtilisin, which reflects enantioselectivity of an enzyme in water and in several organic solvents. The enzymatic reaction studied in water was the hydrolysis of 2-chloroethyl esters of *N*-acetyl-L- and D-amino acids. In organic solvents we examined the subtilisin-catalyzed transesterification reaction between the same esters and propranol. For all nonaqueous solvents tested, this enantioselectivity factor was found to be 10-100 fold lower. Also, there is a linear correlation between the difference in free energy of activiation ($\Delta\Delta$ G#) for subtilisin-catalyze0d transesterification of L- and D-enantiomers of *N*-Ac-Ala-OCH_2CH_2Cl in 13 different organic solvents and the hydrophobicity of the solvent. These results indicate that an enzyme's enantioselectivity decreases as the hydrophobicity of the solvent increases. These data can be rationalized with the following simple model. When a substrate interacts with the enzyme's active center, water must be excluded from between them[20]. In particular, the binding of an *N*-acyl-L-amino acid substrate to the active center of a proteolytic enzyme results in the release of a certain number of water molecules from the hydrophobic binding pocket of the protease; this process is a driving force in the enzyme-substrate interaction[20]. One would expect that the release of water molecules into the reaction medium would become thermodynamically less favorable with increasing hydrophobicity of the solvent. If the D-enantiomer binds to the active center of a protease in the same manner as the *N*-acyl-L-amino acid substrate, then the scissile bond will not face the enzyme's nucleophile, and cleavage cannot occur. In order for the D-isomer to be reactive, it must bind to the enzyme "incorrectly"; this binding will release fewer molecules of water from the hydrophobic binding pocket than that of its L-counterpart. Therefore, the reactivity of the L-enantiomer should be diminished to a greater extent, and consequently the enzyme's enantioselectivity should decrease, when the hydrophobicity of the solvent increases. To test the generality of the observed phenomenon, we investigated the enantioselectivity of five other serine proteases in water and in butyl ether. As seen in Table 4, elastase, α-lytic protease, subtilisin BPN', α-chymotrypsin, and trypsin all exhibit striking enantio-selectivity in water, but not in organic solvents; while the $(k_{cat}/K_m)_L/(k_{cat}/K_m)_D$ in water is on the order of 10^3-10^4, in butyl ether it does not exceed even one order of magnitude. These results demonstrate the predictable and rational control of enantioselectivity of enzymes by changing the reaction medium. The discovered phenomenon allowed us to solve the problem of enzymatic synthesis of peptides containing D-amino acid residues. Using subtilisin Carlsberg as a catayst and *tert*-amyl alcohol as a reaction medium, we synthesized several peptides with D-amino acids (Ala, Phe, Trp, Asn) in the *N*-terminal position (Table 5). One can see the various D-amino acids (Ala, Phe, Trp and Asn) were readily incorporated enzymatically in the *N*-terminal position. It is worth noting that various amino acid derivatives, including both L- and D- isomers and dipeptides, were utilized as nucleophiles. This led us to the first enzymatic synthesis of a dipeptide (*N*-F-D-Ala-D-AlaNH_2) composed of only D-amino acids. Thus, a radically altered sterospecificity of subtilisin in organic solvents affords facile enzymatic preparation of diverse peptides containing D-amino acids.

Table 4. Enantioselectivity of Various Proteolytic Enzymes in Water and in Butyl Ether. Reprinted with permission from T. Sakurai, A.L. Margolin, A.J. Russell and A.M. Klibanov, *J.Am. Chem. Soc.* **110**: 7236 (1988).Copyright 1988 American Chemical Society.

enzyme	amino acid substrate	$(k_{cat}/K_m)_L/(k_{cat}/K_m)_D$ in water	in butyl ether
subtilisin Carlsberg	*N*-Ac-Ala-OCH_2CH_2Cl	1800	4.4
subtilisin Carlsberg	*N*-Ac-Phe-OCH_2CH_2Cl	15000	5.4
elastase	*N*-Ac-Ala-OCH_2CH_2Cl	>1000	4.5
α-lytic protease	*N*-Ac-Ala-OCH_2CH_2Cl	>10000	8.3
subtilisin BPN'	*N*-Ac-Phe-OCH_2CH_2Cl	16000	7.3
α-chymotrypsin	*N*-Ac-Ala-OCH_2CH_2Cl	710	3.2
trypsin	*N*-Ac-Phe-OCH_2CH_2Cl	>4000	3.2

ENZYMES IN THE SYNTHESIS OF CHIRAL DRUGS

The synthesis of chiral drugs is becoming a major challenge in medicinal chemistry. It has long been known that one enantiomer of a racemic drug (or its metabolite) can be much more biologically active than the other[21]. Moreover, the less active enantiomer can have a different type of activity, or be responsible for toxic side effects. Currently, there is a widespread consensus that a racemic drug should be considered, not as an individual compound, but as a combination of drugs[22]. However, due to the difficulty in separating the enantiomers, many synthetic pharmaceuticals are still being developed and marketed as racemates[23]. The development of new methods of asymmetric synthesis and new chiral analytical techniques is rapidly changing the situation.

Enzyme-catalyzed enantioselective resolution, both in aqueous and nonaqueous media, has emerged as a versatile technique for the synthesis of chiral synthons. Enzymatic catalysis in organic solvents has become increasingly popular among organic chemists since it allows for reactions which are impossible in water[24]. Several important chiral building blocks and drugs have been synthesized using this method[25,26].

MDL 28,618A ((+)-**1**), is a new serotonin reuptake inhibitor, 2,3-dihydro-N-methyl-2-[(4-trifluoromethyl)-phenoxy]-1H-indene-2-methanamine hydrochloride)(Fig 5). Preliminary studies have shown that (+)-**1** is at least 10 times more active in inhibiting serotonin uptake both *in vitro* and *in vivo* than the (-)-**1** enantiomer. The development of the resolution of (±-**4**) via lipase-catalyzed acylation of (-)-**4** allowed the synthesis of pure (+)-**1** from the unacylated (+)-**4**. This procedure made possible the facile synthesis of the [^{14}C]-(+)-**1** enantiomer, the first example of the application of enzyme-catalyzed reactions in organic solvents for the synthesis of a radiolabeled compound.

Table 5. Synthesis of Peptides Containing D-Amino Acids Catalyzed by Subtilisin in Anhydrous *tert*-Amyl Alcohol. Reprinted with permission from A.L. Margolin, D.-F. Tai and A.M. Klibanov *J. Am. Chem. Soc.* 109:7885 (1987). Copyright 1987 American Chemical Society.

Substrates (mmol) Amino acid ester	nucleophile	product	isolated yield of the product, %
N-Ac-D-Phe-OCH_2CH_2Cl (3.1)	L-Phe-NH_2 (3.1)	*N*-Ac-D-Phe-L-NH_2	67
N-F-D-Ala-OCH_2CH_2Cl (5.1)	L-Phe-NH_2 (15.6)	*N*-F-D-Ala-L-Phe-NH_2	82
N-Ac-D-Asn-OCH_2CH_2Cl (5.0)	L-Leu-NH_2 (7.5)	*N*-Ac-D-Asn-L-Leu-NH_2	71
N-Ac-D-Trp-OCH_2CH_2Cl (8.4)	L-Phe-NH_2 (8.4)	*N*-Ac-D-Trp-L-Phe-NH_2	47
N-F-D-Ala-OCH_2CH_2Cl (6.0)	D-Ala-NH_2 (6.0)	*N*-F-D-Ala-D-Ala-NH_2	65
N-F-D-Ala-OCH_2CH_2Cl (16.0)	L-Phe-L-Leu-NH_2 (16.0)	*N*-F-D-Ala-L-Phe-L-Leu-NH_2	61
N-CBZ-L-Tyr-OCH_2CH_2Cl (2.5)	D-Ala-NH-$(CH_2)_3$-Ph (7.8)	*N*-CBZ-L-Tyr-D-Ala-NH-$(CH_2)_3$-Ph	54

Fig.5 outlines the synthetic route to (+)-1 [27]. Paraformaldehyde was condensed, in Mannich fashion, with 1-indanone (**2**) and N-benzylmethylamine hydrobromide to form **3**. Reduction of **3** with L-selectride gave **4** (cis/trans; 20/1); after chromatography pure **4** was obtained in 88% yield. A solution of **4** (2.2g) and vinyl acetate (2.3 eq) in tert-butyl methyl ether (t-BME) (50 ml) was treated with lipase from *Pseudomonas fluorescens* (50 mg/ml), and the resulting suspension was stirred at room temperature for 4 days. After silica gel chromatography, 1.23g (43% yield; 97%ee) of (-)-**5** and 1.02g (46% yield, 98%ee) of (+)-**4** was obtained. The excellent optical purity of both enantiomers is due to the high selectivity of the catalyst. Indeed, separate measurements of the initial rates of trans-esterification of (-)-**4** and (+)-**4** gave a ratio of 90. The resolution of (±)-**4** can also be achieved by hydrolysis of its O-acetyl derivative in water, catalyzed by porcine liver esterase. However, the moderate selectivity of this enzyme combined with the instability of (±)-**4** esters in water makes hydrolytic pathways much less attractive.

The saponification of (-)-**5** with sodium hydroxide in ethanol gave (-)-**4** in 96% yield. Both enantiomers of **1** have been prepared from (+)-**4** and (-)-**4** according the procedure depicted in Fig.5. A single crystallization from 2-propanol gave (+)-**1** and (-)-**1** both with excellent optical purity (>99% ee).

The reaction conditions, developed first for the unlabeled synthesis, were employed for the synthesis and resolution of [^{14}C]-(±)-**4**. A total of 4.8 mCi (117 mg) [^{14}C]-(+)-**1** was obtained with a specific activity of 14.6 mCi/mmol and radiochemical purity of 99.8%.

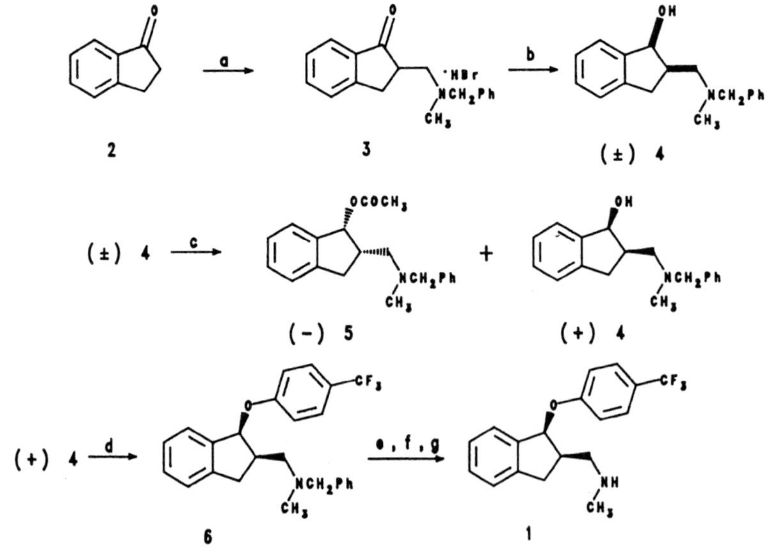

Fig 5 Synthesis of a New Serotonin Uptake Inhibitor (1)
(a) CH_2O, $PhCH_2NHCH_3$•HBr (CH_3CN) reflux, 2h. (b) 2 eq. L-Selectride (THF) 0°C→22°C, 16h. (c) 2 eq H_2C=CHOAc, Lipase P (t-BME) 22°, 4 days. (d) p-$FC_6H_4CF_4$, NaH (DMF) 90°C, 30 min. (e) $ClCO_2CH_2CCl_3$ ($PhCH_3$) reflux, 30 min. (f) Zn (90% HOAc) 23-25°C 30 min. (g) HCl (Et_2O).

REFERENCES

1. A. Zaks and A.M. Klibanov, Substrate Specificity of Enzymes in Organic Solvents vs Water is Reversed, J. Am. Chem. Soc. 108:2767 (1986).
2. A.L. Margolin, D.-F. Tai and A.M. Klibanov, Incorporation of D-Amino Acids into Peptides via Enzymatic Condensation in Organic Solvents, J. Am. Chem. Soc. 109:7885 (1987).
3. T. Sakurai, A.L. Margolin, A.J. Russell and A.M. Klibanov Control of Enzyme Enantioselectivity by the Reaction Medium, J. Am. Chem. Soc. 110:7236 (1988).
4. N. Chinsky, A.L. Margolin and A.M. Klibanov, Chemoselective Enzymatic Monoacylation of Bifunctional Compounds, J. Am. Chem. Soc., 111:386 (1989).
5. M. Kitaguchi, P.A. Fitzpatrick, J.E. Huber and A.M. Klibanov, Enzymatic Resolution of Racemic Amines: Crucial Role of the Solvent, J. Am. Chem. Soc., 111:3094 (1989).
6. G.W.J. Fleet, A. Karpas, R.A. Dwek, L.E. Fellows, A.S. Tyms, S. Petursson, S.K. Namgoong, N.G. Ransden, P.W. Smith, J.C. Son, F. Wilson, D.R. Witty, G.S. Jacob, T. Rademacher, Inhibition of HIV Replication by Amino-Sugar Derivatives, FEBS Lett. 237:128 (1988).
7. B.D. Walker, M. Kowalski, W.C. Goh, K. Kozarsky, M. Krieger, C. Rosen, L. Rohrschneider, W.A. Haseltine, I. Sodroski, Inhibition of Human Immunodeficinecy Virus Synecytium Formation and Virus Replication by Castanospermine, Proc. Natl. Acad. Sci USA 84:8120 (1987).
8. P.S. Sunkara, D.L. Taylor, M.S. Kang, T.L. Bowlin, P.S. Liu, A.S. Tyms and A. Sjoerdsma, Anti-HIV Activity of Castanospermine Analogues, Lancet, 1206 (1989).
9. M. Therisod and A.M. Klibanov, Facile Enzymatic Preparation of Monoacylated Sugars in Pyridine, J. Am. Chem. Soc. 108:5038 (1986).
10. S. Riva, J. Chopineau, A.P.G. Kieboom and A.M. Klibanov Protease-catalyzed Regioselective Esterification of Sugars and Related Compounds in Anhydrous Dimethylformamide, J. Am. Chem. Soc. 110:584 (1988).
11. Y.-F. Wang, J.J. Lalonde, M. Momongan, D.E. Bergbeiter and C.-H. Wong, Lipsase-catalyzed Irreversible Transesterification Using Enol Esters as Acylating Reagents, J. Am. Chem. Soc. 110:7200 (1988).
12. W.J. Hennen, H.M. Sweers, Y-F. Wang and C.-H. Wong, Enzymes in Carbohydrate Synthesis: Lipase-Catalyzed Selective Acylation and Deacylation of Furanose and Pyranose Derivatives, J. Org. Chem. 53:4939 (1988).
13. J-F.Shaw and A.M. Klibanov, Preparation of Various Glucose Esters via Lipase-catalyzed Hydrolysis of Glucose Pentaacetate, Biotechnol. Bioeng. 29:648 (1987).
14. F. Nicotra, S. Riva, F. Secundo and L. Zuccelli, An Interesting Example of Complementary Regioselective Acylation of Secondary Hydroxyl Groups by Different Lipases, Tetrahedron Lett., 30:1703 (1989).
15. M. Therisod and A.M. Klibanov, Regioselective Acylation of Secondary Hydroxyl Groups in Sugars Catalyzed by Lipases in Organic Solvent, J. Am. Chem. Soc. 109:3977 (1987).
16. A.L. Margolin, D.L. Delinck and M.R. Whalon, Enzyme-Catalyzed Regioselective Acylation of Castanospermine, J. Am. Chem. Soc., in press.

17. D.L. Delinck and A.L. Margolin, Enzyme-Catalyzed Acylation of Castanospermine and Deoxynojirimycin, Tetrahedron Lett., submitted.
18. M.R. Ruff. B.M. Martin, E.I. Ginns, W.L. Farrar and C.B. Pert, CD4 Receptor Binding Peptides that Block HIV Infectivity Cause Human Monocyte Chemotaxis, FEBS Lett. 1:17 (1987).
19. J.B. West and C-H. Wong, Enzyme-Catalyzed Irreversible Formation of Peptides Containing D-amino Acids, J. Org. Chem. 51:2728 (1986).
20. A. Fersht, Enzyme Structure and Mechanism, W.H. Freeman, New York, 1985.
21. E.J. Arens, Stereochemistry:A Source of Problems in Medicinal Chemistry, Med. Res. Rev. 6:451 (1986).
22. W.H. DeCamp, The FDA Perspectives on the Development of Stereoisomers, Chirality, 1:2 (1989).
23. E.J. Ariens, E.W. Wuis and E.J. Veringa, Stereoselectivity of Bioactive Xenobiotics, Biochem. Pharmacol, 37:9 (1988).
24. A.M. Klibanov, Enzymatic Catalysis in Anhydrous Organic Solvents, Trends Biochem. Sci. 14:4 (1989).
25. M. DeAmici, C. DeMicheli, G. Carrea and S. Spezia Chemoenzymatic Synthesis of Chiral Isoxazole Derivatives, J. Org. Chem. 54:2646 (1989).
26. D. Bianchi, W. Cabri, P. Cesti, F. Francalanci and F. Rama, Enzymatic Resolution of 2,3-epoxyalcohols, Intermediates in the Synthesis of the Gypsy Moth Sex Pheromone, Tetrahedron Lett. 29:2455 (1988).
27. R.J. Cregge, E.R. Wagner, J. Freedman and A.L. Margolin, Lipase-Catalyzed Transesterification in the Synthesis of a New Chiral Unlabeled and Carbon-14 Labeled Serotonin Uptake Inhibitor, in preparation.

PERSPECTIVES ON THE DISCOVERY OF VITAMIN B_{12}

Karl Folkers

Institute for Biomedical Research
The University of Texas at Austin
Austin, Texas 78712

In high school, my propensity for chemistry was revealed by my receiving the highest refund in money in recognition of my breaking less glassware than anyone else in the class. The magnitude of my refund was more important than that of my grade. In four years at the University of Illinois under Speed Marvel and Roger Adams, I became completely addicted to organic chemistry. During doctoral studies at the University of Wisconsin, Homer Adkins did not approve of my spending three dollars of my fifty dollar monthly income for Gortner's new book on biochemistry. After that, I dared not look at any biochemistry, but during three years at Yale as a postdoc, Professor Treat B. Johnson encouraged me to extrapolate organic chemistry not only to the biological but to the medical. So, when I left Yale I joined the newly established Laboratory for Pure Research at Merck and declined an offer to join a famous company to do research on polymers. At Merck, I thought I should have an M.D., but it was too late. I was married, had a daughter, and commuting to Columbia in New York City from Raway and keeping my job at Merck was not possible. However, 55 years later, I did receive an honorary M.D. In '34, I became susceptible to medical goals and still am after almost five decades. At that early time at Merck, I was naive and yielded to the attraction of trying to purify the antipernicious anemia factor. I had the right boss who allowed me to have my head. At present, I am reading Arthur Kornberg's --"For the Love of Enzymes". On page 28 of the chapter on The Vitamin Hunters, Dr. Kornberg wrote that "nutrition cannot be left to the art of medicine and that the only answer is hard science and sustaining the faith that persistent scientific effort eventually solves most problems, often in a surprisingly novel way".

Certainly, the discovery of vitamin B_{12} resulted from persistent effort and the discovery of B_{12} as a cobalt complex with a cyano group was suprisingly novel. Dr. Kornberg has an excellent perspective. Imagine a vitamin with cyanide! Later I will tell how we found the cyanide.

It will be after 43 years on December 11, 1990, that vitamin B_{12} crystallized from an aqueous medium on December 11, 1947. This event was the crown of the success of the combined efforts of my research team on B_{12} at Merck which consisted of Edward Rickes, Norman Brink, Frank Koniuszy, and Thomas Wood. Our total effort, which was interrupted by the war, spanned about 8 years. It is noteworthy that Professor Martell and Professor Scott should want to include the historical aspects of vitamin B_{12} with the chemical aspects of 1990. In telling these stories of so long ago, I shall try to relate them to research of 1990 in the hope that such history and experience may provide some guidance in 1990. Technically, the isolation of pure and crystalline vitamin B_{12}, constituting the discovery, was very simple when compared to the high resolution techniques of 1990. Four decades ago, chromatography in the the Merck labs was nearly unknown. The intellectual difficulties of trying to isolate B_{12} were formidable, complex, very controversial, and a project of very high risk. Had the levels of approvals been operating four decades ago as they do today, vitamin B_{12} would have remained to be discovered many years later by some biochemist studying enzyme transformations. Imagine a chemist telling his boss today he wants to isolate an unknown vitamin in liver for a very rare disease, and by an assay with human subjects.

The first requirement to conduct research is presumably to select an important project, and this requirement I seem to have met since it appeared to me in the early '40's that the goal of isolating the antipernicious anemia factor was important. I had documented this view in a presentation at a Conference on "The Chemistry and Physiology of Growth" which was in celebration of the Bi-Centennial of Princeton University in 1946, I said:

> "Now that the chemistry of pteroylglutamic acid and its related conjugates is known, it is likely that more rapid progress will be made on the isolation of the liver antipernicious anemia factor(s). The unknown factor or factors for pernicious anemia in liver and other sources are possibly the outstanding unidentified factor(s) of today, if importance is judged by the amount of clinical evidence on therapeutic usefulness."

I had learned from Dr. Major the wisdom of attending periodic medical meetings in addition to meetings of the American Chemical Society. While we were considering the initiation of the research on the isolation of the antipernicious anemia factor, I went to a medical meeting in Atlantic City, and fortuitously, I met Randolph West, M.D., in the lobby of the Claridge Hotel. We had known each other, and he asked me about my on-going research. I replied that I wanted to work on the isolation of the antipernicious anemia factor, and to try chromatography to achieve new purification. He asked me, "what is chromatography?" He offered no judgement on my explanation, but promptly said he would test my fractions. Then, he impressed upon me the hardship in clinical testing I would experience, because there would be so few patients with pernicious anemia even in the metropolitan area of New York City. Dr. West was on the faculty in the Department of Hematology at the College of Physicians and Surgeons of Coumbia University. I was beginning to learn to cooperate with physicians, which I have done during all my career. Back in the lab, I reported my conversation with Dr. West to my boss, Dr. Major, and stated that Dr. West had volunteered to test my chromatographic fractions from the crude liver residue furnished by Dr. Dakin. Henry Dakin had once purified liver. He was English, and most kindly and fatherly, and he was an advisor -- I think at the Board Level of Merck. Dr. Major was satisfied, and approved my sending fractions to Dr. West for clinical evaluation.

Frank Koniuszy and I dissolved a little of the blackish liver residue in an aqueous medium, and passed the solution over a column of alumina. Merck had lots of alumina. The eluate from the column was colorless, because the black "insolubles" of the liver extract had remained at the top of the alumina column. Fearing that the antipernicious anemia factor might not be stable to the heat of evaporation, and for the first time in our laboratory experience, we lyophilized something----the eluate from the alumina column. The lyophilized residue was --"snow-white", and a aliquot was mailed to Dr. West to test on the first available patient with pernicious anemia.

What happened when Dr. West received and looked at the "snow-white" sample was about as follows: he immediately phoned Dr. Dakin and said something like:--"Fantastic, Folkers has already crystallized the antipernicious anemia factor". Such a pronouncement caused Dr. Dakin to phone immediately to Dr. Major for a report, and an explanation on why Dr. Major had not reported such an important event to him. Dr. Major replied that he would get a report from me and within minutes he called me to his office and asked me to explain why I had not informed him--my boss--that I had crystallized the factor, but I detected a "twinkle in his eye". My response was something like: "I have not done any such thing!". Then, I remembered having seen the white ice pattern of the lyophilized residue, and explained that this ice pattern must have been mistakenly viewed by Dr. West as the "crystalline factor".

Obviously, Dr. West promptly tested such a remarkable sample and, indeed, found that it caused remission of the anemia in a patient. Chromatography was then unknown in the Merck labs and we were apparently events was momentous. I realized that the active factor had passed entirely through a column of alumina which indicated that the factor was not a protein as indicated in the literature. It was also promptly evident that the factor could be purified by selecting chromatographic fractions and Dr. West immediately wanted such fractions to test. Both Dr. Dakin and Dr. Major became more enthusiastic and my desire to pursue the isolation to a goal was established, but the difficulties of final the first chemists to use the technique. The intellectual impact of these isolation were as formidable as ever. I went to Chicago and ordered liver extracts equivalent to tons of liver. The Accounting Department threatened to cancel my order which was extraordinarily expensive but Dr. Major again supported me. Dr. West established a clinical network in New York City to find pernicious anemia patients for tests.

Over time, we confirmed that chromatography by various procedures was successful, the factor was stable, and that it could be distributed between certain solvents and that it was of relatively low molecular weight and not a protein. In the lab, we all continually discussed and speculated about identifying a chemical or physical property which could correlate with clinical activity and minimize the need for clinical tests except for confirmation of activity. I studied the literature for strategies. Over time we had a few fractions which were positive in Dr. West's patients as well as a few inactive fractions. Dr. West was strongly not receptive to testing any fraction which was likely to be inactive.

I was also conducting research on pantothenic acid, and one day, I went to the University of Maryland to visit Dr. George Briggs on possible cooperation. I asked him about the regulations of the University and he took a proposal from his desk which he said would illustrate the contractral conditions of the University for cooperation with an industry. The proposal had been written by Dr. Mary Shorb and was based on very limited microbiological data with L. lactis Dorner and a liver extract. She had sought financial aid from two prominent pharmaceutical companies that actually had programs of research on the antipernicious anemia factor as well as from one foundation that funded research in nutrition. Her proposal was rejected. Not only were her data limited but it was so incredible --so very unbelievable-- at that time that there could be a relationship between the growth of an organism and the hemotological response of a patient with pernicious anemia, that the two famous companies and the one foundation declined financial support. Today, we know that closed minds are still inhibiting medical research. I explained antipernicious anemia factor, and could provide her with both active and inactive liver fractions from clinical tests which might or might not indicate a possible correlation of the two activities.

On returning to Merck, I realized that those who approved of university grants could be just as negative. When I saw Dr. Major, I recounted my discussion with Drs. Briggs and Shorb, and I asked for his approval of a sum so small that he might immediately say "yes" to my desire to finance assays by Mary Shorb on our fractions for only a few months. I explained that I could readily tell if there were a promising correlation or not between the growth of her microorganism and the response of the human anemia to my fractions. Both he and I readily visualized the great significance of a possible correlation. True to his open mindedness, and understanding and generosity, he replied, "yes". The sum was about $400. There was no committe action --just two chemists --no medical advice-- just a lab man and his boss. One may contrast the simplicity of this decision with the decision-making of management and committees today.

Her data on our fractions were very promising although her assay was then so highly qualitative that it was more of a test than an assay, but in time we had the red crystals.

The first clinical test of the crystalline vitamin was the greatest of highlights. We did have pharmacology and toxicology at Merck,and we were well aware of the responsibility of clinical safety. It was decided that the clinical dose would be 3 micrograms by injection and that the source would be liver and not fermentation. There was no preliminary testing in mice or rats.

Norm Brink had lyophilized a few drops of solution containing 3 micrograms in an ampule and showed me the ampule, but nothing could be seen,because the 3 micrograms was invisible unless one inspected the ampule with a magnifying lens. I doubted that Dr. West would favorably respond to an "empty ampule" and I asked Norm to put some physiological and sterile saline into the ampule so that Dr. West could at least see a solution. Also, we showed Dr. West data to justify our recommendation that he administer a remarkable dose as small as 3 micrograms to a patient. With Dr. West, we reminisced and contrasted the first white residue to the new red crystals. Norm and I went back to Rahway,and Dr.

West searched for a patient. In a few days, Dr. West called me at my home and described the strong and prompt hemotopoietic response. We had succeded. No one ever questioned the clinical validity of the response of just one patient. Today, one would expect to do safety tests on rodents and even a double blind trial at the clinical level --even for hematology. Dr. West died soon after he tested the crystals, but not before he was honored by the medical profession for his important success.

Several months before the day the vitamin crystallized I had been in Zurich, Switzerland since Dr. Major had sent me there. In the garden of the Baur au Lac Hotel, I read a letter from Norman Brink which stated that medical advice had been so very unfavorable for continuing the research on the isolation of the factor that I could expect the effort to be terminated. Could the growth of this organism really be the property we had speculated on for so long?

I thought I knew the identitity of the physician behind the negative pressure, and he was powerful and influential. Also, commercially, the company had not been sympathetic to my research, because it was not apparent how the company could make a profit from treating a rare disease. My faith and that of Dr. Major in the potential profit value to the company of a new vitamin kept us very willful, and I firmly believed that Dr. Major would not stop my project while I was in Europe at his request, and he did not do so. On my return, Dr. Major said something like--"Take it easy and keep a low profile". In only a few more months, we had the red crystals, and as the old saying goes --"all hell broke loose" -- everyone was excited--even the medical department. Mr. Merck held a news conference in Newark of very large magnitude, over a hundred people.

In retrospect, I believe that we would have isolated the vitamin without the microbiological assay, because we were so conscious of a strategy to correlate some property with activity and this strategy would have become the red band on a column of an extract from liver, in due time. Also, B_{12} crystallizes with the greatest of ease. One executive, examining our data, said that we isolated B_{12} not because of the assay but in spite of it.

I wrote the essential parts of three companion manuscripts for publication in Science. One was on the crystalline B_{12}, a second on the microbiological data with Mary Shorb's name, and a third on the clinical data with Dr. West's name. Before submission of the manuscripts to Science, Dr. Major asked me if a pre-publication copy of the manuscript on the crystalline B_{12} could be sent to Glaxo in England. We had understood that Glaxo had suspended the project, and we knew that by revealing that we had the crystalline vitamin and that it was red in color would make it very easy for Dr. Lester Smith to repeat what we had done and indeed he did so. All he had to do was to take stored fractions from a cabinet, look for a red band on a chromatographic column, separate the band and allow the B_{12} to crystallize.

The timing of the submission of our manuscripts to Science was doubtless facilitated by a fortuitous meeting which I had with Dr. Fritz Lipmann in Atlantic City. He walked up to me on the Boardwalk and asked if indeed we had isolated a new vitamin for pernicious anemia. The "news" was leaking out of Columbia University, and witholding publication any longer was about like trying to hide a big building on fire. Fritz was not a tall man, and he was wearing an overcoat which reached nearly to his ankle which protected him from the wind and cold on the Boardwalk at the time of the Federation meeting. Fritz was a loveable and admirable biochemist. The three papers appeared on April 16, 1948.

I had selected the name --vitamin B_{12}-- with the support of Dr. Crane of Chemical Abstracts, but neither he nor I could have known that Professor Hauge of Purdue University had a manuscript in press with the expression vitamin B_{12}. Professor Hauge told me later that when he saw our papers in Science, he had to change all the twos to threes in the galley of his paper, which subsequently appeared with the expression "vitamin B_{13}". I wondered whether I would have used the designation vitamin B_{13}. For Professor Hauge, this designation was ill-fated--thirteen!

Per Laland of the Nyegaard Company of Oslo was an inspired and dedicated researcher who had previously sought the antipernicious anemia factor. He fractionated liver extracts and his fractions were tested by Jens Dedichen, M.D., on patients. The German invasion of Norway terminated Per's research and almost his life. Had Per been able to continue his fractionation, he might have succeeded. Per was a man of great enthusiasm and kindness. I understood that his data had been provided to Lester Smith of Glaxo because the German invasion terminated Per's fractionation. Lester Smith and L.F.J. Parker described their crystalline red needles at a meeting of the Biochemical Society at Oxford on May 29, 1948. In their Abstract, they stated that --"It appears probable that our substance is the same as that recently described by Rickes, Brink, Koniuszy, Wood, and Folkers with the suggestion that it be given the name vitamin B_{12}".

Smith and Parker not only re-isolated vitamin B_{12} but also a second crystalline entity which was the hydroxycobalamin. The red color and the great ease of the substance to crystallize facilitated subsequent isolations elsewhere in Europe and in the U.S.

We at Merck had 28 publications on B_{12} during 1948-1956. We elucidated the organic chemistry of the benzimidazole, and its riboside, as well the phosphate of the riboside. We correctly identified and synthesized the 1-amino-2-propanol in the correct optically active form. We also had the good fortune to clarify structurally that part of the corrin nucleus which was the only part that was initially ambiguous in the remarkable x-ray crystallography of Dorothy Crowfoot and her associates. Over and above the responsibility of an organic chemist, Norm Brink detected the presence of the cyano group in the vitamin by smell in a structural reaction.

Dorothy Crowfoot finalized the entire organic structure of B_{12} by crystallography, and I recall that Alex Todd said once that if the organic structure of B_{12} had to be solved by classical organic chemistry, it might have taken --"forever".

Before the crystals, we at Merck could never even have dreamed about the forthcoming B_{12}-coenzyme, the biochemistry, or that Woodward and Eschenmoser would synthesize the substance. However, we were alert, at the right time, to the role of vitamin B_{12} as the animal protein factor, which became of great scientific and financial importance. The usefulness of B_{12} to promote animal growth provided the financial return to Merck and vindicated the faith that Dr. Major and I had that a new vitamin would somehow become profitable other than for a rare disease.

In closing, I would like to reflect on my research at Merck of about 50 years ago. The policy, established by George Merck, was --"Pure Research"-- and I was predisposed toward isolation. However, in those years, we did pioneer the chemistry of vitamin B_6, vitamin B_{12}, the synthesis of cortisone and the isolation and chemistry of antibiotics, particularly penicillin and streptomycin, but --and there was a big but --we were acutely aware of the necessity of the potential importance of the goals we sought, regardless of the high risk involved. Even today, I am predisposed toward isolation, because we are presently endeavoring at U.T.A. to isolate a peptide hormone which many endocrinologists believe does not exist. Personally, I give the existence the benefit of the doubt and am well aware of a remark of a company physician who said he would not know what to do with the hormone if I gave it to him in a bottle. We chemists tend to be critical of our physician friends; but we should not be, because the late Dr. Gibson once said to me that I should never forget that the M.D. is not a research degree.

I can also confess that in my career, my first boss and I, Dr. Randolph T. Major, were never primarily attracted to classical medicinal chemistry but we were strongly attracted to vitamins and hormones of the human body. I am only explaining and not apologizing for this aptitude. Clearly, classical medicinal chemistry also has high priority. Since not everyone can excell in all endeavors, one should excell in whatever endeavor one may choose.

In 1990, it is difficult to realize the significance of the achievement of B_{12} as a new vitamin for a rare disease in the intellectual and industrial environment of the mid-forties. The policy of George Merck for Pure Research had paid off.

STERIC COURSE AND MECHANISM OF COENZYME B_{12}-DEPENDENT REARRANGEMENTS

János Rétey

Chair of Biochemistry
University of Karlsruhe
Karlsruhe 1 (Germany)

INTRODUCTION

Vitamin B_{12} is not only one of the most fascinating naturally occurring molecules, its coenzyme form is also a unique co-catalyst. It is unique both in its structure and in its function. By X-ray diffraction studies of Lehnert and Hodgkin a covalent bond has been identified between the central cobalt atom and its axial ligand, 5'-deoxyadenosine. The stability of this metalorganic bond in aqueous solution is surprising and so is its role in a number of enzymic rearrangements. The first of these, the glutamate mutase reaction was discovered 1958 by H.A. Barker et al.[1] (Eqn.1). It was and still is a mechanistically intriguing process without chemical precedence. One of the puzzles for chemists is the specific attack and subsequent substitution at a non-activated methyl group, even though hydrogen atoms in more activated positions are available in the molecule. A similar rearrangement, discovered at about the same time both in bacteria and in mammals[2,3], turned out to be also dependent on coenzyme B_{12} (Eqn.2). In the last few years we focused our attention on this rearrangement catalyzed by the enzyme methylmalonyl-CoA mutase and a considerable portion of my lecture will deal with it. Other coenzyme B_{12}-dependent reactions in which a rearrangement is not immediately obvious are catalyzed by dioldehydratases (Eqn.3) and ammonia lyases (Eqn.4).

In the former the migration of a secondary OH-group has been demonstrated by isotope labelling[4]. This led to the formulation of a general scheme (Eqn.5) valid for about a dozen coenzyme B_{12}-dependent rearrangements. All are characterized by an interchange of a hydrogen atom and a variable group R between adjacent carbon centers. At least in the case of methylmalonyl-CoA mutase the migration of R has been shown to be intramolecular[5,6] and there is no reason to doubt this to be true for all other cases. The migration of the hydrogen atom turned out to be more complex. Frey and Abeles could show first for the dioldehydratase reaction that the migrating hydrogen atom exchanges

Chemical Aspects of Enzyme Biotechnology
Edited by T.O. Baldwin *et al.*, Plenum Press, New York, 1990

Equations

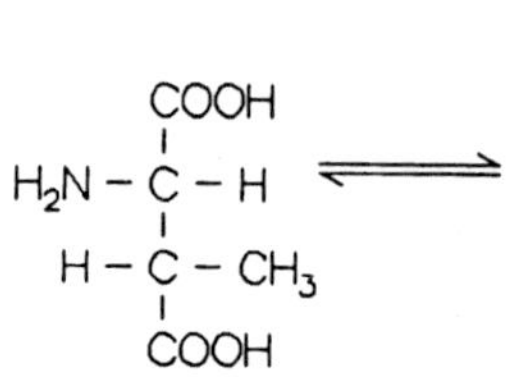

(1)

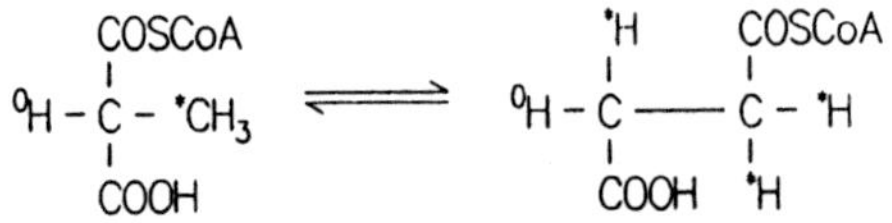

(2)

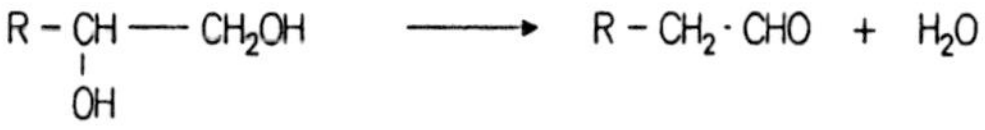

$R = H$ or CH_3

(3)

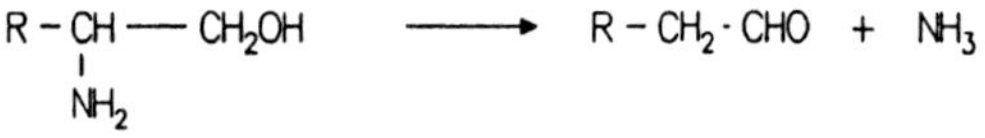

$R = H$ or CH_3

(4)

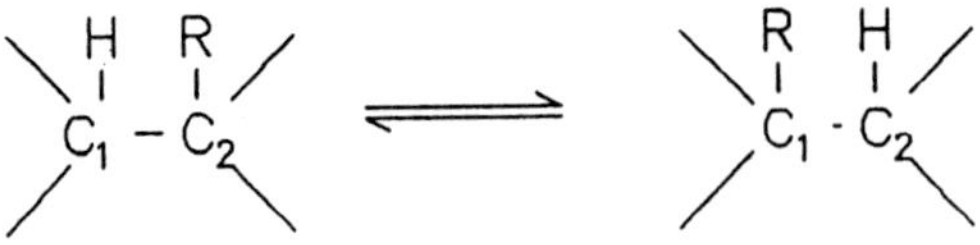

(5)

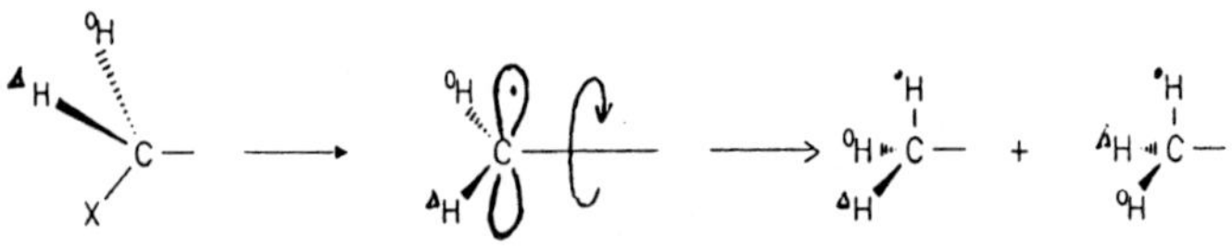

(6)

with those of the cobalt-bound methylene group of coenzyme B_{12}[7]. Subsequently similar hydrogen atom exchanges were demonstrated in the methylmalonyl-CoA mutase[8,9], the ethanolamine ammonia-lyase[10,11] and in the lysine mutase reactions[12]. Since the hydrogen atom migration is not necessarily intramolecular, the intermediacy of a fast rotating 5'-methyl group of 5'-deoxyadenosine has been postulated. This requires the transient cleavage of the cobalt-carbon bond and a further hydrogen atom abstraction from the non-activated 5'-methyl group before the coenzyme can be regenerated. In the cases of dioldehydratase and ethanolamine ammonia-lyase the bond cleavage has been shown to be homolytic by ESR and uv/vis spectroscopy[14-21]. The paramagnetic species were identified as Co(II) and an organic radical. It has been also found[19-20] that the latter are about 12 Å apart from each other in the activated complex. To account for all these results a minimal mechanism can be depicted that is shown in Figure 1.

STERIC COURSE OF THE SUBSTITUTIONS

During the last 20 years considerable work has been devoted to the stereochemical analysis of the coenzyme B_{12}-catalyzed rearrangements. Two substitutions occur in each reaction and in some cases sophisticated methods were required to elucidate their steric courses.

Surprising and somewhat embarrassing stereochemical details have been uncovered in the dioldehydratase and ethanolamine ammonia-lyase reactions. Both enzymes accept a pair of homologous substrates, but treat them in stereochemically distinct ways (Eqn.3,4).

The results and their interpretations have been reviewed[22,23]. In summary, they support the occurrence of radical intermediates previously indicated by ESR spectra[14-21]. Especially, the racemization consistently observed whenever the intermediacy of a methylene radical is assumed implicates a fast rotating radical with indistinguishable homotopic faces (Eqn.6). The product analysis involved an enzymic method developed for the determination of the absolute configuration of chiral methyl groups[24,25].

Substitution of one of the hydrogen atoms by a methyl group might raise the rotational barrier at the binding site as observed with dioldehydratase[26]. In this case the same steric course, i.e. inversion, is observed with both enantiomers of the substrate (propane-1,2-diol) and the configurations of the substrates and of the products will be complementary (Eqn.7).

Just the alternative situation arises when ethanolamine ammonia-lyase reacts with the enantiomers of 2-aminopropanol (Eqn.4, $R=CH_3$). Despite methyl substitution the rotational barrier at the active site remains low enough to establish the more favorable binding mode at the active site. Thus, irrespective of the original configuration of the substrate only a single, stereochemically defined product is formed. The substitution occurs with retention in one case and with inversion in the other[11].

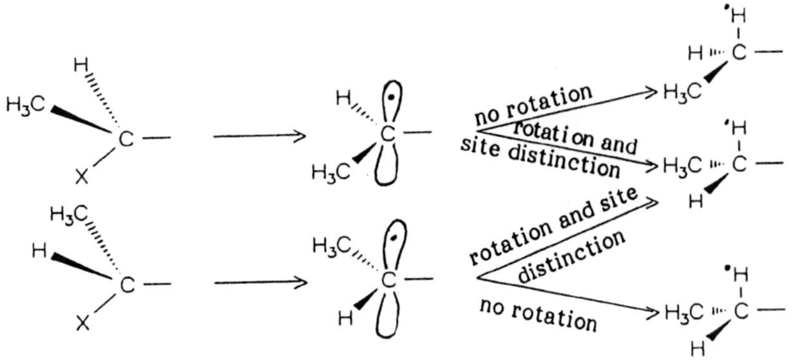

(7)

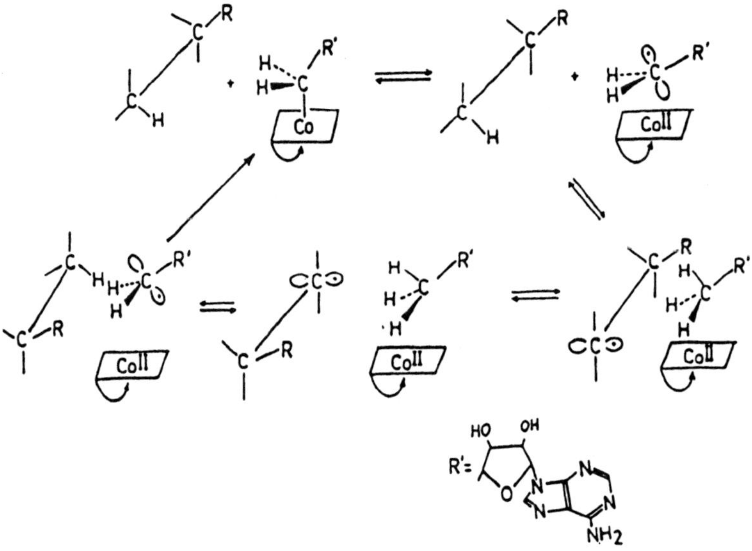

Fig. 1. A minimal mechanism for coenzyme B_{12}-dependent rearrangements

Table 1. Kinetic data of Substrate Analogues for Methylmalonyl-CoA Mutase

Substrate	K_M [mM]	V_{max} [µmol/min · U]
Succinyl-SCoA	0.025	1.0
Succinyl-CH_2CoA	0.136	1.1
Succinyl-CH_2CH_2CoA	2.2	0.012
Succinyl-CH_2SCoA	0.13	0.0048

Stereochemical studies of other coenzyme B_{12}-dependent rearrangements revealed that the substitutions can take place both with inversion and retention, but racemization invariantly occurs when methyl groups are formed or consumed in the reaction. It can be concluded that stereospecific interaction of the enzyme's active site with the intermediate radical is decisive for the stereochemical outcome of the reaction rather than other mechanistic factors important in solution chemistry.

METHYLMALONYL-CoA MUTASE

Among the coenzyme B_{12}-dependent carbon skeleton rearrangements the methylmalonyl-CoA mutase reaction (Eqn.2) has been investigated the most thoroughly. The intramolecular migration of the COSCoA moiety was secured by isotope labelling at a very early stage[5,6].

The steric course of the substitutions at the migrating centers were elucidated using deuterium as a label[27,28] or with substrate analogue ethylmalonyl-SCoA[29]. In the latter the substitutions occur mainly but not exclusively with retention at both migration centers. With the natural substrate also a high degree of retention (ca. 90 %) was observed at the stereogenic center (no determination was attempted at the methyl group), but the produced succinyl-CoA contained not only the expected monodeuterated but also unlabelled and dideuterated molecules (15-20% of each [27,28]). It was surprising since the starting methylmalonyl-SCoA carried deuterium only at the activated α-position (^{o}H = ^{2}H in Eqn.2). Believing in the strict stereospecificity of enzymes these results were first attributed to impure enzyme preparations or experimental errors. Only 20 years later, when highly purified enzyme preparations were available and high resolution NMR spectrometers allowed to monitor enzymic reactions directly in the NMR tube, could it be shown that the surprising deuterium distribution arose not from experimental error but was a consequence of the reaction mechanism[30,31]. Since succinyl-SCoA rapidly hydrolyses under the conditions of the enzymic reaction, the time for NMR observation is limited. To overcome this handicap we synthesized methylmalonyl-carba(dethia)-coenzyme (CH_2CoA) a methylmalonyl-SCoA analogue in which the sulfur is replaced by methylene[32]. Surprisingly, this synthetic analogue turned out to be an excellent substrate for methylmalonyl-CoA mutase (Table 1) and extremely useful in mechanistic studies. Its enzymic conversion to succinyl-CH_2CoA in deuterated buffer could be directly monitored in the NMR tube during days. Even before reaching equilibrium (methylmalonyl: succinyl ca. 1 : 40) about 15 - 18 % of the succinyl species were unlabelled and about the same amount doubly labelled. After establishing equilibrium the ratio of the monodeuterated, dideuterated and unlabelled molecules was roughly 60 : 20 : 20. When methylmalonyl-CoA epimerase was present all unlabelled methylmalonyl-CH_2CoA molecules (formed in the back-reaction) incorporated deuterium into the α-position and the repetition of this process maintained a deuterium leak that continuously increased the deuterium content of the product. After 19 hours trideuterated (ca. 20 %) and tetradeuterated (ca. 10 %)

species were also present. Interestingly, they co-existed with about 6 % unlabelled molecules.

In order to explain the above results we postulate the mechanistic Scheme 1 for the reaction. After homolytic cleavage of the cobalt-carbon bond of the coenzyme the 5′-methylene radical is positioned in a manner that it can abstract a methyl hydrogen of the substrate. Since the CoA portion is geometrically fixed by the complementary binding pocket of the enzyme and an ionic bonding of the carboxylate anion can be also assumed, the activated α-proton is not in reach of the 5′-methylene radical. Thereby one highly reactive radical is generating another one in which now migration of the COXCoA group can occur. The radical center is now in the front (see Scheme 1) and its predominant reaction is the reabstraction of a hydrogen atom from the 5′-methyl group of 5′-deoxyadenosine. This will result in stereospecific formation of monodeuterated succinyl-XCoA as a main product. In order to explain the observed deuterium disproportionation, we have to postulate a 1,2-hydrogen atom shift in intermediate **2** occurring in competition to the "intermolecular" H-transfer. This leads to an inversion at the front center which changes the orientation of the original α-proton (labelled with a circle in Scheme 1). Fast "intermolecular" hydrogen (deuterium) atom transfers can produce unlabelled succinyl-XCoA and 5′-deuterated deoxyadenosine. In the next catalytic cycle the latter can transfer deuterium to the intermediate radical **2** yielding geminally dideuterated succinyl-XCoA. Thus, in the initial stages of the reaction, unlabelled and geminally dideuterated molecules will be formed in equal amounts at the expense of the expected and still predominant monodeuterated species.

When the conversion of α-deuterated methylmalonyl-CH_2CoA was monitored by NMR in the presence of methylmalonyl-CoA mutase but in the strict absence of epimerase, only half of the methylmalonyl-CH_2CoA was converted into succinyl-CoA and the incorporation of solvent deuterium was extremely slow. The distribution of the original α-deuterium within the product molecules obeyed the same law observed in the experiment with epimerase. After 17 h incubation the ratio of monodeuterated, dideuterated and unlabelled species was 1 : 1 : 1 (M. Kunz, W.E. Hull and J. Rétey, unpublished).

Most of the reaction steps postulated in Scheme 1 have no analogy in solution or gas phase chemistry. Although no direct evidence for radical intermediates exists in the methylmalonyl-CoA mutase reaction, the attack at a non-activated methyl group would be difficult to explain otherwise. It is also known that radicals are not prone to rearrange by 1,2-migrations. All organic or metalorganic models for coenzyme B_{12}-dependent rearrangements are either of no radical nature or very low yielding[33-36]. How does methylmalonyl-CoA mutase master these enormous problems? Before attempting to answer this question let us see what we know about the structure of methylmalonyl-CoA mutases? In higher organisms methylmalonyl-CoA mutase is a homodimer of $M_r \approx 150{,}000$. The gene of the enzyme from human liver has been cloned and sequenced[37,38]. The richest source of the mutase is, however, Propionibacterium shermanii[39].

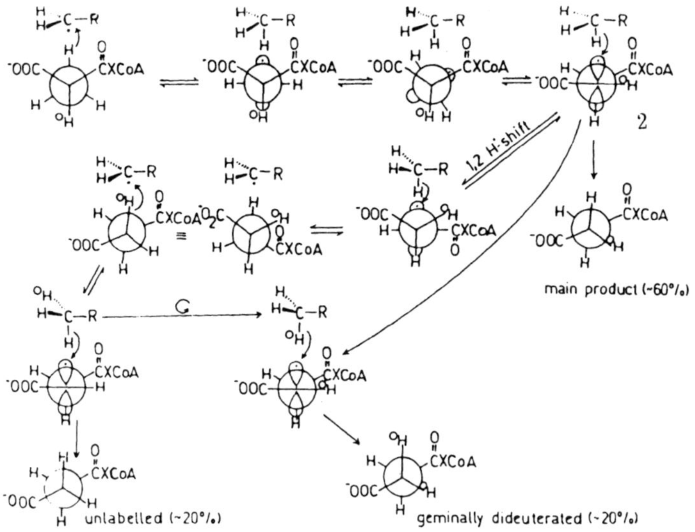

Scheme 1. Detailed mechanism for the methylmalonyl-CoA mutase reaction involving the observed deuterium disproportionation

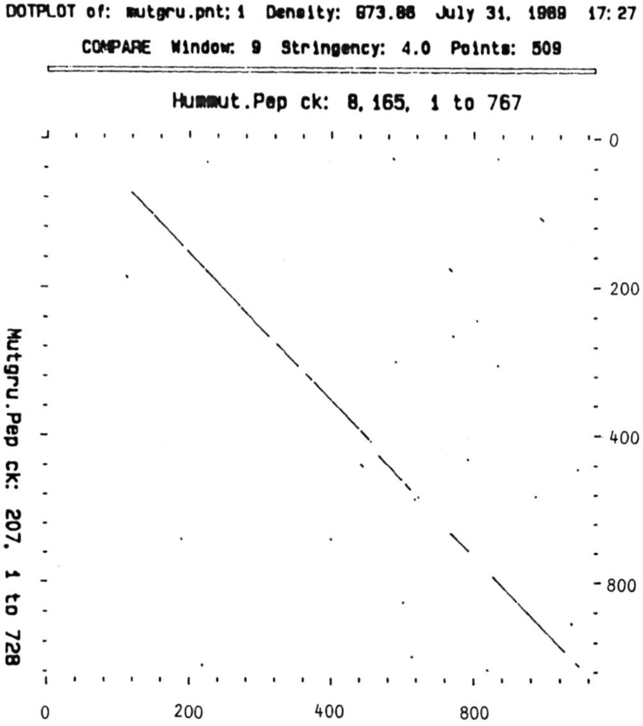

Fig. 2. Dotplot showing homology between the amino acid sequence of human methylmalonyl-CoA mutase from propionibacterium shermanii

The bacterial enzyme is a heterodimer, the two polypeptide chains exhibiting molecular masses of 80,147 and 69,465 Da, respectively. More recently the bacterial mutase gene was also cloned and sequenced[42]. Interestingly, the deduced amino acid sequence of the larger polypeptide chain is highly homologous with that of the human enzyme (in addition to 58 % identical, 11 % similar amino acids, see Fig.2). This is in agreement with earlier findings[43] that antibodies against the larger polypeptide of bacterial mutase strongly cross-react with mammalian and human mutases whereas the smaller subunit is serologically different. Independently, we also cloned and sequenced the bacterial mutase gene from another strain of P. shermanii (I. Burkhart, G. Reck and J. Rétey, unpublished). While the DNA sequence of the smaller subunit differs in eight basis from that published by Leadlay[42], only one silent mutation has been detected in the gene of the larger subunit. This confirms the conservative nature of the larger subunit that harbors the binding site for coenzyme B_{12} and thus the catalytic site for the mutase reaction. What is the role of the smaller subunit? Why it is dispensable in mammalian mutases? Examination of its amino acid sequence suggests that it harbors the binding site for the CoA portion of the substrate. We speculate that in the homodimeric forms of the mutase one subunit can play both roles. As indicated in a speculative model (Fig.3) the necessity of the dimeric form arises because each subunit serves as substrate donor for the other. In particular, the reactive part of the substrate intrudes into the catalytic site of the other subunit. Such a device creates a situation in which the highly reactive radicals are isolated from amino acid residues of the protein but are still fixed by remote binding. The final answer to the question, whether this model is real will be only given by the X-ray structure of an enzym-coenzyme-substrate tertiary complex.

Irrespective of the reality of this model we tested the tolerance of the mutase for the length of the "pike" that reaches the acyl portion of the substrate to the active site. We synthesized a number of methylmalonyl-CoA analogues in which the CoA spacer is shorter or longer than in the natural thioester. Thus we replaced the sulfur by a $-CH_2-$[32], a $-CH_2CH_2-$ and a $-CH_2S-$ group (Y. Zhao and J. Rétey, unpublished). Surprisingly, all these were substrates for methylmalonyl-CoA mutase, as revealed both by NMR and uv monitoring of the reactions and by isolation of the succinyl products. The kinetic data are summarized in Table 1. Successive elongation of the CoA portion continuously decreased V_{max} of the reaction. A positive effect of the sulfur on binding is reflected by the K_M values.

CONCLUSIONS

We now return to the question: Why coenzyme B_{12}-dependent enzymes and, in particular methylmalonyl-CoA mutase, are able to catalyze such chemically difficult reactions? They use coenzyme B_{12} to generate highly reactive intermediates, in which they prevent stabilization reactions typical for the solution or gas phase chemistry of such species. By this "negative catalysis" the life time of the highly reactive intermediates is prolonged giving them the chance to undergo rearrangements that are otherwise suppressed by faster reactions.

The prevention of these reactions may involve isolation from reactive groups, desolvation or fixing a certain conformation. While many enzymes whose mechanisms were the subject of thorough investigations lower the activation energy of the target reaction mainly by transition state binding, the strategy of other enzymes, that generate highly reactive intermediates may lie in preventing undesired reactions and waiting until the target reaction occurs spontaneously. Trapping the desired product may involve regeneration of the high-energy cofactor as in the case of coenzyme B_{12}.

ACKNOWLEDGEMENT

I am grateful to my younger colleagues whose hard work and enthusiasm made my talk possible. The work in the author's laboratory was supported by the Deutsche Forschungsgemeinschaft and by the Fonds der Chemischen Industrie.

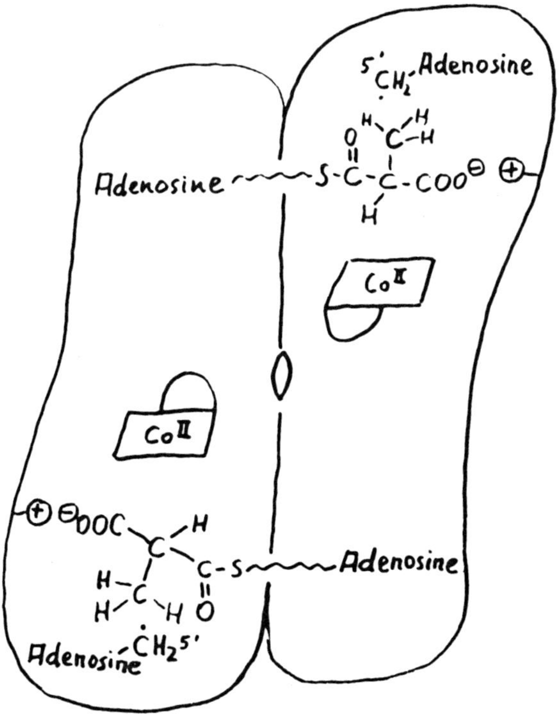

Fig. 3. Speculative model of a homodimeric methylmalonyl-CoA mutase .

REFERENCES

1. H.A. Barker, H. Weissbach and R.D. Smyth, A coenzyme containing pseudo-vitamin B_{12}, Proc.Natl.Acad.Sci. USA 44:1093 (1958).
2. M. Flavin and S. Ochoa, Propionyl coenzyme A carboxylation system, J.Biol.Chem. 229:965 (1957).
3. H. Eggerer, E.R. Stadtman, P. Overath and F. Lynen, Zum Mechanismus der durch Cobalamin-Coenzym katalysierten Umlagerung von Methylmalonyl-CoA in Succinyl-CoA, Biochem. Z. 333:1 (1960).
4. J. Rétey, A. Umani-Ronchi, J. Seibl and D. Arigoni, Zum Mechanismus der Propandioldehydrase-Reaktion, Experientia 22 : 502 (1966).
5. R.W.Kellermeyer and H.G. Wood, Methylmalonyl-Isomerase, A study of the mechanism of isomerization, Biochemistry 1:112 (1962).
6. E.F. Phares, M.V. Long and S.F. Carsen, An intramolecular rearrangement in the methylmalonyl isomerase reaction as demonstrated by positive and negative ion mass analysis of succinic acid, Biochem.Biophys. Res. Commun. 8:142 (1962).
7. P.A. Frey and R.H. Abeles, The Role of the B_{12} Coenzyme in the Conversion of 1,2-Propanediol to Propionaldehyde, J.Biol.Chem. 241:2732 (1966).
8. J. Rétey and D. Arigoni, Coenzym B_{12} als gemeinsamer Wasserstoffüberträger der Dioldehydrase und der Methylmalonyl-CoA-Mutase-Reaktion, Experientia 22:783 (1966).
9. G.J. Cardinale and R.H. Abeles, Mechanistic similarities in the reactions catalyzed by dioldehydrase and methylmalonyl-CoA mutase, Biochim.Biophys.Acta 132:517 (1967).
10. D.A. Weisblat and B.M. Babior, Mechanism of action of ethanolamine ammonia-lyase, a B_{12}-dependent enzyme. VIII. Further studies with compounds labeled with isotopes of hydrogen. Identification and some properties of the rate-limiting step, J.Biol.Chem. 246(19):6064 (1971).
11. P. Diziol, H. Haas, J. Rétey, S.W. Graves and B. Babior, The Substrate-Dependent Steric Course of the Ethanolamine Ammonia-Lyase Reacion, Eur.J.Biochem. 106:211 (1980).
12. J. Rétey, F. Kunz, T.C. Stadtman and D. Arigoni, Zum Mechanismus der β-Lysin-Mutase-Reaktion, Experientia 25:801 (1969).
13. J. Rétey, F. Kunz, D. Arigoni and T.C. Stadtman, Zur Kenntnis der β -Lysin-Mutase-Reaktion: Mechanismus und Sterischer Verlauf, Helv.Chim.Acta 61:2989 (1978).
14. B.M. Babior, T.H. Moss and D.C. Gould, Mechanism of action of ethanolamine ammonia lyase, a B_{12}-dependent enzyme. X. Study of the reaction by electron spin resonance spectrometry, J.Biol.Chem. 247:4389 (1972).
15. M.R. Hollaway, H.A. White, .N. Joblin, A.W. Johnson, M.F. Lappert and O.C. Wallis, Coenzyme-B_{12}-dependent reactions. Part V. A spectrophotometric rapid kinetic study of reactions catalyzed by coenzyme-B_{12}-dependent ethanoloamine ammonia-lyase, Eur.J.Biochem. 82(1):143 (1978).
16. B.M. Babior, T.H. Moss, W.H. Orne-Johnson and H. Beinert, Mechanism of action of ethanolamine ammonia-lyase, a B_{12}-dependent enzyme. 13. Participation of paramagnetic species in the catalytic deamination of 2-aminopropanol,

J.Biol.Chem. 249(14):4537 (1974).
17. O.C. Wallis, R.C. Bray, S. Gutteridge and M.R. Hollaway, The Extents of Formation of Cobalt(II)-Radical Intermediates in the Reaction with Different Substrates Catalysed by the Adenosylcobalamin-Dependent Enzyme Ethanolamine Ammonia-Lyase, Eur.J.Biochem. 125:299 (1982).
18. S.A. Cockle, H.A.O. Hill, R.J.P. Williams, S.P. Davies and M.A. Foster, The Detection of Intermediates During the Conversion of Propane-1,2-diol to Propionaldehyde by Glyceroldehydrase a Coenzyme B_{12}-Dependent Reaction, J.Am. Chem.Soc. 94:275 (1972).
19. K.L. Schepler, W.R. Dunham, R.H. Sands, J.A. Fee and R.H. Abeles, Physical explanation of the EPR spectrum observed during catalysis by enzymes utilizing coenzyme B_{12}, Biochim.Biophys.Acta 397(2):510 (1975).
20. G.R. Buettner and R.E. Coffman, EPR Determination of the Co(II)-Free Radical Magnetic Geometry of the "Doublet" Species Arising in a Coenzyme B_{12} Enzyme Reaction, Biochim.Biophys.Acta 480:495 (1977).
21. J.R. Boas, P.R. Hicks, J.R. Pilbrow and T.D. Smith, Interpretation of Electron Spin Resonance Spectra Due to Some B_{12}-dependent Enzyme Reactions, J. Chem.Soc. Faraday II 74:417 (1978).
22. J. Rétey and J.A. Robinson, Stereospecificity in Organic Chemistry and Enzymology, Verlag Chemie, Weinheim/Deerfield Beach/Basel (1982).
23. J. Rétey, Vitamin B_{12}: Stereochemical aspects of its biological functions and of its biosynthesis, in: "Stereochemistry", C. Tamm, ed. New Comprehensive Biochemistry, Vol. 3, Elsevier Biomedical Press (1982).
24. J.W. Cornforth, J.W. Redmond, H. Eggerer, W. Buckel and C. Gutschow, Asymmetric methyl groups, Nature 221:1212 (1969).
25. J. Lüthy, J. Rétey and D. Arigoni, Preparation and Detection of Chiral Methyl Groups, Nature 221:1213 (1969).
26. J. Rétey, A. Umani-Ronchi and D. Arigoni, Zur Stereochemie der Propandioldehydrase-Reaktion, Experientia 22:72 (1966).
27. M. Sprecher, M.S. Clark and D.B. Sprinson, The Absolute Configuration of Methylmalonyl-Coenzyme A and Stereochemistry of the Methylmalonyl-Coenzyme A Mutase Reaction, J.Biol.Chem. 241:872 (1966).
28. J. Rétey, cited by D. Arigoni and E.L. Eliel, Chirality Due to the Presence of Hydrogen Isotopes at Noncyclic Positions, in: "Top. Stereochemistry", vol. 4, E.L. Eliel and N.L. Allinger, eds. J. Wiley and Sons, New York (1969).
29. J. Rétey, E.H. Smith and B. Zagalak, Investigation of the Mechanism of the Methylmalonyl-CoA Mutase Reaction with the Substrate Analogue: Ethylmalonyl-CoA, Eur.J.Biochemistry 83:437 (1978).
30. K. Wölfle, M. Michenfelder, A. König, W.E. Hull and J. Rétey, On the mechanism of action of methylmalonyl-CoA mutase. Change of the steric course on isotope substitution, Eur.J.Biochem. 156:545 (1986).
31. M. Michenfelder, W.E. Hull and J. Rétey, Quantitative measurement of the error in the cryptic stereospecificity of methylmalonyl-CoA mutase, Eur.J.Biochem. 168:659 (1987).
32. W.E. Hull, M. Michenfelder and J. Rétey, The error

in the cryptic stereospecificity of methylmalonyl-CoA mutase. The use of carba-(dethia)-coenzyme A substrate analogues gives new insight into the enzyme mechanism, Eur.J.Biochem. 173:191 (1988).
33. K. Aeberhard, R. Keese, E. Stamm, V.R. Vögel, W. Lau and J. Kochi, Structure and Chemistry of Malonylmethyl- and Succinyl-Radicals. The Search for Homolytic 1,2-Rearrangements, Helv.Chim.Acta 66:2740 (1983).
34. S. Wollowitz and J. Halpern, 1,2-Migrations in Free Radicals Related to Coenzyme B_{12}-Dependent Rearrangements, J.Am.Chem.Soc. 110:3112 (1988).
35. W. Best and D.A. Widdowson, Reactions Related to Coenzyme B_{12} Dependent Rearrangements: Metal Mediated Free Radical Acyl Migrations in Methyl and Cyclopropyl Substituted Models, Tetrahedron 45:5943 (1989).
36. P. Dowd, The Vitamin B_{12} Promoted Model Rearrangement of Methylmalonate to Succinate is Not a Free Radical Reaction, Contribution to these Proceedings (1990).
37. F.D. Ledley, M. Lumetta, P.N. Nguyen, J.F. Kolhouse and R.H. Allen, Molecular cloning of L-methylmalonyl-CoA Mutase: Gene transfer and analysis of mut cell lines, Proc.Natl.Acad.Sci. USA, 85:3518 (1988).
38. R. Jansen, F. Kalousek, W.A. Fenton, L.E. Rosenberg and F.D. Ledley, Cloning of Full-Length Methylmalonyl-CoA Mutase from a cDNA Library Using the Polymerase Chain Reaction, Genomics 4:198 (1989).
39. R.W. Kellermeyer and H.G. Wood, 2-Methylmalonyl-CoA Mutase from Propionibacterium shermanii (Methylmalonyl-CoA Isomerase), Methods Enzymol. 13:207 (1969).
40. B. Zagalak, J. Rétey and H. Sund, Studies on Methylmalonyl-CoA Mutase from Propionibacterium shermanii, Eur.J.Biochem. 44:529 (1974).
41. F. Francalanci, N.K. Davis, J.K. Fuller, D.M. Murfitt and P.F. Leadlay, The subunit structure of methylmalonyl-CoA mutase from Propionibacterium shermanii, Biochem.J. 236:489 (1986).
42. E.N. Marsh, N. McKie, N.K. Davis and P.F. Leadlay, Cloning and structural characterization of the genes coding for adenosylcobalamin-dependent methylmalonyl-CoA mutase from Propionibacterium shermanii, Biochem.J. 260:345 (1989).
43. I. Burkhart, J. Kleiber, R. Müller, G. Reck and J. Rétey, Investigation of Methylmalonyl-CoA Mutase from Several Sources by Gene Technological Methods, Biol.Chem.Hoppe-Seyler 368:1028 (1987).

ON THE MECHANISM OF ACTION OF VITAMIN B_{12}: A NON-FREE RADICAL MODEL FOR THE METHYLMALONYL-CoA - SUCCINYL-CoA REARRANGEMENT

Paul Dowd,* Guiyong Choi, Boguslawa Wilk, Soo-Chang Choi, Songshen Zhang and Rex E. Shepherd

Department of Chemistry
University of Pittsburgh
Pittsburgh, PA 15260

INTRODUCTION

Vitamin B_{12}, in its coenzyme form, plays an essential role in mammalian metabolism where it serves as an obligatory cofactor for methylmalonyl-CoA mutase, which mediates the carbon skeleton rearrangement of methylmalonyl-CoA to succinyl-CoA[1] (eq 1). The purpose of this enzymic re-

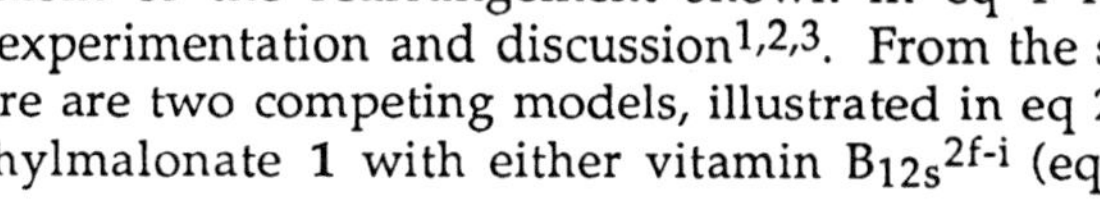

(1)

arrangement is to conduct propionate, by way of methylmalonate and succinate, to the Krebs cycle and the main stream of biochemical metabolism.

The mechanism of the rearrangement shown in eq 1 has been the subject of intense experimentation and discussion[1,2,3]. From the standpoint of model studies, there are two competing models, illustrated in eq 2. Treatment of the bromomethylmalonate **1** with either vitamin B_{12s}[2f-i] (eq 2a) or tri-*n*-

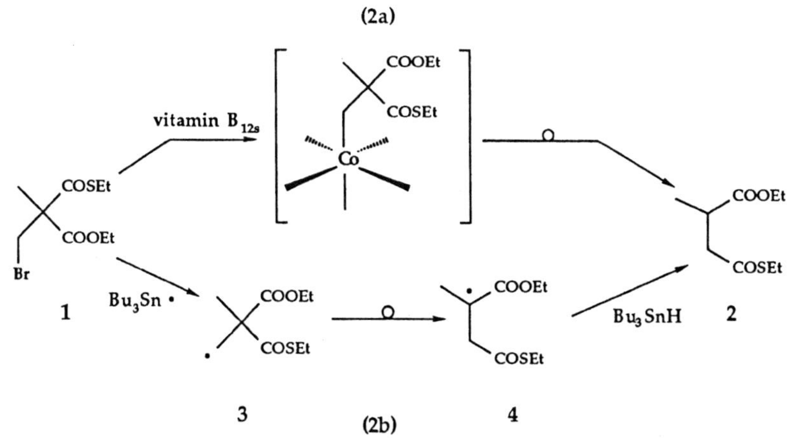

(2)

Chemical Aspects of Enzyme Biotechnology
Edited by T.O. Baldwin *et al.*, Plenum Press, New York, 1990

butyltin hydride[3b,c] (eq 2b) mimics the enzymic transformation in yielding rearranged succinate **2**. The tri-*n*-butyltin hydride promoted rearrangement (eq 2b) proceeds through a free radical mechanism,[3b,c] whereas the vitamin B_{12s} promoted rearrangement does not.[4] At issue is which is the better model for the enzymic carbon skeleton rearrangement.

RESULTS AND DISCUSSION

Wollowitz and Halpern[3b,c] showed that rearrangement of the methylmalonyl radical **3** to the succinyl radical **4** (eq 2b) proceeds with migration of the thioester and is characterized by a rate constant for rearrangement k_{rearr} = 2.5 s^{-1} (30 °C). The vitamin B_{12s} promoted rearrangement (eq 2a) also proceeds with exclusive thioester migration,[2f-i] but the rate of that rearrangement is unknown.

We recently demonstrated, by the absence of intramolecular trapping, that the vitamin B_{12s} promoted rearrangement of methylmalonate **1** to succinate **2** (eq 2a) is *not* a free radical reaction.[4] Thus, when the pentenylmalonate **5** is treated with vitamin B_{12s} (eq 3), the succinate **6** is formed[5] in 40-60%

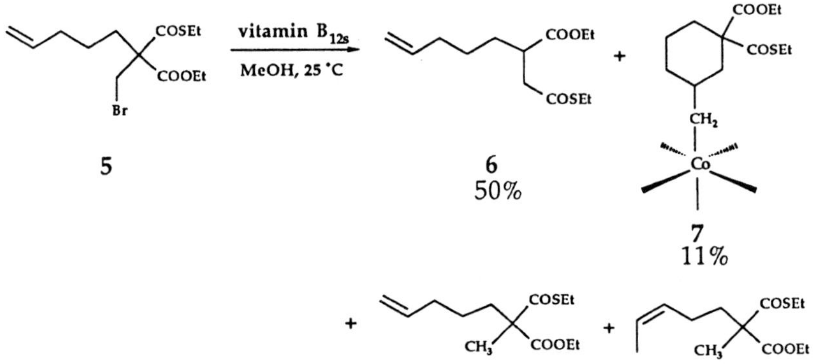

5 **6** 50% **7** 11% **8** 15% **9** 15% (3)

yield together with the cobalamin **7**[5] (11%), in addition to the reduction and hydrogen abstraction products **8** and **9**.[4,5] Were this solely a free radical reaction, no succinate would be observed. With k_{rearr} = 2.5 s^{-1}, the rate of free radical rearrangement to succinate[3b,c] is too slow to compete with intramolecular radical addition to the double bond. In the parent 6-heptenyl radical, cyclization takes place with a rate constant k_{cycl} = 6.7 x 10^3 s^{-1} (extrapolated to 30 °C).[6]

There must also be a minor radical component to the reaction of eq 3 to account for the cyclized product **7**. Indeed, the presence of **7** should permit a further evaluation of the rate of the vitamin B_{12s} promoted reaction (eq 2a) and lead to a useful comparison of the latter with the free radical model (eq 2b). The present series of experiments is intended to lay the groundwork for that determination. Thus, it was important, instead of relying on the rates of cyclization of the parent 6-heptenyl and 5-hexenyl radicals, to establish absolute rates for the cyclization of the radicals **10** and **14**.

The distribution of products in the model reaction of eq 3 can be understood if the first step involves electron transfer and cleavage of the carbon-bromine bond.[7] The primary radical **10** (Scheme 1) closes to the cyclohexylmethyl

Scheme 1

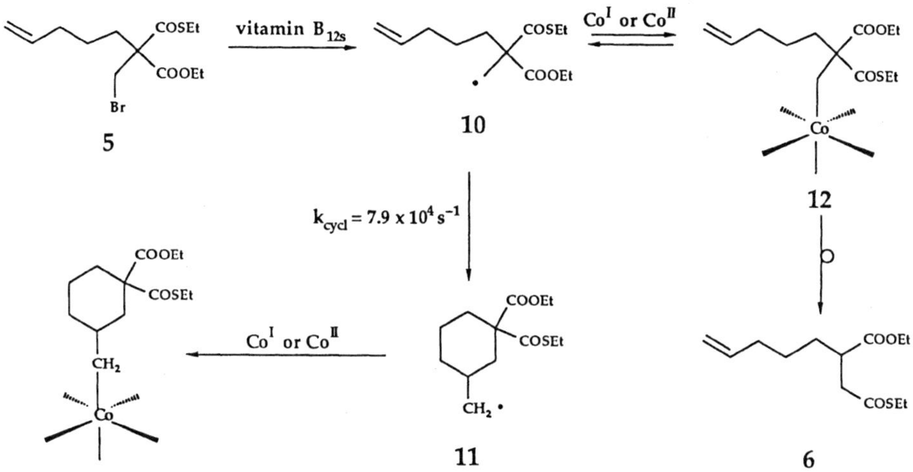

5 10 12 7 11 6

radical **11** with a rate constant $k_{cycl} = 7.9 \times 10^4\ s^{-1}$ (30 °C) (Table 1). The cyclized radical product **11** is then trapped by vitamin B_{12r} (Co^{II}) or vitamin B_{12s} (Co^{I})[8] in a diffusion controlled step leading to the vitamin B_{12} adduct **7**.

In the vitamin B_{12s}-promoted reaction (Scheme 1), the primary radical **10** undergoes diffusion controlled, reversible electron transfer or carbon-cobalt bond formation to the B_{12} adduct **12** competitive with ring closure to **11**. The B_{12}-trapping step is followed by rearrangement to the succinate **6**.

Table 1

Temperature Dependence of the Rate of Cyclization of the 6-Heptenyl Radical 10. The rate of cyclization of the 6-heptenyl radical **10** was measured according to Scheme i. The bromide **5** was treated with tri-*n*-butyltin hydride and AIBN in benzene at

Scheme i

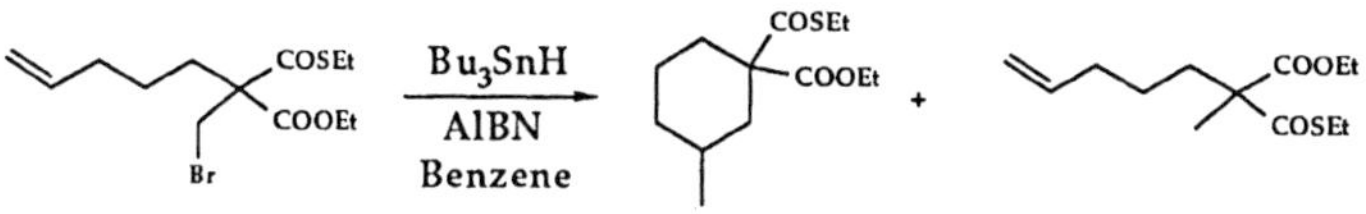

5 Cyclized Reduced

various temperatures. The absolute rate constant for reduction of primary radicals is well established[10] to be: log k_{red}=9.07-3.69/2.3RT. Therefore the rate constant for cyclization, k_{cycl}, will be given by:

$$\frac{[Red]}{[Cycl]} = \frac{k_{red}[Bu_3SnH]}{k_{cycl}}$$

For fast radical cyclizations, such as that in Scheme i, high ratios of Bu_3SnH can be used to obtain pseudo first order conditions, the concentration of Bu_3SnH remaining approximately constant. Thus, the rate constant for cyclization, k_{cycl}, can be calculated from the relative peak areas of the cyclized and reduced products as analyzed by gas chromatography.

For the analysis an HP-5890A gas chromatograph fitted with an HP-5 (5% diphenyl- and 95% dimethylpolysiloxane) 25 m x 0.2 mm x 0.33 µm capillary column was used with 40 ml/min He gas flow and temperature programming from 120-180 °C at a rate of 25 °C/min. Under these conditions the reduced product had retention time 11.7 min and the cyclized product 13.6 and 14.0 min (pair of diastereomers). Injection of appropriate standard samples was used to establish that no correction on the relative peak areas of the two products was required.

The observed rate constants are shown below.

T, °C [a]	$10^2 \times k_{cycl}/k_{red}$ [b]	$10^5 \times k_{cycl}$ [b]
70	7.06 ± 0.9	3.67 ± 0.5
60	5.95 ± 0.5	2.62 ± 0.2
50	4.59 ± 0.1	1.70 ± 0.04
42	4.18 ± 0.3	1.34 ± 0.1
30		0.79[c]

(a) The kinetic runs were conducted in a Neslab exacal High Temperature Bath with the associated temperature controller estimated to be accurate to 0.1 °C.
(b) Each point is the average of at least three determinations with the concentration of Bu_3SnH ranging from 0.4 to 0.02 M. Rate constants in s^{-1}.
(c) Extrapolated value according to: log k_{cycl} = 10.6-7.9/2.3 RT.

In a parallel set of experiments, the butenylbromomethylmalonate **13** was treated with vitamin B_{12s} (Scheme 2). In this instance, the first inter-

Scheme 2

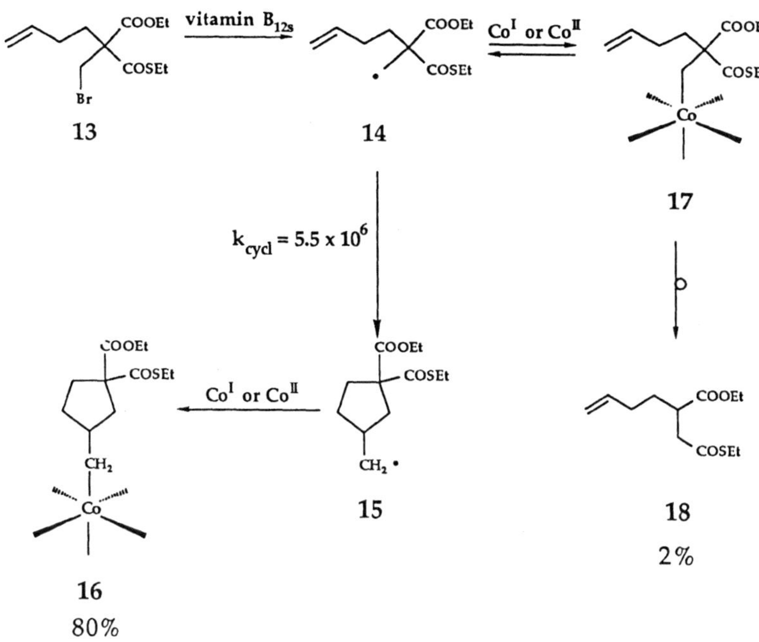

mediate product of electron transfer is the 5-hexenyl radical **14**, which closes to the cyclopentylmethyl radical **15** with a rate constant k_{cycl} = 5.5 x 10^6 (30 °C) (Table 2). Since **14** is a 5-hexenyl radical, it cyclizes much faster than the 6-heptenyl radical **10**[9]. Accordingly, the vitamin B_{12s} reaction now consists principally of cyclized product **16**[5] (72%) (Scheme 2) with a minor amount of succinate rearrangement product **18**[5] (1-2%).

Table 2

Temperature Dependence of the Rate of Cyclization of the 5-hexenyl Radical. Following the procedure outlined in Table 1, the rate constant for cyclization, k_{cycl}, of the 5-hexenyl radical **14** was determined according to Scheme ii.

Scheme ii

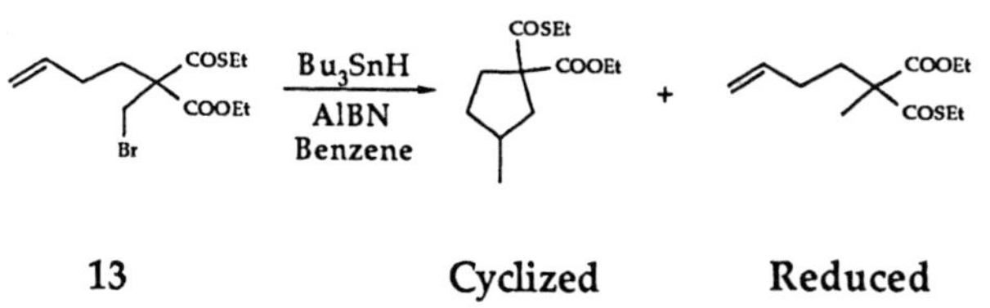

Table 2 (continued)

The observed rate constants are shown below:

T, °C	$k_{cycl.}/k_{red}$ [a]	$10^7 \times k_{cycl}$ [a]
70	5.7 ± 0.7	3.0 ± 0.4
60	4.8 ± 0.7	2.1 ± 0.3
50	3.51 ± 0.3	1.3 ± 0.1
40	2.92 ± 0.2	0.91 ± 0.07
30		0.55 [b]

(a) Each point is the average of at least three determinations with Bu_3SnH concentration ranging from 1.0 to 0.2 M. Rate constants in s^{-1}.
(b) Extrapolated value according to: $\log k_{cycl} = 12.95 - 8.6/2.3\,RT$.

MECHANISM OF REARRANGEMENT

It is suggested in Schemes 1 and 2 that free radical intermediates **10** and **14** are formed initially by electron transfer from vitamin B_{12s} to the starting bromides **5** and **13**. However, the rearrangment step leading to succinate cannot be a free radical process. Therefore, the most likely intermediate for the rearrangement is a carbanion. We suggest than an $S_{RN}1$ mechanistic sequence, with the initially formed radicals being trapped by vitamin B_{12s}, yields an anion radical intermediate (Scheme 3). The latter can then rearrange

Scheme 3

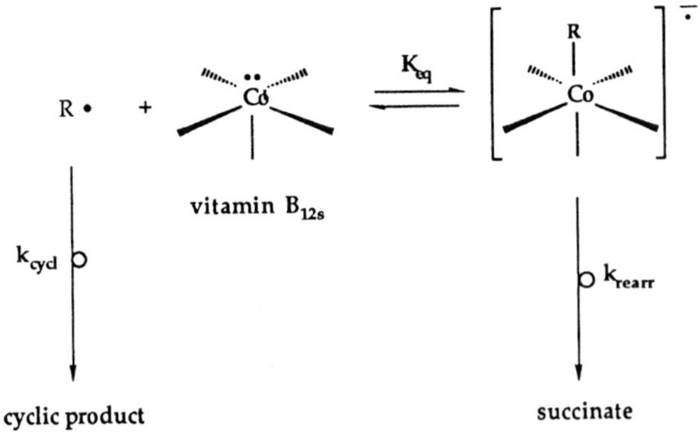

to the product succinate, possibly proceeding through a bound anionic intermediate.

The ratio of the rate of formation of succinate to the rate of cyclization may be represented by eq 4, and the equilibrium constant, K_{eq}, for the transformation in Scheme 3 is given by eq 5. Substitution of eq 5 into eq 4 yields

$$\frac{\text{rate succinate}}{\text{rate cyclic product}} = \frac{k_{rearr} \cdot [\text{R-Co}^{\overline{\cdot}}]}{k_{cycl} \cdot [\text{R}\cdot]} \qquad (4)$$

$$K_{eq} = \frac{[\text{R-Co}^{\overline{\cdot}}]}{[\text{R}\cdot][\ddot{\text{Co}}^{\text{I}}]} \qquad (5)$$

the expression in eq 6. Eq 6 shows that the ratio of the yields of succinate and

$$\frac{\text{rate succinate}}{\text{rate cyclic product}} = \frac{k_{rearr}}{k_{cycl}} \cdot K_{eq} \cdot [\ddot{\text{Co}}^{\text{I}}] \qquad (6)$$

cyclized product depends linearly on the concentration of vitamin B_{12s}. Indeed, as shown in Figure 1 and Table 3 increasing the concentration of

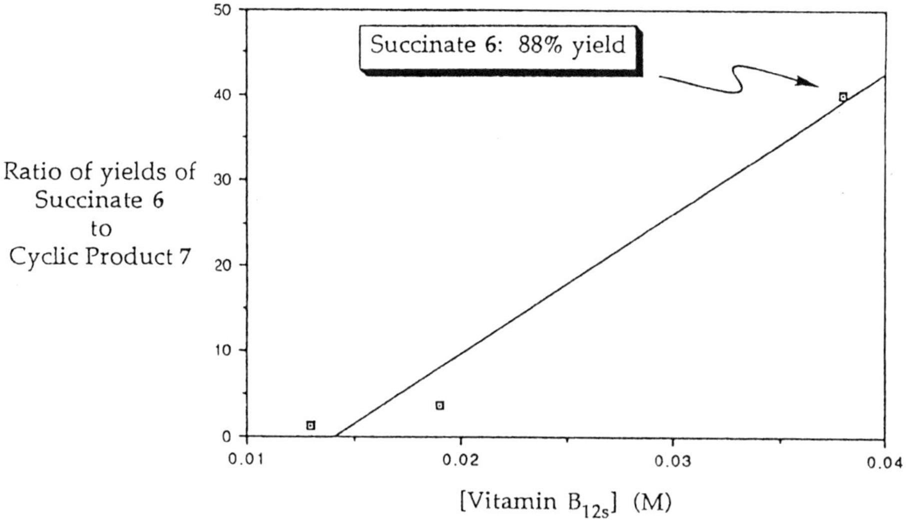

Figure 1. Plot of Succinate **6**/Cyclic Product **7** yields versus Vitamin B_{12s} Concentration

vitamin B_{12s} sharply increases the yield of succinate relative to cyclized product.

Table 3

$[Co^{I}]$, M	Succinate **6**	Cyclic Product **7**
0.013 (1 eq)	27%	22%
0.019 (1.5 eq)	50%	11%
0.038 (3 eq)	87%	<5%

Although, it would be desirable, it is not yet possible to evaluate the rate constant k_{rearr} for rearrangement to succinate from the expression in eq 6 because K_{eq} is not known.

The absence of radical involvement in the methylmalonate to succinate model rearrangement is consistent with the mechanistic hypothesis for the coenzyme B_{12} dependent rearrangement shown in Scheme 4. If coenzyme B_{12}

Scheme 4

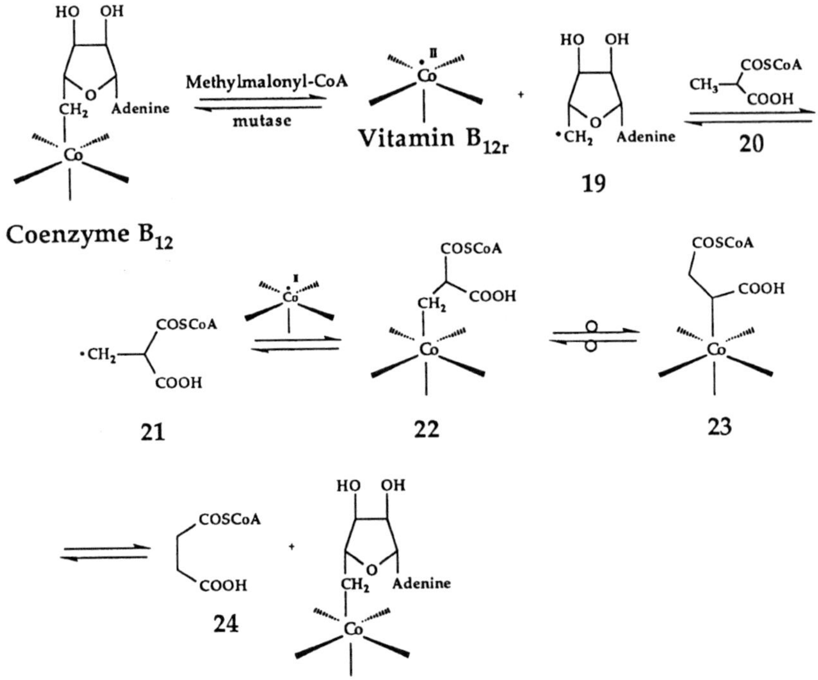

undergoes homolytic cleavage to vitamin B_{12r} and the deoxyadenosyl radical **19**, the latter can abstract hydrogen from the substrate methylmalonyl-CoA (**20**). The intermediate methylmalonyl radical **21** will reform the carbon-cobalt bond at a diffusion controlled rate yielding the adduct **22**, then rearrange to the succinyl cobalamin **23** through an enzymic-induced "carbanionic" intermediate. Reformation of coenzyme B_{12} and product succinyl-CoA (**24**) completes the catalytic cycle.[13]

SUMMARY AND CONCLUSIONS

The model experiments with the alkenyl sidechain have demonstrated that free radicals play no role in the vitamin B_{12s}-based model for the methylmalonyl-CoA rearrangement. This is the reason it is attractive to suggest that the reactive intermediate in the carbon-skeleton rearrangement is a carbanion. However, there is a very interesting comparison to be made. When the bromide **5** is treated with sodium naphthalenide (eq 7), a mixture of ester **25** and

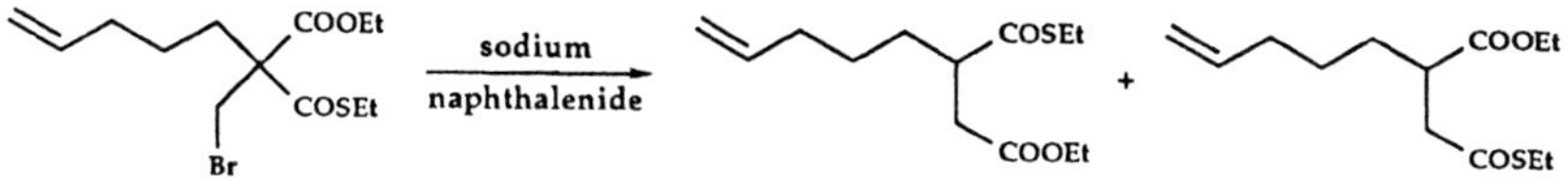

5 25 40% 26 40% (7)

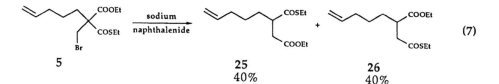

thioester **26** rearrangement products is produced in excellent yield (eq 7). We observed *none* of the ester migration product **25** in the vitamin B_{12s} promoted rearrangement (eq 3). Moreover, we observed *none* of the interesting hydrogen abstraction product **9** in the sodium naphthalenide rearrangement (eq 7). Thus, although the carbanion is an attractive alternative to the free radical, the experiment above emphasizes that one is not dealing with uncomplexed anions and that the cobalt may be playing a decisive role, particularly in facilitating the hydrogen abstraction leading to **9**. We are most interested in exploring the origin of **9** because its production does not involve an active hydrogen and abstraction of unactivated hydrogens is one of the unsolved problems of the coenzyme B_{12}-dependent rearrangements.

REFERENCES AND NOTES

(1) J. Katz and I. L. Chaikoff, *J. Am. Chem. Soc.* **77**, 2659 (1955); M. Flavin and S. Ochoa, *J. Biol. Chem.*, **229**, 965 (1957); E. R. Stadtman, P. Overath, H. Eggerer and F. Lynen, *Biochem. Biophys. Res. Commun.*, **2**, 1 (1960); J. R. Stern,and D. L. Friedman, *Biochem. Biophys. Res. Commun.*, **2**, 82 (1960); H. Eggerer, P. Overath, F. Lynen and E. R. Stadtman, *Biochem. Biophys. Res. Commun.*, **82**, 2643 (1960); S. Gurnani, S. P. Mistry, and B. C. Johnson, *Biochim. Biophys. Acta*, **38**, 187 (1960); R. Stjernholm and H. G. Wood, *Proc. Nat. Acad. Sci., USA*, **47**, 303 (1961).

(2) For cobalt-based methylmalonate models see: (a) P. Dowd and M. Shapiro, *Tetrahedron*, **40**, 3063 (1984); (b) P. Dowd and M. Shapiro,*J. Am. Chem. Soc.*, **98**, 3724 (1976); (c) G. Bidlingmaier, H. Flohr, U. M. Kempf, T. Krebs and J. Rétey, *Angew. Chem. Int. Ed. Engl.*, **15**, 613 (1976); (d) H. Flohr, W. Pannhorst and J. Rétey, *Helv. Chim. Acta.*, **61**, 1565 (1978); (e) J. Rétey, in *Vitamin B_{12}*, B. Zagalak, W. Friedrich, Eds.; Walter de Gruyter, Berlin, pp 439-460 (1979); (f) A. I. Scott and K. Kang, *J. Am. Chem. Soc.*, **99**, 1997 (1977); (g) A. I. Scott, J. Kang, D. Dalton and S. K. Chung, *J. Am. Chem. Soc.*, **100**, 3603 (1978); (h) A. I. Scott, J. Kang, P. Dowd and B. K. Trivedi, *Bioorganic Chem.*, **9**, 426 (1980); (i) A. I. Scott, J. B. Hansen and S. K. Chung, *J. Chem. Soc., Chem. Comm.*, 388 (1980).

(3) (a) J. Halpern, *Science*, (Washington, D. C.) **227**, 869 (1985); (b) S. Wollowitz and J. Halpern, *J. Am. Chem. Soc.*, **110**, 3112 (1988); (c) S. Wollowitz and J. Halpern, *J. Am. Chem. Soc.*, **106**, 8319 (1984).

(4) G. Choi, S.-C. Choi, A. Galan, B. Wilk and P. Dowd, *Proc. Nat. Acad Sci. USA*, in press.

(5) The structure of this product was established by spectroscopic comparison with an independently synthesized authentic sample.

(6) See: (a) A. L. J. Beckwith, *Tetrahedron*, **37**, 3078 (1981); (b) A. L. J. Beckwith and C. H. Schlesser, *Tetrahedron*, **41**, 3930 (1985); (c) A. L. J. Beckwith and G. Moad, *J. Chem. Soc., Chem. Commun.*,472 (1974) .

(7) See, for example, R. Breslow and P. L. Khanna, *J. Am. Chem. Soc.*, **98**, 1297 (1976).

(8) A. Ghosez, T, Göbel and B, Giese, *Chem. Ber.*, **121**, 1807 (1988).

(9) At 25 °C. the parent 5-hexenyl radical cyclizes 40 times faster than the parent 6-heptenyl radical.[6b]

(10) C. Chatgillaloglu, K. U. Ingold, and J. C. Scaiano, *J. Am. Chem. Soc.*, **103**, 7739 (1981).

(11) J. J. B. Cannata, A. Focasi, Jr., R. Mazumder, R. C. Warner, and S. Ochoa, *J. Biol. Chem.*, **240**, 3249 (1965).

(12) W. W. Miller and J. H. Richards, *J. Am. Chem. Soc.*, **91**, 1498 (1969).

(13) This research was generously supported by the National Institute for General Medical Sciences of the National Institutes of Health under Grant GM 19906.

VITAMIN B_{12}: THE BIOSYNTHESIS OF THE TETRAPYRROLE RING:

MECHANISM AND MOLECULAR BIOLOGY

Peter M. Jordan

School of Biological Sciences
Queen Mary and Westfield College, University of London
Mile End Road, London, E1 4NS, U.K.

INTRODUCTION

Vitamin B_{12} is the precursor of coenzyme B_{12}, an essential component of a group of enzymes responsible for the catalysis of various novel molecular rearrangements in which a hydrogen and one of several R groups are exchanged positionally as follows $-CHR-CH_2X \longrightarrow -CH_2-CHRX$.

Scheme 1. Structures of vitamin B_{12} and coenzyme B_{12}

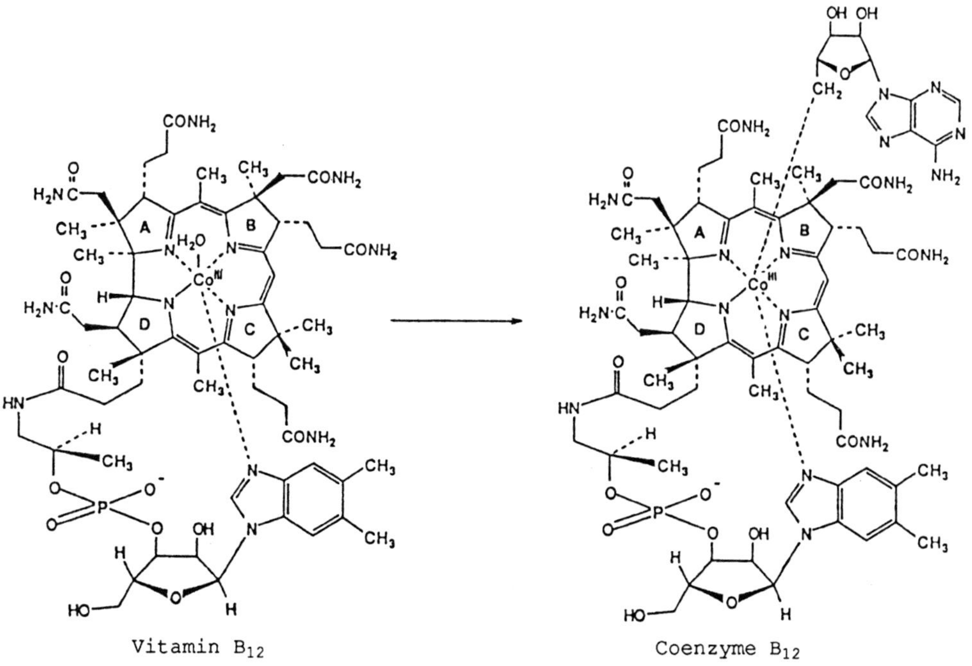

The vitamin B_{12} ring system originates from a porphyrinogen which has been modified by methylation and contracted by the elimination of C-20 to give the corrin macrocycle. The four nitrogen atoms of the corrrin system chelate with spectacular tenacity the cobalt ion which, once incorporated, is virtually impossible to remove except by destruction of the ring itself. The corrin ring system resists reduction, unlike a metalloporphyrin, permitting the fission of the cobalt-carbon bond present in the coenzyme and the remarkable reactions which follow.

The initial stages of vitamin B_{12} biosynthesis are similar to those for the biosynthesis pathways of heme, chlorophyll and other tetrapyrroles, with four common intermediates, 5-aminolevulinic acid, porphobilinogen, preuroporphyrinogen and uroporphyrinogen III. In organisms such a *Rhodobacter spheriodes*, which elaborates heme, bacteriochlorophyll and corrin systems, uroporphyrinogen III may either be methylated and embark upon the route to vitamin B_{12} or decarboxylate to coproporphyrinogen III and follow the pathway to heme and bacteriochlorophyll.

Scheme 2. The biosynthesis of uroporphyrinogen III from 5-aminolevulinic acid

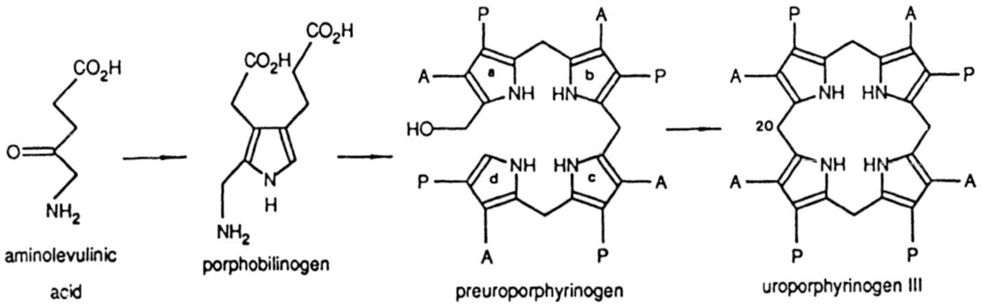

This account summarises research carried out over a period of two decades by the author and collaborators on the biosynthesis of these early tetrapyrrole pathway intermediates and the enzymes responsible for their formation. Some of the earlier work was carried out in conjunction with Professor Akhtar.

THE BIOSYNTHESIS OF 5-AMINOLEVULINIC ACID

The biosynthetic route to 5-aminolevulinic acid differs according to the organism with glycine and succinyl-CoA as the precursors in the purple non-sulfur bacteria such as *Rhodobacter spheroides* and animals. In many bacteria such as *Clostridium tetanomorphum*, *Salmonella typhimurium* and in eukaryotes including algae and higher plants, the sole precursor of 5-aminolevulinic acid is glutamic acid which provides the entire carbon skeleton for the tetrapyrrole ring.

The involvement of glutamate was first recogised by Beale and Castelfranco from experiments with higher plants and the broad details of the enzymic stages were elucidated by Kannangara and his associates using greening barley leaves. The most remarkable aspect of the glutamate pathway is the initial activation of glutamate by the formation of glutamyl-tRNA. The activated carboxyl group of the glutamyl-tRNA is then reduced to an intermediate with the aldehyde oxidation state which is transaminated to yield 5-aminolevulinic acid. This reaction is unusual in that no other amino-donor appears to be involved.

The essential features of the glycine and glutamate pathways are shown in schemes 3(a) and 3(b) respectively.

Scheme 3. The biosynthesis of 5-aminolevulinic acid

(a)

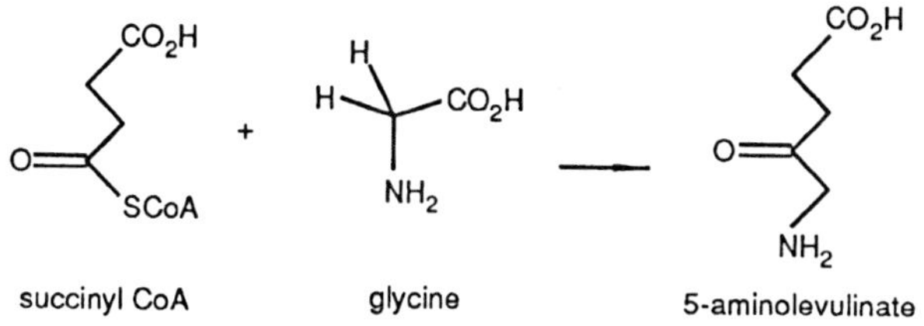

(b)

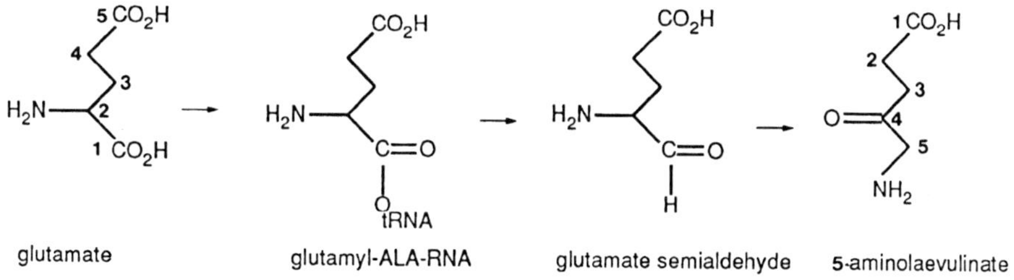

(c)

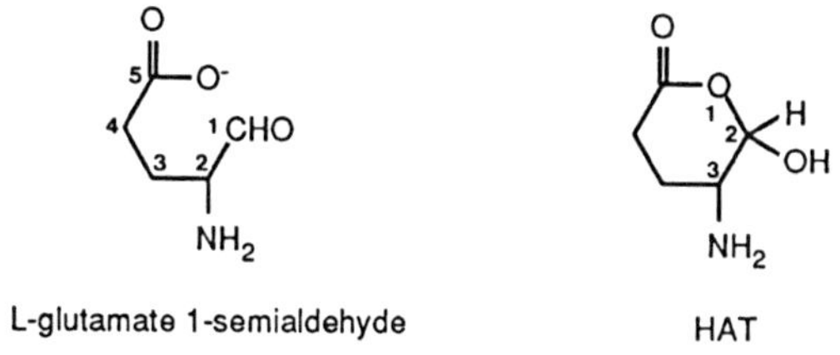

In scheme 3(a), glycine condenses with succinyl-CoA in a pyridoxal 5′-phosphate-dependent reaction catalysed by 5-aminolevulinic acid synthase. This pathway was thought, until relatively recently, to be the sole route to 5-aminolevulinic acid but now the glutamate pathway, scheme 3(b), is considered to be more abundant. The glutamate pathway is particularly prominent amongst organisms which biosynthesise vitamin B_{12} or siroheme.

Studies have focussed, in our laboratory, on the precise nature of the intermediate(s) between glutamyl-tRNA and 5-aminolevulinic acid (Jordan, 1989). On chemical grounds, the existence of an α-amino-aldehyde such as glutamate 1-semialdehyde seemed improbable since such compounds are highly unstable and have not been isolated in pure form. Accordingly we investigated the nature of the proposed aldehyde intermediate by ^{1}H nmr and natural abundance ^{13}C nmr. The proton nmr spectrum indicated that the protons at C-3 were non-equivalent, as found in glutamic-anhydride, suggesting that a ring system was present.

Furthermore there was no evidence of an aldehyde proton. When the ^{13}C spectrum was examined, the existence of a ^{13}C signal at δ=91.8ppm suggested strongly that the C-1 position was attached to two oxygen atoms. Thus it would appear that glutamate 1-semialdehyde occurs in a cyclic form as 2-hydroxy-3-aminotetrahydropyran-1-one (HAT) as indicated in scheme 3(c). The cyclic form accounts for the unexpected stability of this intermediate and poses interesting questions about its enzymic synthesis and further conversion into 5-aminolevulinic acid.

The mechanism and steric course of 5-aminolevulinic acid synthase has been studied in *Rhodobacter spheroides* by the use of *2(R)*-[^{3}H]- and *2(S)*-[^{3}H]-glycines (Jordan and Akhtar, 1969; Zaman et al., 1973) In common with other pyridoxal 5′-phosphate-catalysed enzyme reactions, the first event in the enzymic synthesis of 5-aminolevulinic acid involves the binding of glycine to the pyridoxal 5′-phosphate enzyme complex, (i) in scheme 4, to form a Schiff base intermediate (ii). Stereospecific loss of the hydrogen with the *pro-(R)*-configuration (iii) generates a stabilized carbanion intermediate which condenses with succinyl-CoA resulting in an enzyme-pyridoxal 5′-phosphate-2-amino-3-ketoadipic acid complex (iv). Decarboxylation to the planar intermediate (v) followed by stereospecific reprotonation, in the *pro-(R)*-configuration, generates the 5-aminolevulinic acid-pyridoxal 5′-phosphate-Schiff base (vi), hydrolysis of which releases 5-aminolevulinic acid from the enzyme-pyridoxal 5′-phosphate complex. Thus overall the *pro-(S)*-hydrogen at C-2 of glycine (●) is retained and occupies the *pro-(S)*-configuration at C-5 in the 5-aminolevulinic acid.

Scheme 4. Mechanism and steric course of the 5-aminolevulinic acid synthase reaction

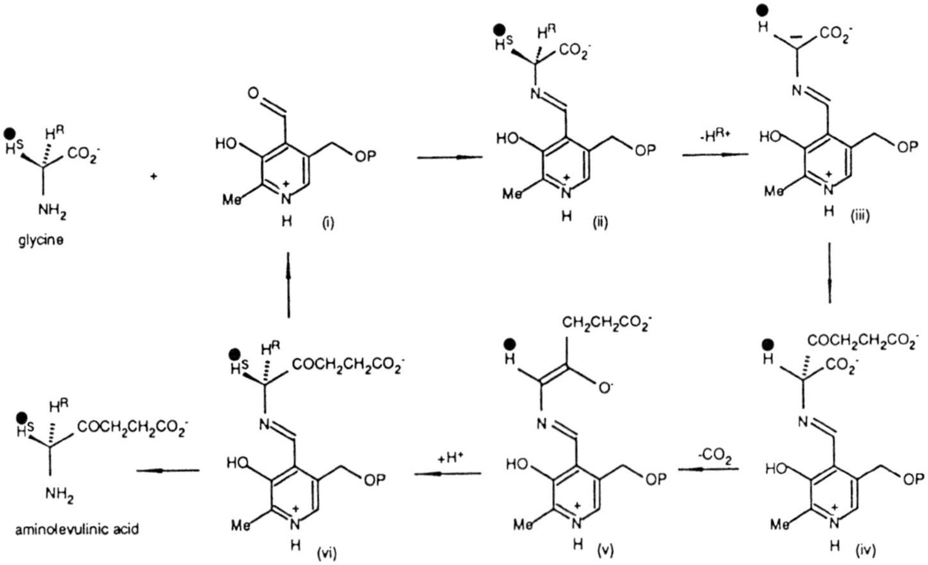

Highly purified 5-aminolevulinic acid synthase from *Rhodobacter spheroides* catalyses partial reactions involving the stereospecific exchange of the *pro-(R)*-hydrogen at the C-2 position of glycine (Laghai and Jordan, 1976) and the *pro-(R)*-hydrogen at the C-5 position of 5-aminolevulinic acid (Laghai and Jordan, 1977). These findings, together with those described above, suggest that the *pro-(R)*-hydrogens of both glycine and 5-aminolevulinic acid are handled by the same enzymic base at the active site of 5-aminolevulinic acid synthase. The overall enzymic reaction follows a cryptic stereochemical course in which there is overall inversion of configuration (Abboud et al., 1974). The inversion could, in principle, occur at either the condensation or the decarboxylation step although the latter is favored. A knowledge of the steric course of 5-aminolevulinic acid synthase has permitted the synthesis of 5-amino-*5(S)*-[^{3}H]levulinic acid for studies on the mechanistic and steric course of several enzymic reactions in the heme, chlorophyll and corrin pathways some of which will be discussed in a later section of this account.

5-AMINOLEVULINIC ACID DEHYDRATASE

5-Aminolevulinic acid dehydratase (porphobilinogen synthase) catalyses the dimerization of two molecules of 5-aminolevulinic acid to give porphobilinogen as shown in scheme 5.

Scheme 5. Biosynthesis of porphobilinogen from 5-aminolevulinic acid

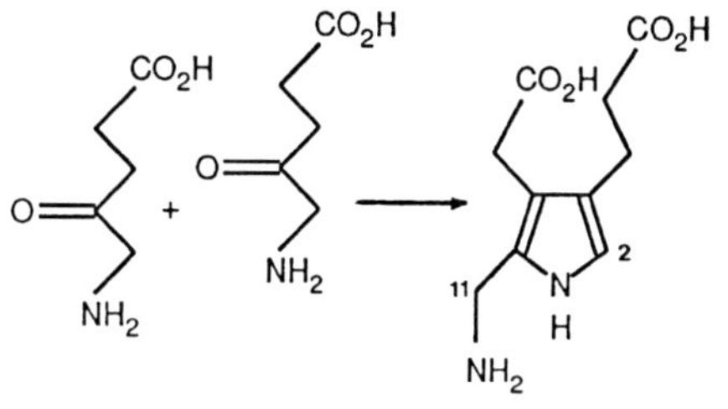

The elucidation of the mechanism of any enzymic reaction requires a knowledge of the order of binding of the substrate molecules. For a reaction such as that catalyzed by 5-aminolevulinic acid dehydratase, which involves the incorporation of two identical substrates, a single turnover reaction study using isotopic labelling to distinguish between the two substrate molecules is the only feasible approach. This technique involves an initial rapid incubation of a limiting amount of 5-amino[^{13}C]levulinic acid, or 5-amino[^{14}C]levulinic acid, with the enzyme with the anticipation that only one of the two substrate binding sites will become occupied preferentially. Addition of unlabelled substrate completes the "turn over" of the labelled substrate which is incorporated regiospecifically into the product. Analysis of the product by nmr or by chemical degredation determines the position of the label and hence the order of substrate addition to the enzyme.

When porphobilinogen was generated from a single turnover experiment in which 5-amino[5-^{13}C]levulinic acid was incubated for 100msec with bovine liver dehydratase followed by the addition of unlabelled substrate, ^{13}C nmr analysis showed a single resonance at δ=117ppm indicating that the label had been incorporated regiospecifically into the pyrrole ring at C-2. In the control experiment, in which excess labelled substrate was added, the C-2 and C-11 positions were, as expected, equally labelled as shown by resonances at δ=117ppm and 36.5ppm respectively.

The nmr results were confirmed by single turnover experiments using 5-amino[5-^{14}C]levulinic acid with the dehydratases isolated from *Rhodobacter spheroides*, bovine liver and human erythrocytes. Chemical degradation of the resulting porphobilinogen was performed to determine the location of the label. In all cases the labelled substrate was incorporated preferentially into C-2 of porphobilinogen.

Table 1. Labelling in porphobilinogen at ring (C-2) and side chain (C-11) positions from single turnover experiments with 5-amino[5-^{14}C]levulinic acid

Source of dehydratase	%label at C-2	%label at C-11
Rhodobacter spheroides	60	40
Bovine liver	86	14
Human erythrocytes	77	23
control (bovine)	49	51

The results from these investigations indicated that of the two 5-aminolevulinic acid molecules, the one initially bound to the enzyme provides the "propionic acid side" of porphobilinogen (Jordan and Seehra, 1980a, 1980b; Jordan and Gibbs, 1985). Additional experiments established that the first molecule of substrate to bind to the enzyme does so through a Schiff base (Gibbs and Jordan, 1986). On the basis of these results, a mechanism may be written for the 5-aminolevulinic acid dehydratase reaction which is shown in scheme 6.

Scheme 6. Mechanism of action of 5-aminolevulinic acid dehydratase

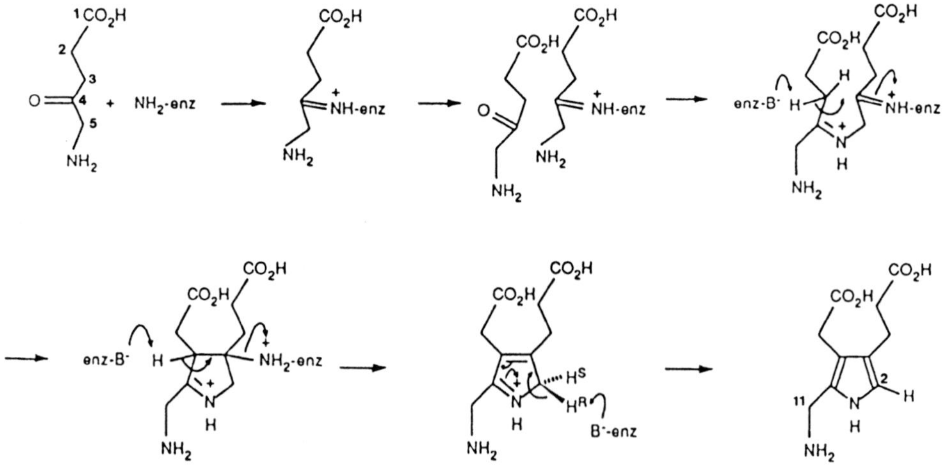

In this mechanism the first substrate to bind at the catalytic site does so in the form of a Schiff base. The second substrate can either form a second Schiff base with the amino group of the first substrate thus facilitating deprotonation at C-3 and permitting aldol condensation or, alternatively, aldol condensation may precede Schiff base formation. The deprotonation at C-3 is almost certainly facilitated by the zinc ion which is an essential part of the catalytic site (Gibbs and Jordan,

1981), probably acting as an enzymic base in the form of zinc hydroxide, $Zn^{II}OH^-$. Finally, the loss of the *pro-(R)*-hydrogen from the 2-position occurs stereospecifically during aromatization. This was demonstrated by incorporating 5-amino-*(5S)*-[^{3}H]levulinic acid, synthesized from *(2S)*-[^{3}H]glycine using 5-aminolevulinic acid synthase, into porphobilinogen. During the enzymic synthesis of porphobilinogen both labels are retained (Chaudhry and Jordan, 1976) since the chiral centre which becomes C-11 in porphobilinogen is undisturbed during the reaction.

Studies in our laboratory have established the identity of the lysine residue responsible for the formation of the Schiff base between the enzyme and the first-bound molecule of 5-aminolevulinic acid (Gibbs and Jordan, 1986). In the presence of 5-amino[^{14}C]levulinic acid, the active site lysine was labelled in both the human and bovine dehydratases by reduction with borohydride and the active site cyanogen bromide peptides from both enzymes were isolated and sequenced. The sequence of amino acids in the vicinity of the lysine appears to be highly conserved as shown in figure 1.

	*
Human active site peptide	M-V-K-P-G-M
	*
Bovine active site peptide	M-V-K-P-G-R-P-Y-L-D-L-V-R-E-

Figure 1. Sequence of the active site peptides carrying the reactive lysine (*) from human erythrocyte and bovine liver dehydratases

The active site peptide sequence data obtained in our laboratory, coupled with the protein sequence data of the dehydratase derived from the human cDNA by Professor Desnick and his coworkers, have allowed us to assign the the exact location of the active site lysine at position-252 in the human enzyme and, by analogy, to position-247 in the *Escherichia coli* enzyme. In view of the similarities in primary structure between the different dehydratases investigated, it is likely that the three dimensional structures are also closely related.

UROPORPHYRINOGEN III BIOSYNTHESIS

The classical experiments of Bogorad in the 1950s established that the biosynthesis of the tetrapyrrole uroporphyrinogen III requires two enzymes, porphobilinogen deaminase and uroporphyrinogen cosynthase. In the absence of cosynthase, uroporphyrinogen I is the product.

Scheme 7. Uroporphyrinogens I and III

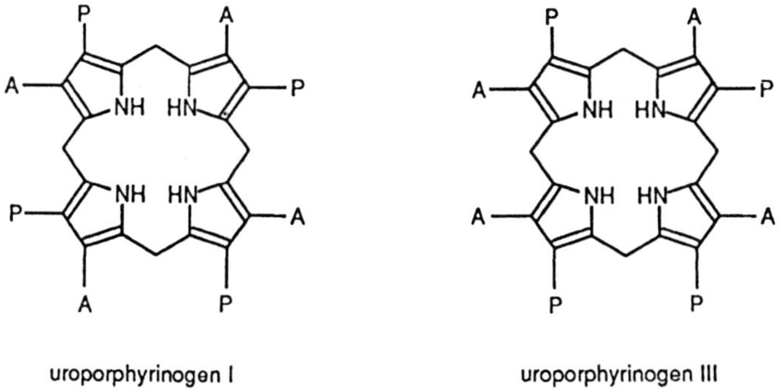

a) PORPHOBILINOGEN DEAMINASE

Much of our knowledge about porphobilinogen deaminase has resulted from studies on the enzyme isolated from *Rhodobacter spheroides* (Jordan and Shemin, 1974). This versatile bacterium is able to elaborate heme, bacteriochlorophyll and corrin when grown anaerobically in the light. The role of porphobilinogen deaminase was first determined in the laboratory of Professor Scott at Texas A & M University when highly purified porphobilinogen deaminase was incubated with 11-[^{13}C]porphobilinogen in the nmr tube. Under these conditions a transient tetrapyrrole intermediate was discovered which exhibited nmr signals around δ=23ppm and 57ppm integrating in the ratio 3:1 respectively (Burton et al., 1979). Most significantly, preuroporphyrinogen was shown to act as the substrate for the second enzyme uroporphyrinogen III cosynthase, in the absence of deaminase, yielding uroporphyrinogen III (Jordan et al., 1979). In the absence of cosynthase, preuroporphyrinogen cyclizes non-enzymically to yield uroporphyrinogen I. These experiments demonstrated for the first time that porphobilinogen deaminase and uroporphyrinogen III cosynthase act sequentially and independently as shown in scheme 8. Preuroporphyrinogen, an unrearranged 1-hydroxymethylbilane, was later synthesized in the laboratory of Professor Battersby at Cambridge and shown to have identical structural and enzymic properties to the intermediate generated in the experiments described above. Preuroporphyrinogen was shown subsequently to act as the substrate for the cosynthases isolated from plants and animals establishing it as a universal precursor for uroporphyrinogen III (Jordan and Berry, 1980). Uroporphyrinogen III is the common tetrapyrrole precursor for haem, chlorophylls, corrins and all other tetrapyrroles.

Scheme 8. The role of preuroporphyrinogen as an intermediate in the synthesis of uroporphyrinogen III

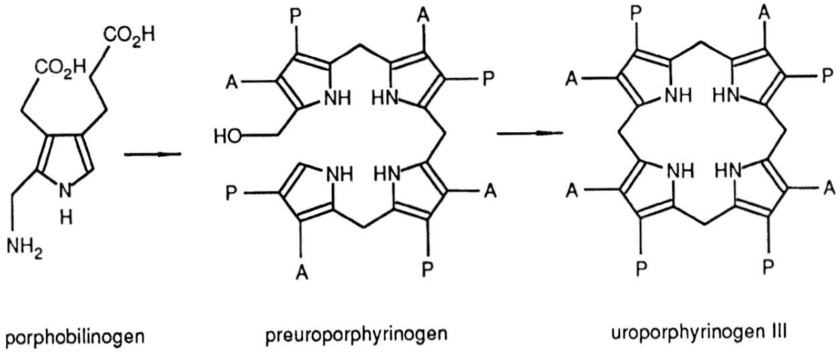

With the knowledge that the unrearranged 1-hydroxymethylbilane, preuroporphyrinogen, is the product of the deaminase reaction it was important to determine the order in which the four porphobilinogen units are incorporated during the assembly of the tetrapyrrole. Single turnover experiments with porphobilinogen deaminase, similar to those described for 5-aminolevulinic acid dehydratase in the previous section, were thus employed. Highly purified deaminase was treated initially with [^{14}C]-porphobilinogen and the enzyme-bound label was then incorporated into uroporphyrinogen III by the addition of unlabelled substrate in the presence of cosynthase. The uroporphyrinogen III was transformed enzymically into protoporphyrin IX and degraded by oxidation to determine the position of the [^{14}C] label. The results showed that the deaminase catalyses the assembly of the tetrapyrrole in a "clockwise" manner from ring "a" to "d" as shown in scheme 9 (Jordan & Seehra, 1979; Seehra and Jordan 1980). Similar findings were obtained simultaneously using a ^{13}C approach by Battersby and his coworkers in Cambridge.

Scheme 9. Sequential assembly of the tetrapyrrole from porphobilinogen

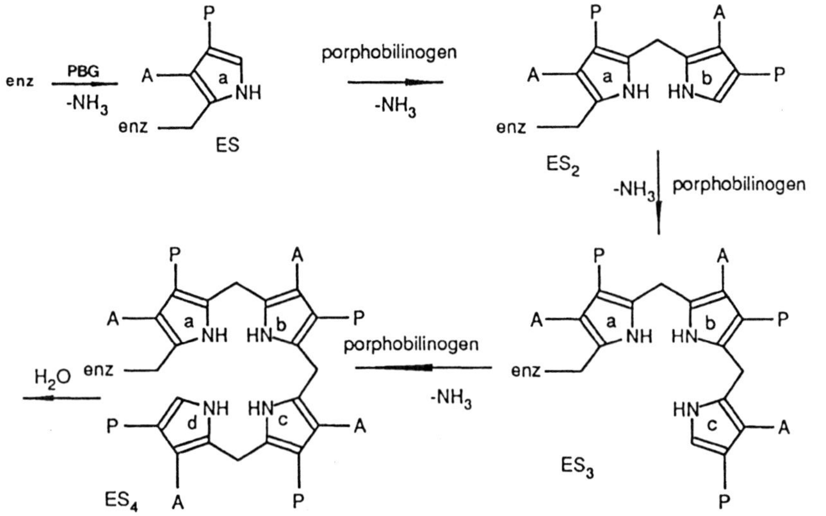

Further experiments with the deaminase from *Rhodobacter spheroides*, established that porphobilinogen could be "titrated" into the enzyme and that by using electrophoresis, the isolation of individual enzyme-intermediate complexes could be achieved. Complexes with one, two and three molecules of the "substrate" attached to the enzyme and termed ES, ES_2 and ES_3 respectively were characterized (Jordan & Berry, 1981; Berry et al., 1981). The complex ES_4, however, is too unstable to isolate and rapidly releases the tetrapyrrole product.

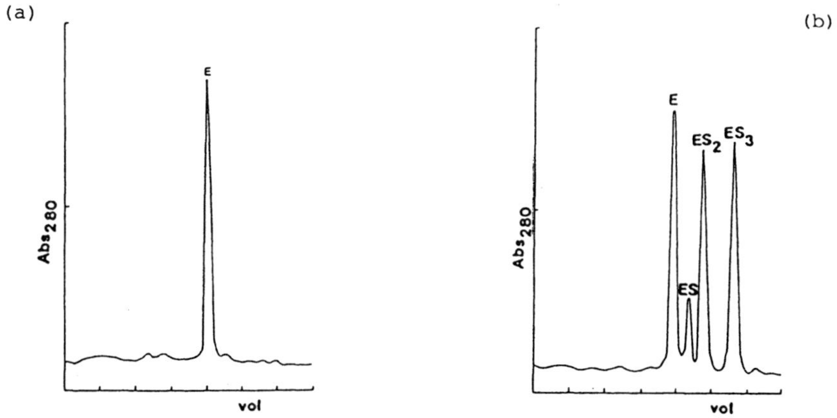

Figure 2. Porphobilinogen deaminase from *Escherichia coli* (a) before and (b) after the addition of substrate showing the separation of enzyme-intermediate complexes ES, ES_2 and ES_3 on f.p.l.c.

The enzyme-intermediate complexes may be conveniently isolated by f.p.l.c. Figure 2(a) shows a peak of porphobilinogen deaminase isolated from *Escherichia coli* (E). On the addition of porphobilinogen the enzyme-intermediate complexes ES, ES_2 and ES_3 may be isolated (figure 2(b)) using high resolution anion exchange chromatography (Warren and Jordan, 1988c). The properties of the enzyme-intermediate complexes from *Escherichia coli* have been investigated in detail (Warren and Jordan, 1988c). In the absence of further substrate, each enzyme-intermediate complex hydrolyses to hydroxy-porphobilinogen which also acts as a reasonable substrate for the enzyme. In the presence of ammonia (or hydroxylamine), porphobilinogen is displaced from the complexes. The enzyme can thus catalyse the removal or addition of NH_3 and H_2O as shown in scheme 10.

Scheme 10. Porphobilinogen deaminase catalysed reactions

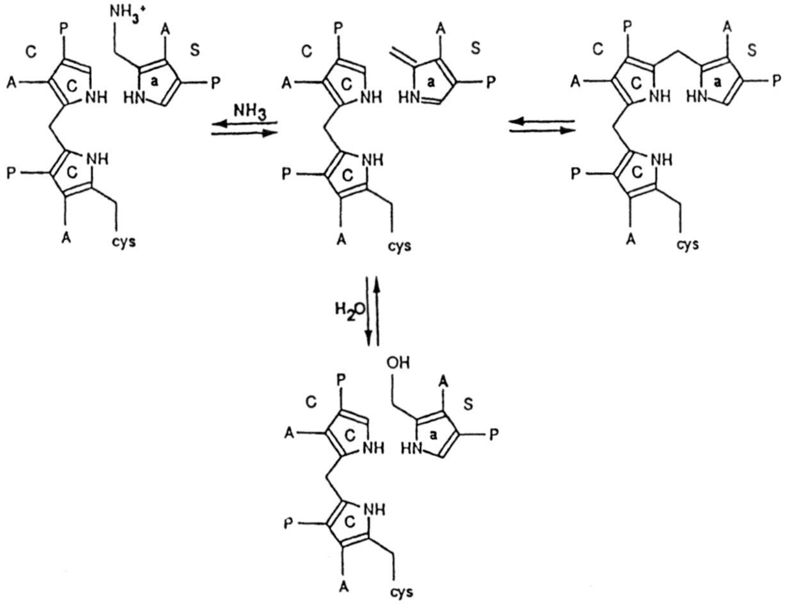

The discovery of the dipyrromethane cofactor as a unique structural component of the porphobilinogen deaminase enzyme first stemmed from the observation that when the enzyme from *Rhodobacter spheroides* was digested with proteolytic enzymes, a peptide containing a pink chromophore was generated with spectral properties similar to those of a dipyrromethene (Berry and Jordan, 1984, unpublished data). The identification of the gene, *hemC*, encoding porphobilinogen deaminase from *Escherichia coli*, the determination of its nucleotide sequence and the over-expression of the enzyme from strains containing high copy number recombinant plasmids (Thomas and Jordan, 1986) permitted the isolation of milligram quantities of the *E. coli* enzyme and allowed the above observations to be extended. Firstly when the purified *E. coli* deaminase was treated with acid the same pink dipyrromethene chromophore also appeared, together with uroporphyrin I, suggesting that the native enzyme contained the building units necessary for the formation of a tetrapyrrole (Jordan & Warren, 1987). When treatment of the native enzyme with Ehrlich's reagent yielded a characteristic test for a dipyrromethane it suggested that a resident dipyrrole system was present as part of the native enzyme structure (Jordan and Warren, 1987). The stoichiometry of two pyrrole rings was further reinforced when it was shown that the ES_2 complex yielded twice as much uroporphyrin on acid treatment as the native enzyme. The role of the enzyme-bound dipyrromethane was also investigated using ES_2 since when this complex was treated with Ehrlich's reagent a reaction characteristic of a linear tetrapyrrole (bilane) was observed. This provided strong evidence for the fact that the two substrate units were linked directly to the dipyrromethane and established quite clearly that its role was to act as a prosthetic group or "cofactor". The name of the **dipyrromethane cofactor** was therefore introduced (Jordan and Warren, 1987). Professor Battersby and his coworkers have also provided evidence for the presence of a dipyrromethane in the *E. coli* enzyme.

If the native deaminase contained a resident and functional dipyrromethane cofactor, made up of two porphobilinogen molecules linked together, it was reasoned that the activity of the deaminase enzyme would depend on the availability of the precursor, porphobilinogen. Thus a *hemA*⁻ mutant, unable to synthesise the first intermediate of the tetrapyrrole biosynthesis pathway, 5-aminolaevulinic acid, and grown in the absence of 5-aminolevulinic acid, had almost no porphobilinogen deaminase activity. However, when the mutant was grown in the presence of 5-aminolevulinic acid, completely normal deaminase activity was restored (Warren and Jordan, 1988a; 1988c). Comparable results were obtained when the *hemA*⁻ mutant was transformed with a plasmid (pST48) containing the *hemC* gene. These findings proved unambiguously that the activity of the deaminase depended on a supply of porphobilinogen for the synthesis of the dipyrromethane cofactor and added to the wealth of evidence discussed above. These results are summarized in Table 2.

Table 2. Requirement of 5-aminolevulinic acid for the formation of active porphobilinogen deaminase in a *hemA*⁻ mutant of *E. coli*

bacterial strain	5-aminolevulinic acid in growth medium	deaminase units/mg
wild type	-	15
wild type	+	14
hemA⁻	-	1
hemA⁻	+	15
hemA⁻/pST48	-	4
hemA⁻/pST48	+	125

These fndings set the stage for two types of experiment with isotopically labelled 5-aminolevulinic acid. Firstly, 5-amino-[5-^{14}C]levulinic acid was incubated with a strain of *Escherichia coli* constructed from the *hemA*⁻ mutant but containing a recombinant plasmid carrying the *hemC* gene. This meant that the cofactor precursor, porphobilinogen, would be synthesised exclusively, without dilution, from the added label. The resulting deaminase was isolated and found to be highly labelled with [^{14}C]-radioactivity. The availability of the [^{14}C]-enzyme permitted a crucial experiment to be carried out to confirm further the role of the dipyrromethane. If the dipyrromethane was indeed acting as a resident cofactor then it should remain attached to the enzyme during catalytic turnover. To test this proposal, two types of ES complex were prepared, one in which the cofactor was labelled as described above (E^*S) and another prepared with unlabelled enzyme which had been incubated with [^{14}C]-porphobilinogen (ES^*). When the E^*S complex was incubated with excess unlabelled porphobilinogen, no label was incorporated into the product and **all** the radioactivity remained associated with the deaminase protein (Jordan and Warren, 1987; Warren & Jordan, 1988a,c). Treatment of the complex with NH_2OH also failed to release label. Only treatment with strong acid caused the removal of any label from the deaminase. In sharp contrast, reaction of ES^* with either substrate or NH_2OH caused the liberation of essentially all the radioactivity from the enzyme complex. These results, which are shown in table 3, established that the dipyrromethane cofactor was permanently attached to the deaminase and that, although made up of two substrate-like pyrrole rings, it was functionally **totally** distinct from the substrate.

Table 3. [^{14}C] label in deaminase protein after incubation of (E^*S) and (ES^*) with porphobilinogen and NH_2OH.

intermediate complex used	initial [^{14}C] label in complex	[^{14}C]-label after catalytic turnover	[^{14}C]-label after NH_2OH
E^*S	3,204dpm	3,197dpm	3,174dpm
ES^*	3,001dpm	8dpm	5dpm

The second type of experiment with labelled 5-aminolevulinic acid involved growing a recombinant *hemA*$^-$ strain in the presence of 5-amino[5-^{13}C]levulinic acid followed by the determination of the nmr spectrum of the purified [^{13}C]-enzyme. If the structure of the cofactor was indeed a dipyrromethane, four resonances should arise as predicted in scheme 11.

Scheme 11. Incorporation of 5-amino[5-^{13}C]levulinic acid into porphobilinogen and thence into the dipyrromethane cofactor

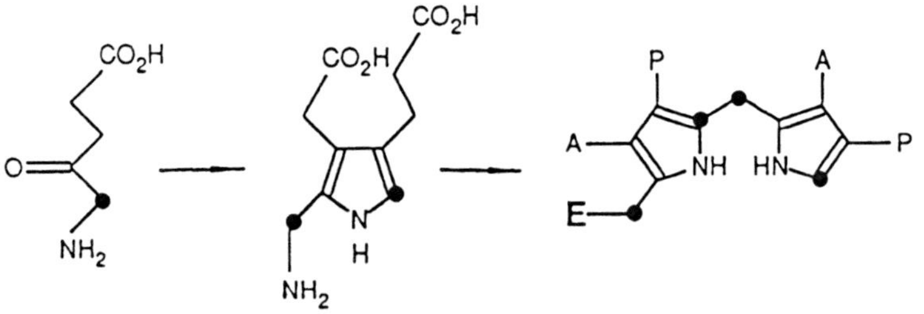

The difference spectrum, determined in the laboratory of Professor Ian Scott at Texas A & M University, revealed four broad resonances, two of which arose from aromatic carbon atoms at δ=116ppm and 128ppm and two others assigned to labelled methylene carbon atoms which overlapped between δ=25-30ppm. The spectrum shown in figure 3 is thus entirely consistent with the dipyrromethane structure proposed originally (Jordan and Warren, 1987). Close examination of the ^{13}C nmr spectrum revealed that the methylene resonance with a chemical shift at about δ=27ppm was very broad in the native enzyme indicating that it may be directly attached to the enzyme protein. Denaturation of the labelled enzyme at pH 12 simplified the spectrum in this area and visualized the two individual resonances (figure 3). The chemical shifts strongly suggested that one of these labelled methylene groups was at the *meso*-position between the pyrrole rings but that the other was linked covalently to a sulphur atom of an enzymic cysteine residue (Jordan et al, 1988c).

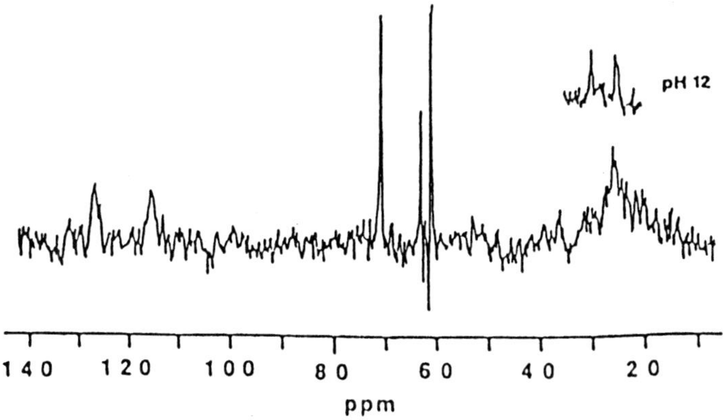

Figure 3. ^{13}C Nmr difference spectrum of porphobilinogen deaminase isolated from *E. coli* grown in the presence of 5-amino[5-^{13}C]levulinic acid

Results from protein chemistry experiments and site-directed mutagenesis established that cysteine-242 is the amino acid residue in the *E. coli* enzyme which is linked to the dipyrromethane cofactor. Cysteine-242 is found in the conserved sequence leu-glu-gly-gly-**cys**-gln-val-pro (Jordan et al, 1988c) as shown in scheme 12.

Scheme 12. The dipyrromethane cofactor attached to cysteine-242 in the *E. coli* porphobilinogen deaminase

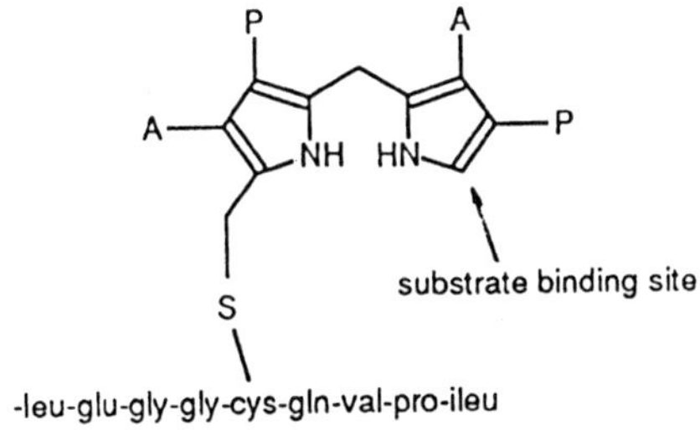

Further evidence for the role of the dipyrromethane cofactor was obtained from the results of experiments with the "chain termination" inhibitor α-bromoporphobilinogen which was synthesized in the laboratory of Professor Scott. This inhibitor is deaminated by the enzyme, like the substrate, and becomes covalently linked to the deaminase as an enzyme-intermediate analog. However, once incorporated, the substituted α-position blocks the linking of further substrate (or inhibitor) units to the enzyme. Thus reaction of E, ES, ES_2 or ES_3 with α-bromoporphobilinogen (B) gives EB, ESB, ES_2B and EB respectively. The EB formed in the reaction of ES_3 with α-bromoporphobilinogen arises from the reaction $ES_3 + B = ES_3B$. Since ES_3B is an ES_4 analog, it rapidly hydrolyses to HO-S_3B (α-bromopreuroporphyrinogen) and free enzyme, E, which reacts with more inhibitor to yield EB. Most importantly, the native enzyme which had been inactivated with α-bromoporphobilinogen exhibits a very weak Ehrlich's reaction indicating that the chain terminating inhibitor is linked covalently to the dipyrromethane cofactor thus blocking the reaction with the reagent

(Warren & Jordan, 1988c). This provides strong evidence that the substrate analog is covalently linked to the dipyrromethane cofactor. The terminated enzyme-inhibitor complexes are shown in scheme 13.

Scheme 13. Termination complexes arising from the treatment of enzyme and intermediate complexes with α-bromoporphobilinogen

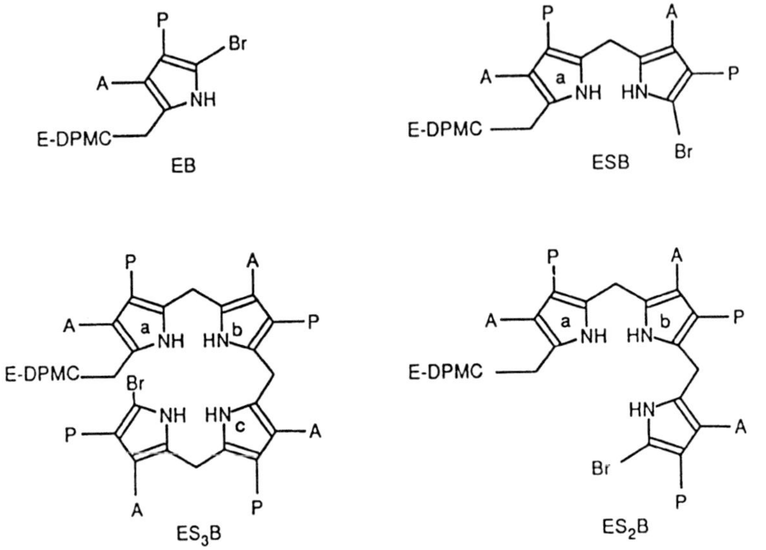

Having established the structure, role and location of the cofactor it was important to investigate other porphobilinogen deaminases to determine whether the dipyrromethane is a universal cofactor for all deaminases. The remarkable similarity in the primary amino acid sequence of three deaminases (*Escherichia. coli*, human and rat) suggested they all are likely to have a conserved structure and, by implication, a similar mechanism of action. Thus enzymes from human erythrocytes, barley chloroplasts and spinach leaves were all highly purified. When these enzymes were treated with Ehrlich's reagent the characteristic dipyrromethane reaction was observed in all cases (Warren & Jordan, 1988b). The deaminase from *R. sphaeroides* was also reisolated and the presence of the dipyrromethane was reconfirmed. These results established that the dipyrromethane cofactor is a **universal** substrate recognition and attachment site in the deaminases from plants, animals and bacteria.

The question of how the dipyrromethane cofactor is itself assembled and linked to the deaminase "apoenzyme" was addressed by incubating porphobilinogen with apodeaminase obtained from the recombinant *hemA*⁻ mutant. Although relatively low yields (15%) of active enzyme were obtained, the experiments indicated that the deaminase was able to catalyse the self-assembly of the cofactor from porphobilinogen, albeit at a slower rate than the polymerization reaction (Warren & Jordan, 1988c). Furthermore, it was established that once converted into the holoenzyme, the deaminase assumes remarkable stability and can tolerate temperatures of 65°C with little effect on activity. Evidence for a substantial conformational change during the apoenzyme - holoenzyme transition was obtained. The complete polymerization process catalysed by the porphobilinogen deaminase is shown in scheme 14.

Scheme 14. The mechanism of cofactor assembly, porphobilinogen polymerization and preuroporphyrinogen release

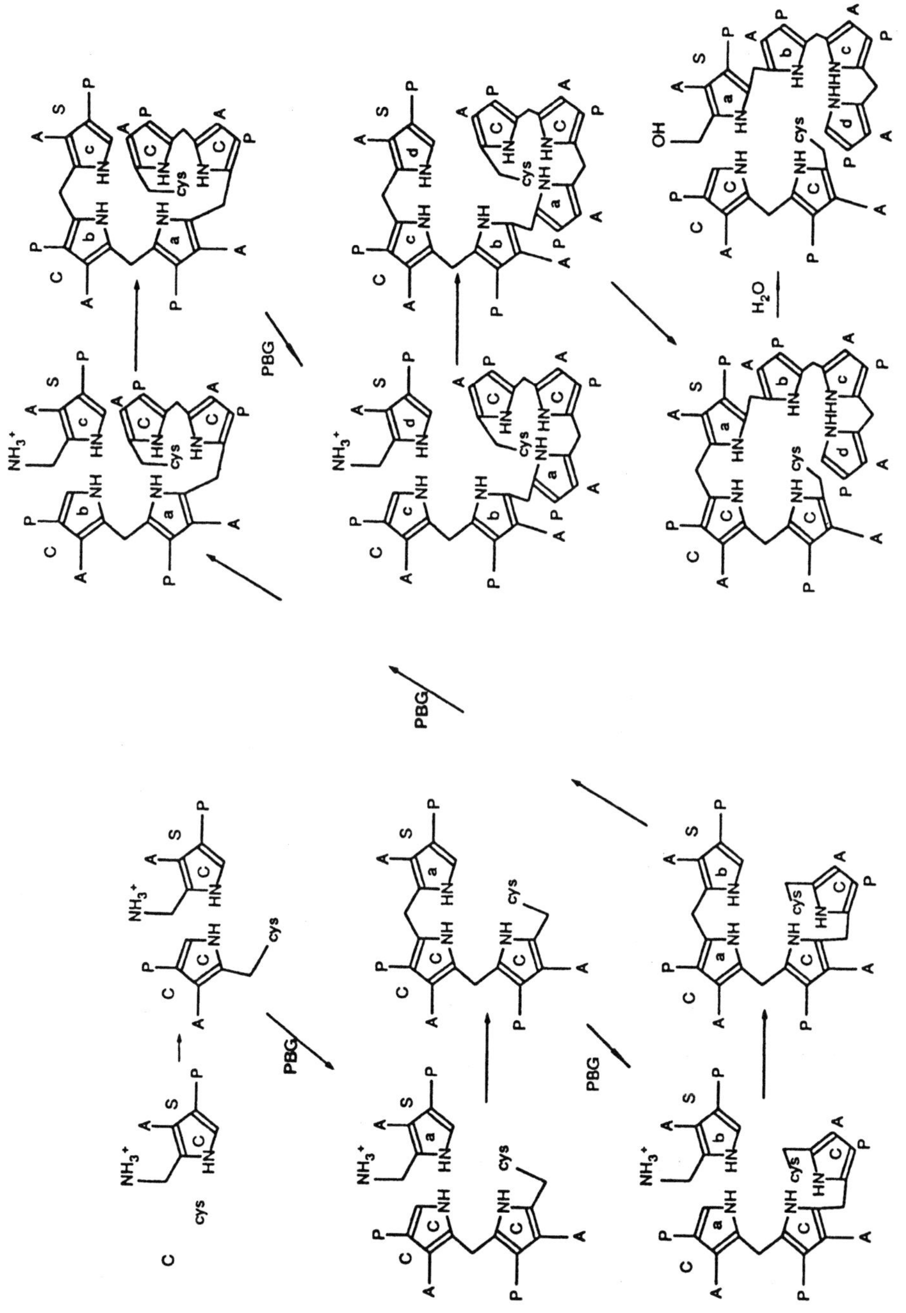

Polymerases have evolved several unique mechanisms. For instance, glycogen synthase uses a glycogen "primer" to initiate the reaction; DNA and RNA polymerases employ a "template" as a crucial part of the substrate recognition site; fatty acid synthase and polyketide synthases use a 4-phosphopantetheine "moving arm" to deliver the bound substrate to several catalytic sites, etc. Porphobilinogen deaminase falls into the category of enzymes which use a primer in order to initiate and propogate the polymerisation reaction. By virtue of its being covalently attached to the deaminase, the cofactor may also act as a chain length regulator to ensure that the polymerization is confined to the addition of only the **four** molecules of substrate required for the biosynthesis of the tetrapyrrole ring.

Our study of the porphobilinogen deaminase enzyme has been greatly facilitated as a result of the identification, cloning and sequencing of the porphobilinogen deaminase gene *hemC* from *Escherichia coli* (Thomas and Jordan, 1986). This has provided a DNA-derived protein sequence, figure 4, and permitted the design of many new experiments. Comparison of the *E. coli* deaminase sequence with that of the human enzyme reveals a remarkable 60% amino acid similarity between the proteins.

```
      Δ   Δ                                           Δ   Δ
MLDNVLRIATRQSPLALWQAHYVKDKLMASHPGLVVELVPMVTRGDVILDTPLAKVGGKG   60

   Δ                  Δ                 Δ
LFVKELEVALLENRADIAVHSMKDVPVEFPQGLGLVTICEREDPRDAFVSNNYDSLDALP  120

          ΔΔ                Δ     Δ  Δ                Δ
AGSIVGTSSLRRQCQLAERRPDLIIRSLRGNVGTRLSKLDNGEYDAIILAVAGLKRLGLE  180

 Δ                       Δ                         Δ
SRIRAALPPEISLPAVGQGAVGIECRLDDSRTRELLAALNHHETALRVTAERAMNTRLEG  240

 ▲                              Δ   Δ
GCQVPIGSYAELIDGEIWLRALVGAPDGSQIIRGERRGAPQDAEQMGISLAEELLNNGAR  300

EILAEVYNGDAPA-COOH 313
```

Figure 4. DNA-derived primary structure of *E. coli* porphobilinogen deaminase showing conserved arginine and lysine residues (Δ) and the cofactor attachment site (▲)

The cloning of the gene into high copy number plasmids has permitted the isolation of milligrams of the homogeneous deaminase and permitted the initial crystallization of the enzyme (Jordan et al, 1988b). X-Ray studies, in collaboration with Dr Wood and Professor Blundell at Birkbeck College, London University, have resulted in large and reproduceable single orthorhombic crystals (0.7 x 0.5 x 0.3 mm). The crystals have a space group $P2_12_12$ and cell dimensions of a=88.01A, b=75.86A, c=50.53A. with $\alpha=\beta=\gamma=90°$ indicating one molecule in the asymmetric unit. The crystals are stable in the X-ray beam for about 40 hours. X-Ray data for the native enzyme and three heavy atom heavy metal derivatives have been collected to 3A and a low resolution contour map has been produced. Trial oscillation photographs at the Daresbury Synchrotron Laboratory show that the native crystals diffract to about 1.7A indicating that a high resolution structure refinement should be possible in the immediate future.

In addition to the enzyme groups responsible for the catalytic reaction, other groups must be involved with the translocation of the polypyrrole chain as new substrate molecules are added to the enzyme. Comparison between the protein sequences of four deaminases indicates that a large number of positively charged amino acids are conserved (see (Δ) in figure 4) possibly to bind the 12 carboxylate side chains of the tetrapyrrole and dipyrromethane cofactor. In this context site directed mutants have been generated to study these groups and to explore the mechanism of chain elongation.

b) UROPORPHYRINOGEN COSYNTHASE

Uroporphyrinogen cosynthase (synthase) catalyses the rearrangement of the "d" ring of preuroporphyrinogen and the final cyclization to give uroporphyrinogen III as shown in scheme 15.

Scheme 15. The reaction catalyzed by uroporphyrinogen III cosynthase

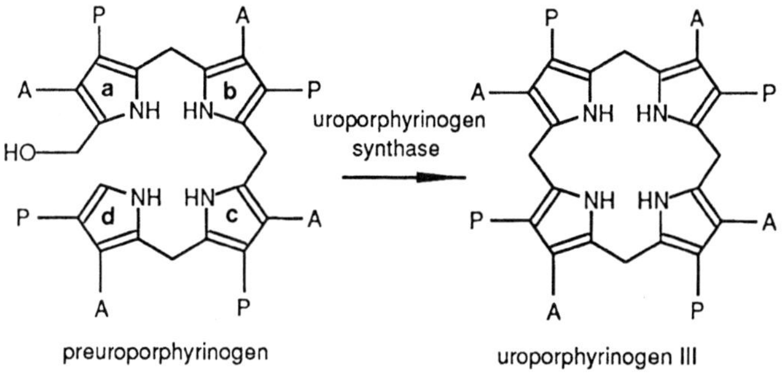

The discovery of preuroporphyrinogen as the substrate for the cosynthase permitted the development of a rapid assay for the enzyme (Jordan 1982) and greatly assisted further studies.

During the sequencing of the *hemC* gene encoding the *E. coli* porphobilinogen deaminase (Thomas and Jordan, 1986), an open reading frame was discovered overlapping the TGA stop codon of the *hemC* gene. Further investigation revealed that this reading frame specifies the *hemD* gene encoding uroporphyrinogen III cosynthase. The *hemD* gene was sequenced (Jordan et al., 1987; Jordan et al., 1988a) and the enzyme was isolated in milligram amounts from a bacterial strain transformed with a multi-copy plasmid containing the *hemD* gene (Alwan et al., 1989). The *hemD* gene appears to be under the control of the *hemC* promoter and is thus the second gene in a *hem* operon. Other genes (*hem3*, *hem4*, *hem5*, etc) appear to be present in the *hem* operon (figure 5) and have been sequenced in our laboratory.

cyaA	P	P	hemC	hemD	hem3	hem4	hem5	hem6?

Figure 5. The *hem* operon at 85 minutes on the *E coli* chromosome

The availabilty of high expression strains for the cosynthase should assist our investigations into the mechanism by which the enzyme catalyses the remarkable rearrangement and cyclisation process during the transformation of preuroporphyrinogen into uroporphyrinogen III. Currently the most favored mechanism is that originally proposed by Mathewson and Corwin about thirty years ago.

The steric course of the reactions catalysed by porphobilinogen deaminase and uroporphyrinogen cosynthase have been studied using randomly and stereospecifically tritiated porphobilinogen. Incorporation of 5-amino-*5(R,S)*-[$^3H_2$]levulinic acid into porphobilinogen results in the labelleing of both hydrogens at the 11-position. Incorporation of the randomly tritiated porphobilinogen into uroporphyrinogen III and thence into protoporphyrin IX leads to the loss of exaxtly 50% of the tritium label indicating that during the overall transformations the *meso*-hydrogens in preuroporphyrinogen, uroporphyrinogen III, and protoporphyrinogen have been manipulated stereospecifically. The steric course of the reaction was investigated further by synthesizing *11(S)*-[^3H]-porphobilinogen from *2(S)*-[^3H]-glycine as described earlier. The labelled porphobilinogen species was transformed into heme which was degraded to the four biliverdin isomers for analysis. The results showed that tritium label was only incorporated into the *meso*-position at C-10. In view of our current understanding about the mechanism of action of porphobilinogen deaminase and uroporphyrinogen cosynthase and our knowledge that such substitution reactions almost certainly occur with retention of configuration, one must conclude that during the oxidation of protoporphyrinogen the three labelled hydrogens are removed in oxidative processes from one face of the macrocycle (at C-5, C-15 and at C-20) and that the fourth hydrogen (■) is lost as a proton from C-10 (Jones et al., 1984).

Scheme 16. Steric course of incorporation of *11(S)*-[^3H]porphobilinogen into uroporphyrinogen III and thence into protoporphyrin IX

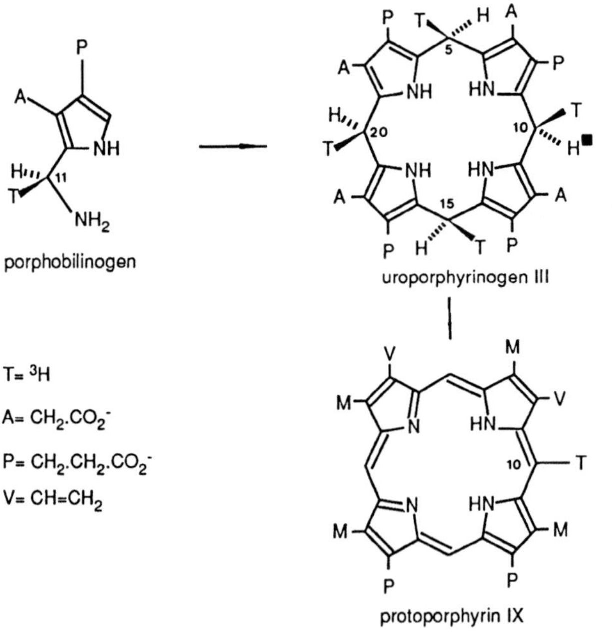

This approach has been extended to the investigation of the events at the C-10 position during the biosynthesis of cobyrinic acid from porphobilinogen in *Propionobacterium shermanii*. When *11(S)*-[^3H] porphobilinogen was incorporated into cobyrinic acid, there was a retention of label consistent with the presence of one tritium (Jordan, in preparation). Since the *meso*-positions at C-5 and C-15 are methylated and the C-20 position has been eliminated the remaining label is located at C-10.

Scheme 17 shows the labelling pattern in uroporphyrinogen and cobyrinic acid. The stereospecific loss of the C-10 hydrogen, arising from the *pro-R*-configuration at C-11 of porphobilinogen, which is deduced from this result, could occur either during methylation at C-12 or alternatively as a result of tautomeric shifts. If the former is the case this may allow one to suggest the steric course of related methylations which occur during the formation of dihydro-factor II. In the absence of detailed knowledge about the precise order of the methylation reactions, these possiblities cannot yet be distinguished with certainty.

Scheme 17. Labelling of uroporphyrinogen III and cobyrinic acid synthesized from *11(S)*-[^{3}H]porphobilinogen in extracts of *Propionobacterium shermanii*

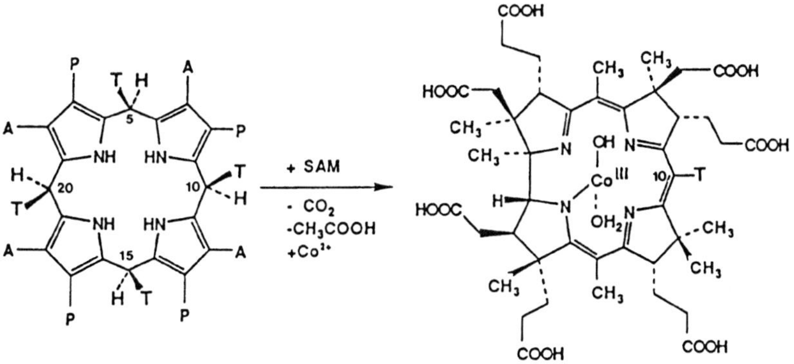

The isolation of the enzymes in B_{12} synthesising organisms has suffered from the fact that only small quantities of the proteins are produced. In the immediate future, the application of molecular biology will greatly assist the bioorganic chemist in unravelling the details at each step in the biosynthesis of this important vitamin.

ACKNOWLEDGEMENTS

This account is dedicated to Professor M. Akhtar, FRS who initiated the study of tetrapyrrole biosynthesis with the author in 1969 at Southampton University and who is a valued colleauge and friend.

This work was largely supported by the Science and Engineering Research Council.

REFERENCES

Abboud, M.M., Jordan, P.M., and Akhtar, M., (1974), J. Chem.Soc. Chem.Comm. 643-644.

Alwan, A.F., Mgbeje, B.I.A., and Jordan, P.M., (1989), Biochem. J. *264*, 397-402.

Berry, A., Jordan, P.M., and Seehra, J.S., (1981), FEBS Letters *129*, 220-224.

Burton, G., Fagerness, P., Hozozawa, S., Jordan, P.M., and Scott, A.I., (1979), J. Chem. Soc., Chem. Comm., 202-204.

Chaudhry, A.G., and Jordan, P.M., (1976), Biochem. Soc. Trans. *4*, 760-761.

Gibbs, P.N.B., and Jordan, P.M., (1981), Biochem. Soc. Trans. *9*, 232-233.

Gibbs, P.N.B., and Jordan, P.M., (1986), Biochem.J. *236*, 447 -451.
Jones, C., Jordan P.M., and Akhtar, M.A., (1984), J. Chem. Soc. Perkin Trans. I, 2625-2633.
Jordan P.M., and Akhtar, M.A., (1969), Tet. Letters, 875-877
Jordan, P.M., and Shemin, D., (1973), J. Biol. Chem. *248*, 1019-1024.
Jordan, P.M., Burton, G., Nordlov, H., Schneider, M.M., Pryde, L., and Scott, A.I., (1979), J. Chem. Soc., Chem. Commun. 204-205.
Jordan, P.M., and Seehra, J.S., (1979), FEBS Letters *104*, 364-366.
Jordan, P.M., and Berry, A., (1980), FEBS Letters, *112*, 86-88.
Jordan, P.M., and Seehra, J.S., (1980a), FEBS Letters, *114*, 283-286.
Jordan, P.M., and Seehra, J.S., (1980b), J. Chem. Soc. Chem. Comm. 240-242.
Jordan, P.M., and Berry, A., (1981), Biochem. J. *195*, 177-181.
Jordan, P.M., (1982), Enzyme *28*, 158-169.
Jordan, P.M., and Gibbs, P.N.B., (1985), Biochem. J. *227*, 1015-1020.
Jordan, P.M., Mgbeje, B.I.A., Alwan, A.F., and Thomas, S.D., (1987), Nucl. Acids Res., *15*, 10538.
Jordan, P.M., and Warren, M.J., (1987), FEBS Letters, *225*, 87-92.
Jordan, P.M., Mgbeje, B.I.A., Thomas, S.D., and Alwan, A.F., (1988a), Biochem. J., *249*, 613-616.
Jordan, P.M., Thomas, S.D., and Warren, M.J., (1988b), Biochem. J., *254*, 427-435.
Jordan, P.M., Warren, M.J., Williams, H.J., Stolowich, N.J., Roessner, C.A., Grant, S.K., and Scott, A.I., (1988c), FEBS Letters, *235*, 189-193.
Jordan, P.M., (1989), in "The Biosynthesis and Regulation of Heme and Chlorophyll Biosynthesis", ed. Dailey, H.A., Wiley.
Laghai, A., and Jordan, P.M., (1976), Biochem. Soc. Trans. *4*, 52-53.
Laghai, A., and Jordan, P.M., (1977), Biochem. Soc. Trans. *5*, 299-301.
Seehra, J.S., and Jordan, P.M., (1980), J. Amer. Chem. Soc. *102*, 6841-6846.
Thomas, S.D., and Jordan, P.M., (1986), Nucl. Acids Res., *14*, 6215-6226.
Warren, M.J., and Jordan, P.M., (1988a), Biochem. Soc. Trans., *16*, 962-963.
Warren, M.J., and Jordan, P.M., (1988b), Biochem. Soc. Trans., *16* 963-965.
Warren, M.J., and Jordan, P.M., (1988c), Biochemistry *27*, 9020-9030.
Zaman, A., Jordan, P.M., and Akhtar, M., (1973), Biochem. J. *135*, 257-263.

BIOSYNTHESIS OF VITAMIN B_{12}: BIOSYNTHETIC AND SYNTHETIC RESEARCHES

Alan R. Battersby

University Chemical Laboratory
Lensfield Road
Cambridge, CB2 1EW
U.K.

INTRODUCTION

The problem posed by the biosynthesis of vitamin B_{12} is one of the most formidable and challenging in the whole area of biosynthetic research. What I hope to illustrate is how such a problem can be attacked using the traditional strengths of organic chemistry, provided one is also able to draw upon many other related areas of knowledge and expertise. The ones of particular importance in our work on B_{12} have been those involving sophisticated n.m.r. spectroscopy combined with isotopic labelling, enzymology, kinetics, genetics and molecular biology.[1]

The first part of the lecture focuses on the early part of the pathway to B_{12} and the second part deals with our studies of the late biosynthetic stages. The structure of vitamin B_{12} is shown in Scheme 1, it was known from the very early work[1] that B_{12} is formed from cobyrinic acid and this in turn is biosynthesised from uro'gen III. Uro'gen III is also the precursor of all the other Pigments of Life including heme and chlorophyll so a massive effort in Cambridge has been aimed at elucidating the biosynthetic pathway to uro'gen III, the key parent substance.[2]

BIOSYNTHESIS OF URO'GEN III: DEAMINASE

It was known at the outset of our studies in 1968 that uro'gen III is built from four molecules of porphobilinogen PBG, (top left, Scheme 2) and that two enzymes

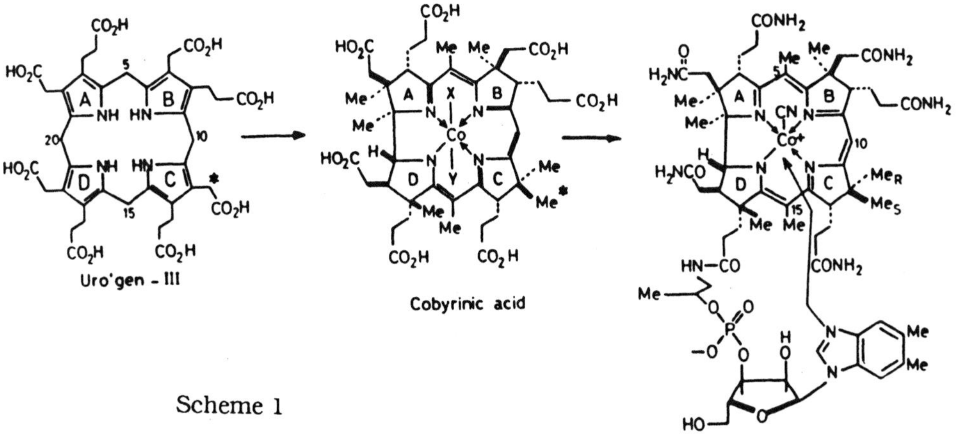

Scheme 1

hydroxymethylbilane synthase (also called deaminase) and uroporphyrinogen III synthase (also called cosynthetase) are required to carry out this remarkable transformation. This knowledge came mainly from the work of Bogorad, Granick, Neuberger, Rimington and Shemin.

Our studies of the biosynthesis of uro'gen III over the past 22 years have focussed on the problem of how living organisms build this unexpected structure from 4 PBG units. The structure one would expect to be formed by assembling 4 PBG molecules into a large ring would have the acetate and propionate residues running in sequence around the periphery of the macrocycle. Inspection of uro'gen III shows that this is not so and therefore rearrangement or several rearrangements must have occurred. The experiments logically probed the mechanism of this building process and the main results which are summarised in Scheme 2 are as follows:

(*a*) Only one rearrangement is involved, it is intramolecular and affects ring D of uro'gen III. The other 3 PBG units are built in as intact units.[3]

(*b*) The role of deaminase is to build an open-chain linear tetrapyrrole[4] (a bilane); deaminase is not (as had been thought) an enzyme which constructs a macrocycle of one sort or another. Deaminase is a head-to-tail assembly enzyme and it is only *after* the linear tetrapyrrole has been built that rearrangement occurs.[4]

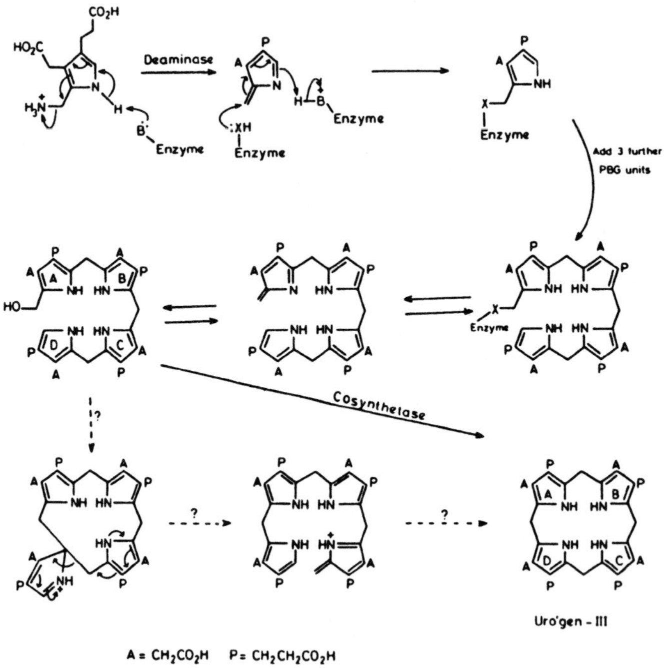

Scheme 2

(*c*) The linear bilane produced by deaminase is the hydroxymethylbilane[5] (centre left of Scheme 2) which in the absence of the second enzyme, cosynthetase, is released into the medium. This bilane is built starting with ring A, adding ring B followed by rings C and D in that order.[6]

(*d*) Cosynthetase accepts the hydroxymethylbilane as its substrate and rapidly ring-closes it, with a single intramolecular rearrangement of the terminal ring, to form uro'gen III.[5]

(*e*) The hydroxymethylbilane undergoes spontaneous (*i.e.* non-enzymic) ring-closure at an appreciable rate at *ca.* pH 8 and the product is uro'gen I. This is isomeric with uro'gen III but now the acetate and propionate side-chains *are* arranged sequentially around the periphery of the macrocycle. This instability means that it is not easy by enzymic means to accumulate large quantities of hydroxymethylbilane (but see later) and it is fortunate that an effective synthesis of this valuable substance has been developed.[5]

(*f*) A solid foundation of knowledge has been laid concerning the kinetic properties of deaminase, its competitive inhibition by its product, hydroxymethylbilane, and its relationship to cosynthetase.[7] Though the two enzymes may be closely associated in the natural state, each one can catalyse the formation of its product independently of the other enzyme.[8]

Many of these findings have been built into the biosynthetic pathway illustrated in Scheme 2 which also shows that the first PBG to be bound to the enzyme is connected covalently[7-9] to the protein through some group X; we shall discuss the nature of X later in the lecture.

There had been strong indications from earlier work[10] that a very reactive tetrapyrrolic intermediate is formed in the assembly process which can be trapped by small nucleophiles such as ammonia. This reactive intermediate is shown as the pyrrolenine at the centre of Scheme 2 and extensive support for its existence was provided by isotopic labelling, n.m.r. spectroscopic and trapping experiments.[7] The possibility has occasionally been raised[11] that this azafulvene is the normal product from deaminase which is then directly cyclised by cosynthetase. If this is true, then the hydroxymethylbilane is a side-product formed by water trapping the released azafulvene. Any stereochemistry due to isotopic labelling of the methylene attached to the X-group would be lost in such a process. In fact, synthesis of (11*R*)- and (11*S*)-[11-$^{3}H_{1}$]PBG followed by their conversion by deaminase into hydroxymethylbilane each gave stereoselectively labelled bilane with *overall retention* of configuration.[12] A workable quantity of the labile hydroxymethylbilane was obtained by passing a solution of PBG through a column to which deaminase had been bound and stabilising the bilane by collecting it in alkali at pH >12. Trapping the azafulvene with ammonia gave the same stereochemical results.[13] These stereochemical studies do not, by themselves, rigorously confirm that hydroxymethylbilane is a true biosynthetic intermediate. But when they are considered with all the other knowledge of deaminase and cosynthetase, the evidence for that view far outweighs the few indications which are claimed to point against it.

Scheme 3

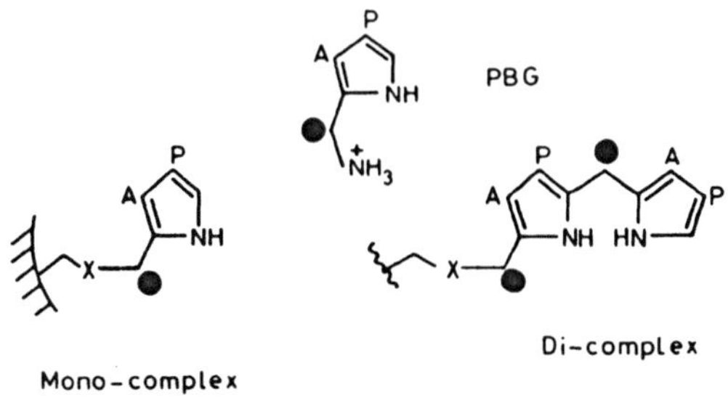

NATURE OF THE X-GROUP

Our experimental approach for studying the X-group involved binding [11-^{13}C]PBG to deaminase (see Scheme 3) when it should be possible in principle to prove what X is from the chemical shift of the ^{13}C-signal from the carbon bound to X. In practice, the enzyme partially loaded with labelled PBG gave no recognisable ^{13}C-signals and furthermore was unstable showing continual loss of pyrrolic material during handling; the results from these early experiments have been reviewed.[14] The way to overcome these problems was found by synthesising model peptides carrying a CH_2-pyrrole group attached in one example to the ε-nitrogen of a lysyl residue and in another to the sulphur of a cysteine residue. Such molecules were found to be sufficiently stable at pH 12-14 to allow acquisition of a ^{13}C-n.m.r. spectrum.[15]

It has been known for some years that the covalently bound mono-, di-, tri- and tetrapyrrolic complexes on deaminase are catalytically competent and can be separated analytically.[16,7-9] Further progress depended on the introduction of two new techniques. *(a)* The deaminase gene in *E. coli* was sequenced,[17-19] cloned and over-expressed. *(b)* Fast protein liquid chromatography (f.p.l.c.) allowed separation of the various enzyme-PBG complexes on a preparative (*ca.* 10 mg) scale. The overexpression achieved in Cambridge was initially 200-fold[20] and has now been increased to about 1000-fold.

These approaches and knowledge from the synthetic work above allowed us to determine the nature of the X-group.[21] It was a most surprising discovery which changed the way everyone viewed deaminase.

A large quantity of pure deaminase was isolated and it was shown to be unloaded with respect to building the mono-, di- etc. PBG complexes; this unloaded form is called *holoenzyme*. It was loaded with a limited quantity of [11-^{13}C]PBG and the enzyme-(^{13}C-PBG)$_1$-complex and enzyme -(^{13}C-PBG)$_2$-complex were isolated by f.p.l.c.. The ^{13}C-n.m.r. spectrum from the former complex run at pH >12 showed sharp ^{13}C-signals (Figure 1) as did the natural abundance ^{13}C-spectrum of unlabelled enzyme-(^{12}C-PBG)$_1$-complex. The difference between these two spectra showed one strong sharp signal at δ24.6 from the ^{13}C-enriched CH_2 group attached to the X-group (Figure 2). This shift corresponded exactly to a pyrrole-$^{13}CH_2$-pyrrole system and showed for the first time that the residue in deaminase to which the first PBG unit covalently binds is a tightly bound *pyrrole residue* already present in the enzyme.[21] The ^{13}C-n.m.r. spectrum of purified enzyme-(^{13}C-PBG)$_2$-complex was also recorded and the difference spectrum as before showed one signal at δ24.6 which was nearly twice the size of that from the mono-complex thus confirming the presence of a previously unsuspected and novel pyrrolic cofactor firmly attached to the protein of deaminase to which the first PBG unit becomes covalently bound.[21]

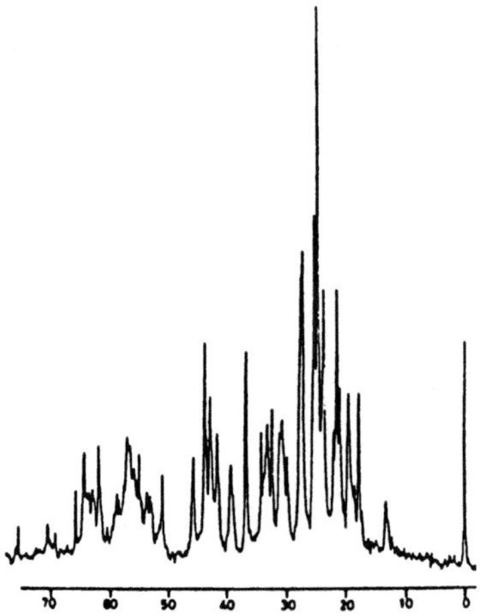

FIGURE 1

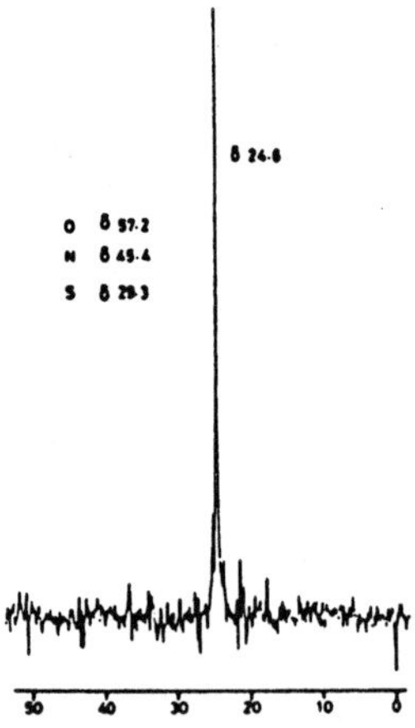

FIGURE 2

By turning over 11 moles of [9-^{14}C]PBG per mole of deaminase, it was found that the subsequently isolated *holoenzyme* was radioinactive. Thus, deaminase as normally isolated carries no PBG substrate and the pyrrolic cofactor attached to the protein *does not turn over* during the assembly of hydroxymethylbilane.

Further understanding came from two important observations: *(a)* treatment of deaminase *holoenzyme* with formic acid gave mainly uroporphyrin I formed by air oxidation of uro'gen I which is initially generated. The amount produced corresponded to the presence of not one but *two pyrrole residues i.e.* a dipyrromethane which because of the structure of uro'gen I, must be formed from two PBG units *(b)*, treatment of the *holoenzyme* with Ehrlich's reagent (*p*-dimethylaminobenzaldehyde) caused striking spectroscopic changes which by comparison with the changes observed with synthetic systems[21-23] confirmed that the cofactor is a dipyrromethane. The cofactor can at this stage be shown as in Scheme 4 attached to the protein through some group Y.

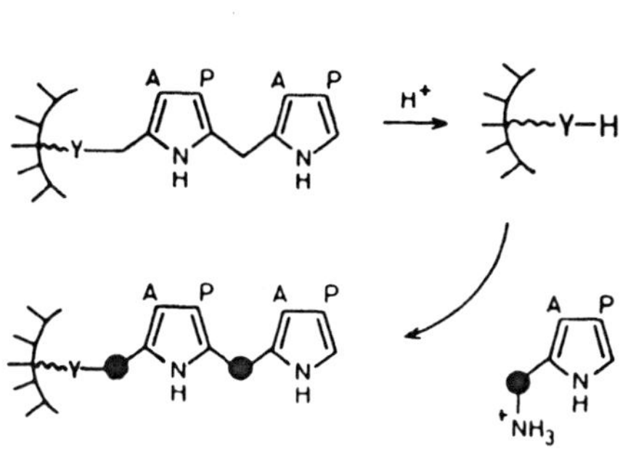

Scheme 4

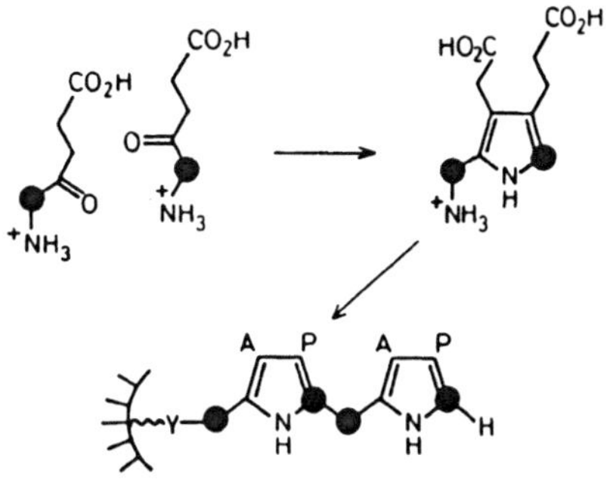

Scheme 5

Careful acidic cleavage of the cofactor from the *holoenzyme* gave, as shown in Scheme 4, the cofactor-free *apoenzyme* which had lost its catalytic activity. However, by incubating the *apoenzyme* with [11-^{13}C]PBG, most of the enzymic activity was restored and the cofactor was reconstructed with a ^{13}C atom directly bound to the Y-group. ^{13}C-N.m.r. difference spectroscopy then established that Y is *sulphur* and so the cofactor is attached *via* the sulphur atom of a cysteine residue.[24] Labelled deaminase was also produced in the natural way by growing the overproducing *E. coli* cells in the presence of [5-^{13}C]aminolaevulinic acid (the precursor of PBG) to generate the labelling pattern in Scheme 5. Four clear signals appeared in the ^{13}C-n.m.r. difference spectrum one of which confirmed unambiguously the attachment to sulphur and the others showed the correct chemical shift (and couplings) for the remaining 3 labelled sites,[25] see Scheme 5. Finally, this labelled deaminase was converted into enzyme-(^{13}C-PBG)$_1$-complex using [11-^{13}C]PBG to confirm rigorously by the pattern and couplings of the signals the attachment of the first PBG unit in the building process to the free α-position of the dipyrromethane cofactor.[22]

There remained one question. To which of the 4 cysteine residues in deaminase is the dipyrromethane cofactor attached? To solve this problem, deaminase was produced with its cofactor ^{14}C-labelled. This was specifically cleaved with endoproteinase Glu-C, the resultant cofactor-carrying peptide was isolated in pure state and sequenced as shown in Scheme 6. The results [26] unambiguously located the attachment of the cofactor to cysteine-242.

All the foregoing studies established that deaminase is a unique enzyme[27] using a novel dipyrromethane cofactor to which four more PBG units are attached one after another. This assembles a hexapyrrolic chain (Scheme 7) from which the tetrapyrrolic hydroxymethylbilane is specifically cleaved as the product, probably by protonation at the arrowed site. The *apoenzyme* possesses its own catalytic activity to construct and bind the cofactor through cysteine-242. The bound cofactor acts as an anchor for assembly of 4 more PBG units and does not itself turn over.

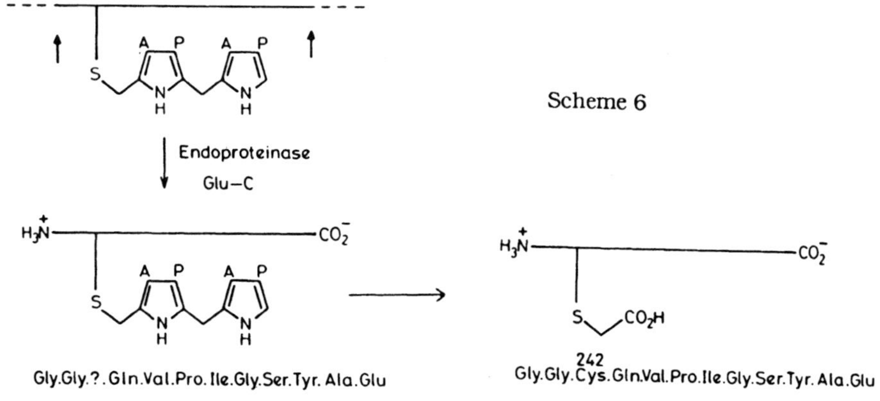

Scheme 6

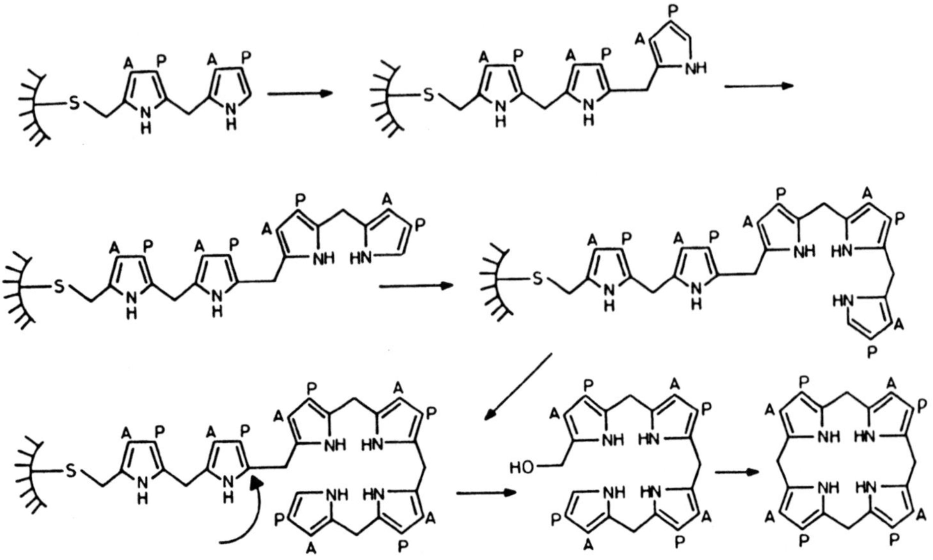

Scheme 7

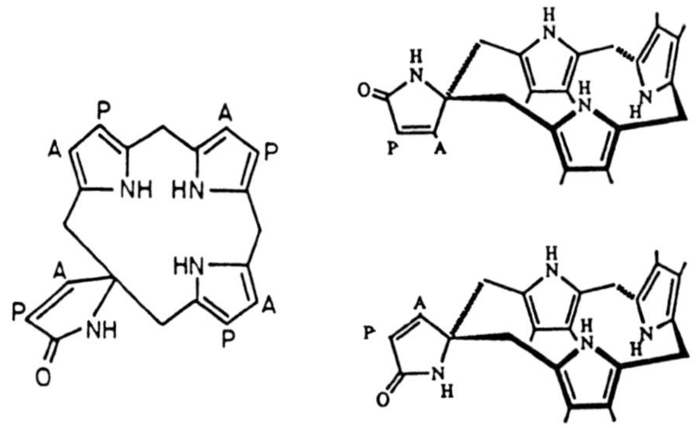

Scheme 8

Following first description[21] from Cambridge of the discovery and role of the dipyrromethane cofactor of deaminase, there were developments from other groups. The independent detection of the cofactor was reported[28] and the following year, its role was confirmed.[29] Results in agreement with ours on cysteine-242 were added[30] and valuable studies by ^{13}C-n.m.r. and mutagenesis appeared[31] which supported several of the structural features discussed above.

BIOSYNTHESIS OF URO'GEN III: COSYNTHETASE

One attractive mechanism for the formation of uro'gen III from the hydroxymethylbilane was shown in Scheme 2. This involves the intermediacy of a spiro-pyrrolenine fused to a tripyrrolic macrocycle (bottom left of Scheme 2). When this type of mechanism was first proposed,[32] it was thought that the macrocycle of the illustrated spiro-intermediate would be too strained and a protonated form of it was suggested to increase flexibility. We showed this to be unnecessary by synthesising several compounds containing this macrocycle[33,34] but a strongly puckered conformation was necessary to relieve strain.[33] This locked conformation led to the existence of two atropisomers of the important spiro-lactam system which was synthesised; [34] the planar representation of the two atropisomers are shown in Scheme 8. These isomers were separable and the corresponding octa-acids were tested with cosynthetase. One had no effect on the enzyme but the other was a very strong inhibitor with a K_i value of *ca.* 1μM which is an order of magnitude lower than the K_m for the substrate, hydroxymethylbilane. This outcome of one positive and one negative affect on cosynthetase could not have been better and gives very strong evidence[34] in support of the spiro-mechanism shown in Scheme 2.

By combining the steps illustrated in Schemes 2 and 7, the detailed biosynthetic pathway from PBG to uro'gen III can be assembled. In addition much knowledge has been gained of the two key enzymes, deaminase and cosynthetase, knowledge which is sure to grow over the next few years.

BIOSYNTHESIS OF VITAMIN B_{12}: THE METHYLATION STEPS

More than 20 years of research by the Cambridge group have led to the following main advances, knowledge of which in brief summarised form is needed by the reader in order to appreciate the latest developments. Naturally, other groups have also been involved in several areas discussed below and any of their valuable contributions not specifically referenced are collected in ref. 1.

(*a*) It was stated at the outset of this lecture that cobyrinic acid, and so vitamin B_{12} also, are formed from uro'gen III.[35-37]. At some stage in the biosynthesis, the 12-acetate residue must be decarboxylated to form the 12β-methyl group of cobyrinic acid. It was shown that this step is *not* the next one after uro'gen III has been built;[36] rather it is *C*-methylation which causes the pathway to branch towards vitamin B_{12}.

(*b*) Two new pigments were isolated from organisms which produce B_{12}. We[38-40] and others[1] worked out their structures but that work is outside the scope of this lecture; it is sufficient to show the structures and names in Scheme 9. Suitable labelling experiments demonstrated that both these isobacteriochlorins are converted into cobyrinic acid by a broken cell enzyme preparation from *Propionibacterium shermanii.*

(*c*) At this stage there was an inconsistency. Uro'gen III contains 8 double bonds whereas the substances in Scheme 9 have 9; but *C*-methylation does not affect the oxidation level. So the enzymic methylation of uro'gen III was carried out at <5 vpm O_2 and then the product was shown[41] to be the yellow dihydro system in Scheme 10. This is the true biosynthetic intermediate, called precorrin-2, and the product previously isolated had been formed by air oxidation. There are many strong indications that the trimethylated intermediate is produced and further transformed at this same dihydro oxidation level. The trimethylated system, precorrin-3, is also shown in Scheme 10.

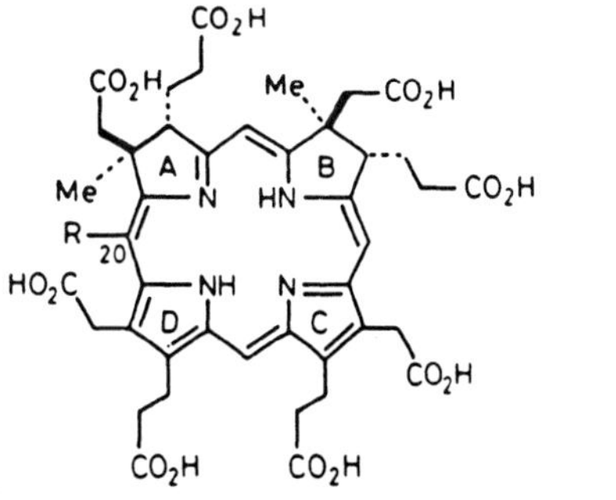

R = H Sirohydrochlorin
R = Me Trimethylisobacteriochlorin

Scheme 9

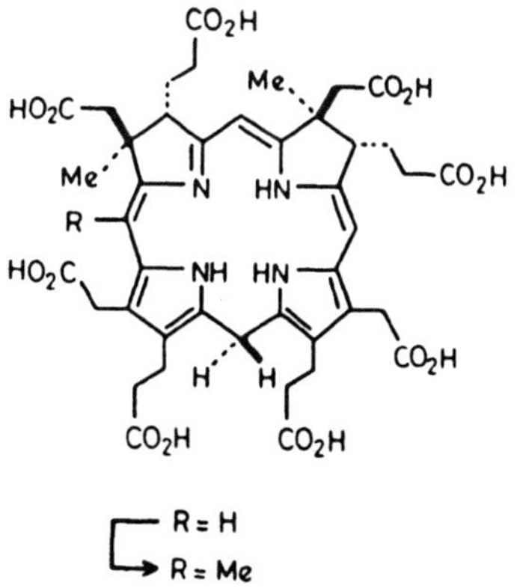

Scheme 10

(*d*) By the time the biosynthetic pathway reaches cobyrinic acid, a direct link has been formed between rings A and D. What has happened to C-20 and the attached *C*-methyl group which are present in precorrin-3? Careful experiments demonstrated that these two carbons are extruded into the medium as acetic acid[42,43] with the original C-20 methyl forming the methyl of acetic acid and C-20 itself appearing as the carboxyl group.

(*e*) 8 *C*-Methyl groups are introduced from *S*-adenosylmethionine (SAM) into uro'gen III during the biosynthesis of cobyrinic acid. When the structure of the first methylation product,[44,45] precorrin-1, is considered with those of precorrin-2 and precorrin-3, Scheme 10, one can deduce the order of introduction of the first 3 methyl groups as being first at C-2 then at C-7 followed by that at C-20. Despite enormous efforts, no more fully methylated intermediates beyond precorrin-3 have been isolated.

(*f*) The problem of order of *C*-methylation was therefore approached in a different way using pulse-labelling. The development of an appropriate methodology based on use of [^{13}C-*methyl*]SAM in combination with ^{13}C-n.m.r. spectroscopy[46] allowed us to show in 1982 that the fourth *C*-methyl group is inserted[46] at C-17 of precorrin-3, a step which will generate the pyrrocorphin shown in Scheme 11. The pulse labelling approach was subsequently extended[47,48] to provide knowledge of the order of introduction of the remaining 4 methyl groups but these late stages are outside the scope of today's lecture.

(*g*) There is, in principle, a possible alternative to direct C-17 methylation of precorrin-3. Precorrin-3 might first undergo decarboxylation at C-12 *before* methylation occurs at C-17. However, suitable incorporation experiments[49,48b] showed that this did not occur and therefore the pyrrocorphin in Scheme 11 and also its 12-decarboxylated analogue are of considerable biosynthetic interest.

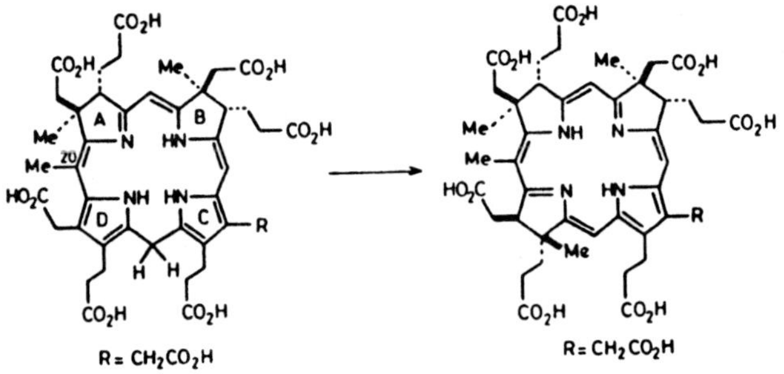

Scheme 11

BIOSYNTHESIS OF VITAMIN B_{12}: SYNTHETIC STUDIES

For experiments on the stages of the pathway beyond precorrin-3, it is important to have precorrin-2 and especially precorrin-3 available in specifically labelled forms and, if possible, in good quantity. Accordingly a synthetic programme was initiated in the late 1970's aimed at the construction of the aromatised forms of precorrin-1,[50] precorrin-2[51] and both precorrin-3[52] and its 12-decarboxylated analogue all of which have been successfully completed. The way we synthesised[53] the last of this group will now be outlined.

The synthesis depended on the photochemical antarafacial cyclisation of an 18π open-chain seco-system.[54] This was to be constructed using the imide synthesised in Scheme 12 as the key chiral starting material, the initial steps depending on the work of Koga *et al.*[55] It was then possible by appropriate use of Lawesson's reagent, to generate the two isomeric monothioimides used in Schemes 13 and 14.

The coupling step in Scheme 13, using the phosphonium salt for the pyrrolic portion, proceeded smoothly and the nitrile function (which was essential for success of the coupling reaction) was removed as illustrated. This involved reduction to an aminomethyl group followed by a reverse Mannich reaction to yield the required eastern block.

The route to the western block was similar and is shown in Scheme 14 but now the nitrile residue must be converted into a methyl group. This proved to be a real stumbling block but success was eventually achieved as summarised in Scheme 14. It remained to add the additional carbon atom for C-5 by Eschenmoser's sulphur extrusion chemistry.

Also, the eastern block was modified by formylation and conversion of the thiolactam into the thioimino ether to yield the two building blocks at the start of Scheme 15. Now we were ready to join these two blocks to form the seco-system, Scheme 15, which was then cyclised photochemically. The 2,7,12,20-tetramethylisobacteriochlorin so synthesised[53] was identical with the small amount of authentic material which had been produced enzymically.[49]

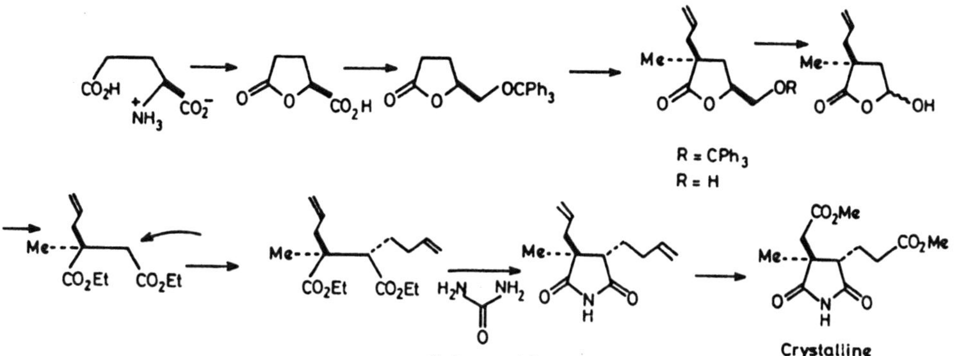

Scheme 12

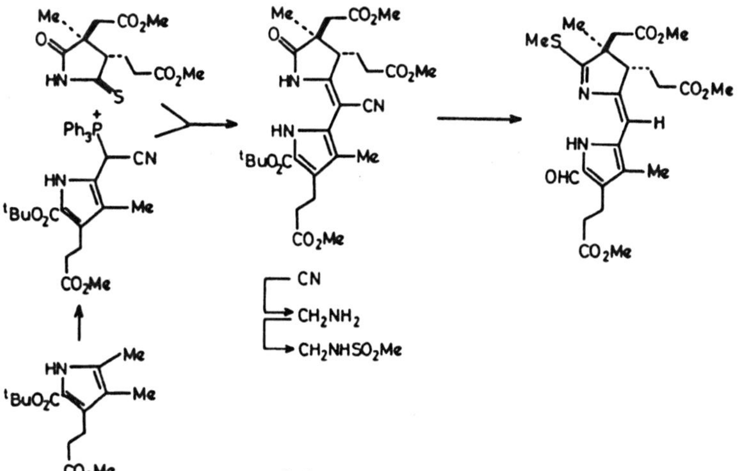

Scheme 13

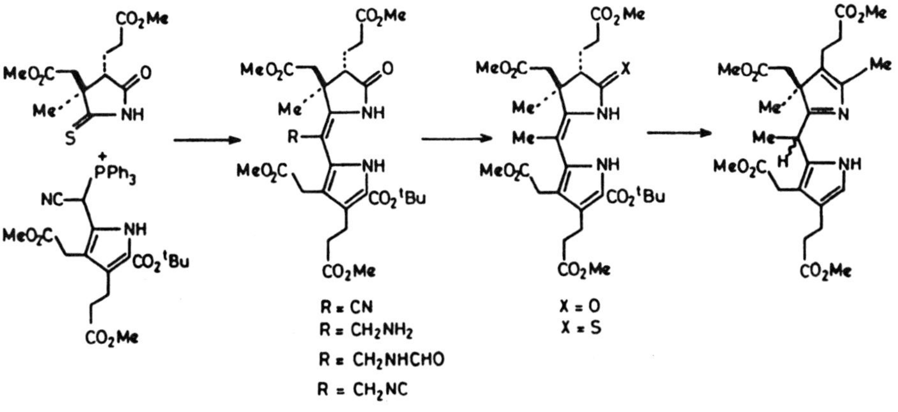

Scheme 14

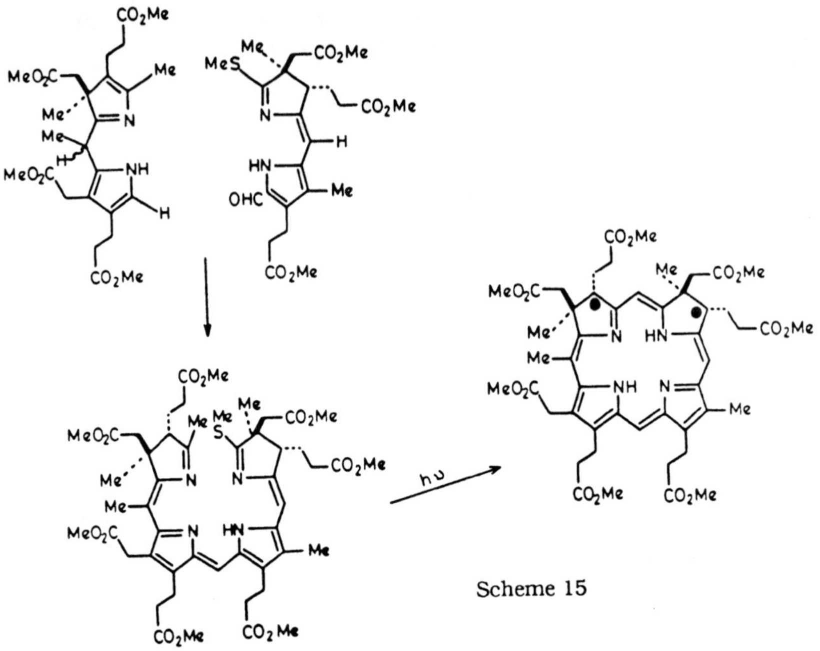

Scheme 15

BIOSYNTHESIS OF VITAMIN B_{12}: EXPERIMENTS ON PYRROCORPHINS

One of the driving forces for the foregoing synthetic work on isobacteriochlorins was that we aimed to prepare from them the putative pyrrocorphin intermediate(s) referred to earlier. Scheme 16 shows one example of the outstanding work of Eschenmoser *et al*[56] in which a dihydroisobacteriochlorin was isomerised and *C*-methylated with preference for C-17. We planned to extend this chemistry eventually into the natural series. There are, however, two major problems to overcome before this approach can be successfully used to prepare, for example, the pyrrocorphin with R = CH_2CO_2Me in Scheme 17. Firstly, the conditions involving a Grignard reagent in Scheme 16 cannot be used in the presence of ester side-chains and secondly, precorrin-3 ester is chiral, so stereochemical problems are added to those of controlling the regio-chemistry; 16 isomers can arise.

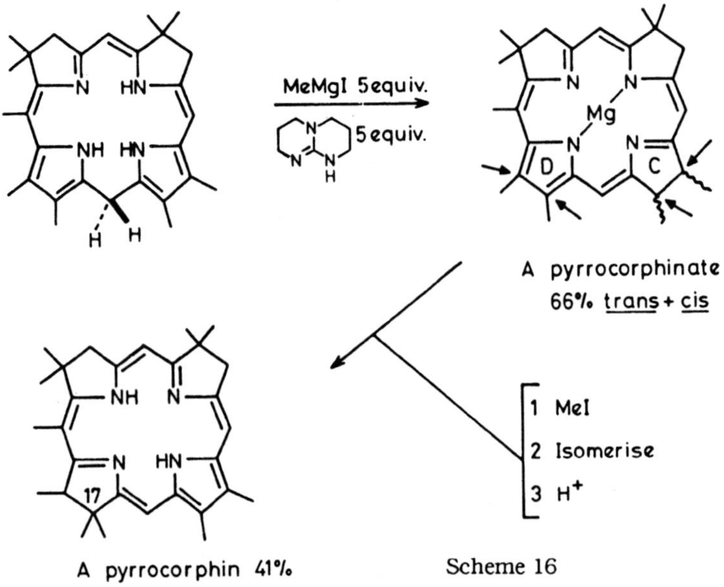

Scheme 16

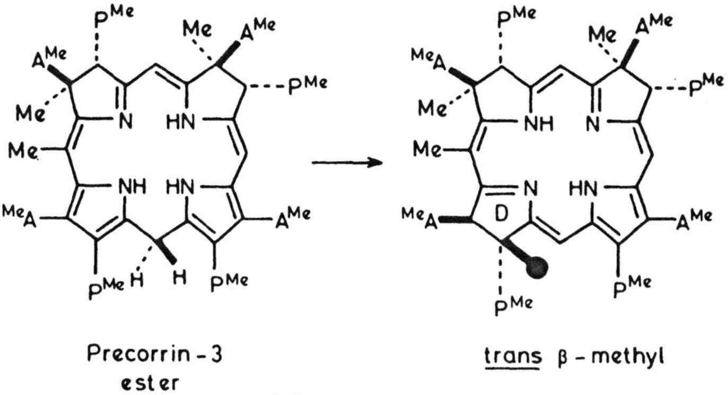

Scheme 17

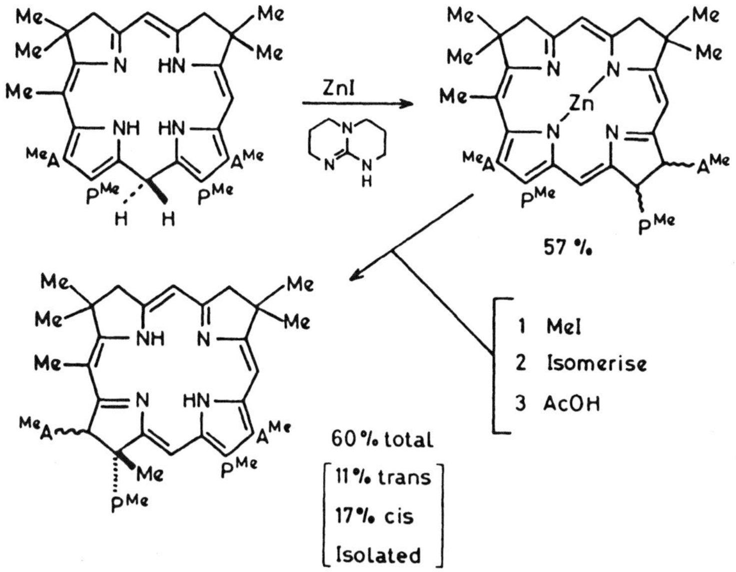

Scheme 18

Conditions were eventually found to allow pyrrocorphins with ester side-chains to be prepared[57] and these were applied to the C-20 methylated system shown in Scheme 18. The major products were *C*-methylated at C-17 and the *cis*- and *trans*-pyrrocorphins were isolated together with isomers methylated at C-13. All these structures were rigorously established[58] by the combined use of uv-visible and n.m.r. spectroscopy (especially n.O.e. difference methods) in combination with mass spectrometry.

These successes opened the way for us to apply the same methods to the ester of precorrin-3, Scheme 17. Indeed, this has now been done[59] but that story makes almost another complete lecture and cannot be covered today.

Researches on the biosynthesis of vitamin B_{12} are now at a most fascinating stage. As you will have seen from this lecture, the necessary skills in labelling and spectroscopy, enzymology, molecular biology and genetics, synthesis and structure determination have all been finely honed. Added to that is the knowledge we now have of how to handle successfully such labile substances as dihydroisobacteriochlorins and pyrrocorphins with acetate and propionate side chains. The coming years should be exciting ones.

ACKNOWLEDGEMENTS

I have been extremely fortunate during the last 22 years of research in this area to work with a continuing series of courageous and imaginative young scientists. The entire list is too long to give here but all the names of those from about the last 12 years are given in the reference list, the senior Cambridge colleagues during that period being Drs. Chris Abell, Chris Fookes, Graham Hart, Finian Leeper and George Matcham and I also want to include our senior French colleagues, Francis Blanche and Joel Crouzet. These in turn have stood on the shoulders of many earlier group members and we must not forget their crucial contributions. Only by this joint effort was it possible to make the progress I have outlined and I am glad to have this chance to record my great debt to all my colleagues.

REFERENCES

1. For reviews see A. R. Battersby and E. McDonald in "Vitamin B_{12}", ed. D. Dolphin, Wiley, New York, p. 107 (1982); A. R. Battersby, Biosynthesis of Vitamin B_{12}. *Accs. Chem. Res.*, *19*, 147 (1986); A. R. Battersby, Synthetic and Biosynthetic Studies on Vitamin B_{12}. *J. Nat. Prod.*, *51*, 643 (1988).
2. A. R. Battersby, C. J. R. Fookes, G. W. J. Matcham and E. McDonald, Biosynthesis of the Pigments of Life: Formation of the Macrocycle, *Nature*, *285*, 17, (1980).
3. A. R. Battersby, E. Hunt and E. McDonald, Biosynthesis of Type-III Porphyrins: Nature of the Rearrangement Process, *J. Chem. Soc., Chem. Commun.*, 442 (1973); A. R. Battersby, G. L. Hodgson, E. Hunt, E. McDonald and J. Saunders, Nature of the Rearrangement Process Leading to the Natural Type III Porphyrins, *J. Chem. Soc., Perkin Trans. 1*, 273 (1976).
4. A. R. Battersby, C. J. R. Fookes, M. J. Meegan, E. McDonald and H. K. W. Wurziger, Proof that the Single Intramolecular Rearrangement Leading to Natural Porphyrins (Type-III) occurs at the Tetrapyrrole Level, *J. Chem. Soc., Perkin Trans. 1*, 2786 (1981); A. R. Battersby, C. J. R. Fookes, K. E. Gustafson-Potter, E. McDonald and G. W. J. Matcham, Chemical and Enzymic Transformation of Isomeric Aminomethylbilanes into Uroporphyrinogens: Proof that Unrearranged Bilane is the Preferred Enzymic Substrate and Detection of a Transient Intermediate, *J. Chem. Soc., Perkin Trans. 1*, 2413 (1982).
5. A. R. Battersby, C. J. R. Fookes, K. E. Gustafson-Potter, E. McDonald and G. W. J. Matcham, Proof by Spectroscopy and Synthesis that Unrearranged Hydroxymethylbilane is the Product from Deaminase and the Substrate for Cosynthetase in the Biosynthesis of Uroporphyrinogen-III, *J. Chem. Soc., Perkin Trans. 1*, 2427 (1982).
6. A. R. Battersby, C. J. R. Fookes, G. W. J. Matcham and E. McDonald, Order of Assembly of the Four Pyrrole Rings during Biosynthesis of the Natural Porphyrins, *J. Chem. Soc., Chem. Commun.*, 539 (1979).
7. A. R. Battersby, C. J. R. Fookes, G. W. J. Matcham, E. McDonald and R. Hollenstein, Purification of Deaminase and Studies on its Mode of Action, *J. Chem. Soc., Perkin Trans. 1*, 3031 (1983).
8. A. R. Battersby, C. J. R. Fookes, G. J. Hart, G. W. J. Matcham and P. S. Pandey, The Interaction of Deaminase and its Product (Hydroxymethylbilane) and the Relationship between Deaminase and Cosynthetase, *J. Chem. Soc., Perkin Trans. 1*, 3041 (1983).
9. P. M. Jordan and A. Berry, Mechanism of action of porphobilinogen deaminase, *Biochem. J.*, *195*, 177 (1981).
10. R. Radmer and L. Bogorad, A Tetrapyrrylmethane Intermediate in the Enzymatic Synthesis of Uroporphyrinogen, *Biochemistry*, *11*, 904 (1972); R. C. Davies and A. Neuberger, Polypyrroles Formed from Porphobilinogen and Amines by Uroporphyrinogen Synthetase of *Rhodopseudomonas spheroides*, *133*, 471 (1973).

11. For recent instances see A. I. Scott, C. A. Roessner, N. J. Stolowich, P. Karuso, H. J. Williams, S. K. Grant, M. D. Gonzalez and T. Hoshino, Site-Directed Mutagenesis and High-Resolution NMR Spectroscopy of the Active Site of Porphobilinogen Deaminase, *Biochemistry*, *27*, 7984 (1988); S. Rosé, R. B. Frydman, C. de los Santos, A. Sburlati, A. Valasinas and B. Frydman, Spectroscopic Evidence for a Porphobilinogen Deaminase-Tetrapyrrole Complex that is an Intermediate in the Biosynthesis of Uroporphyrinogen III, *Biochemistry*, *27*, 4871 (1988).

12. J.-R. Schauder, S. Jendrzjewski, A. Abell, G. J. Hart and A. R. Battersby, , Stereochemistry of Formation of the Hydroxymethyl Group of Hydroxymethylbilane, the Precursor of Uro'gen-III, *J. Chem. Soc., Chem. Commun.*, 436 (1987).

13. W. Neidhart, P. C. Anderson, G. J. Hart and A. R. Battersby, Synthesis of (11*S*)- and (11*R*)-[11-$^2H_1$]Porphobilinogen; Stereochemical Studies on Hydroxymethylbilane Synthase (PBG Deaminase), *J. Chem. Soc., Chem.Commun.*, 924 (1985).

14. A. R. Battersby, Biosynthesis of the Pigments of Life, *Ann. N. Y. Acad. Sci.*, Vol. 471, p. 138 (1986).

15. A. D. Miller, F. J. Leeper and A. R. Battersby, Synthesis and Properties of *S*-Pyrrolylmethylcysteinyl and ε-*N*-Pyrrolylmethyllysyl Peptides, *J. Chem. Soc. Perkin Trans. 1*, 1943 (1989).

16. P. M. Anderson and R. J. Desnick, Purification and Properties of Uroporphyrinogen I Synthase from Human Erythrocytes, *J. Biol. Chem.*, *255*, 1993 (1980).

17. S. D. Thomas and P. J. Jordan, Nucleotide Sequence of a Hem-C locus encoding porphobilinogen deaminase of *Escherichia Coli* K-12, *Nucleic Acids Res.*, *14*, 6215 (1986).

18. A. Sasarman, A. Nepveu, Y. Echelard, J. Dymetryszyn, M. Drolet and C. Goyer, Molecular Cloning and Sequencing of the Hem-D Gene of *Escherichia-Coli* K-12 and Preliminary Data on the Uro Operon, *J. Bacteriol.*, *169*, 4257 (1987).

19. P. R. Alefounder, C. Abell and A. R. Battersby, The sequence of *hem*C, *hem*D and two additional *E. coli* genes, *Nucleic Acids Res.*, *16*, 9871 (1988).

20. A. D. Miller, L. C. Packman, G. J. Hart, P. R. Alefounder, C. Abell, and A. R. Battersby, Evidence that pyridoxal phosphate modification of lysine residues (Lys-55 and Lys-59) causes inactivation of hydroxymethylbilane synthase (porphobilinogen deaminase), *Biochem. J.*, *262*, 119 (1989).

21. G. J. Hart, A. D. Miller, F. J. Leeper and A. R. Battersby, Proof that Hydroxymethylbilane Synthase (PBG Deaminase) uses a Novel Binding Group in its Catalytic Action, *J. Chem. Soc., Chem. Commun.*, 1762 (1987).

22. G. J. Hart, A. D. Miller, U. Beifuss, F. J. Leeper and A. R. Battersby, Discovery of a Novel Dipyrrolic Cofactor Essential for the Catalytic Action of Hydroxymethylbilane Synthase (PBG Deaminase), *J. Chem. Soc. Perkin Trans. 1* (1990), in press.

23. J. Pluscec and L. Bogorad, A Dipyrrylmethane Intermediate in the Enzymatic Synthesis of Uroporphyrinogen, *Biochemistry*, *9*, 4736 (1970).

24. G. J. Hart A. D. Miller and A. R. Battersby, Evidence that the pyrromethane cofactor of hydroxymethylbilane synthase (PBG deaminase) is bound through the sulphur atom of a cysteine residue, *Biochem. J.*, *252*, 909 (1988).

25. U. Beifuss, G. J. Hart, A. D. Miller and A. R. Battersby, ^{13}C-N.M.R. Studies on the Pyrromethane Cofactor of Hydroxymethylbilane Synthase, *Tetrahedron Letts.*, 2591 (1988).

26. A. D. Miller, G. J. Hart, L. C. Packman,and A. R. Battersby, Evidence that the pyrromethane cofactor of hydroxymethylbilane synthase (porphobilinogen deaminase) is bound to the protein through the sulphur atom of cysteine-242, *Biochem. J.*, *254*, 915 (1988).

27. P. R. Alefounder, G. J. Hart, A. D. Miller, U. Beifuss, C. Abell, F. J. Leeper and A. R. Battersby, Biosynthesis of the Pigments of Life: Structure and Mode of Action of a Novel Enzymatic Cofactor, *Bioorganic Chemistry*, *17*, 121 (1989).

28. P. M. Jordan and M. J. Warren, Evidence for a Dipyrromethane Cofactor at the Catalytic Site of *Escherichia-Coli* Porphobilinogen Deaminase, *FEBS Lett.*, *225*, 87 (1987).

29. P. M. Jordan, S. D. Thomas, and M. J. Warren, Purification Crystallization and Properties of Porphobilinogen Deaminase from a Recombinant Strain of *Escherichia-Coli* K12, *Biochem. J.*, *254*, 427 (1988).

30. P. M. Jordan, M. J. Warren, H. J. Williams, N. J. Stolowich, C. A. Roessner, S. K. Grant, and A. I. Scott, Identification of a Cysteine Residue as the Binding Site for the Dipyrromethane Cofactor at the Active Site of *Escherichia-Coli* Porphobilinogen Deaminase, *FEBS Lett.*, *235*, 189 (1988).

31 A. I. Scott, N. J. Stolowich, H. J. Williams, M. D. Gonzalez, C. A. Roessner, S. K. Grant, and C. Pichon, Concerning the Catalytic Site of Porphobilinogen Deaminase, *J. Amer. Chem. Soc.*, *110*, 5898 (1988); A. I. Scott, C. A. Roessner, N. J. Stolowich, P. Karuso, H. J. Williams, S. K. Grant, M. D. Gonzalez, and T. Hoshino, Site-Directed Mutagenesis and High-Resolution NMR Spectroscopy of the Active Site of Porphobilinogen Deaminase, *Biochemistry*, *27*, 7984 (1988).

32. J. H. Mathewson and A. H. Corwin, Biosynthesis of Pyrrole Pigments: A Mechanism for Porphobilinogen Polymerization, *J. Amer. Chem. Soc.*, *83*, 135 (1961).

33. W. M. Stark, M. G. Baker, P. R. Raithby, F. J. Leeper and A. R. Battersby, The Spiro Intermediate Proposed for Biosynthesis of the Natural Porphyrins: Synthesis and Properties of Its Macrocycle, *J. Chem. Soc., Chem. Commun.*, 1294 (1985); W. M. Stark, M. G. Baker, F. J. Leeper, P. R. Raithby and A. R. Battersby, Synthesis of the Macrocycle of the Spiro System Proposed as an Intermediate Generated by Cosynthetase., *J. Chem. Soc. Perkin Trans. 1*, 1187 (1988).

34. W. M. Stark, G. J. Hart and A. R. Battersby, Synthetic Studies on the Proposed Spiro Intermediate for Biosynthesis of the Natural Porphyrins: Inhibition of Cosynthetase, *J. Chem. Soc., Chem. Commun.*, 465 (1986).

35. A. I. Scott, C. A. Townsend, K. Okada and M. Kajiwara, Biosynthesis of Corrins. I. Experiments with [^{14}C]Porphobilinogen and [^{14}C]Uroporphyrinogens, *J. Amer. Chem. Soc.*, *96*, 8054 (1974); A. I. Scott, N. Georgopapadakou, K. S. Ho, S. Klioze, E. Lee, S. L. Lee, G. H. Jemme III, C. A. Townsend and I. A. Armitage, Concerning the Intermediacy of Uro'gen III and of a Heptacarboxylic Uro'gen in Corrinoid Biosynthesis, *ibid*, *97*, 2548 (1975).

36. A. R. Battersby, M. Ihara, E. McDonald, F. Satoh and D. C. Williams, Derivation of Cobyrinic Acid from Uroporphyrinogen-III, *J. Chem. Soc., Chem. Commun.*, 436 (1975); A. R. Battersby, E. McDonald, R. Hollenstein, M. Corresponding Ring-C Methyl Heptacarboxylic Porphyrinogen and Proof of Seven Intact Methyl Transfers, *J. Chem. Soc. Perkin 1*, 166 (1977)..

37. H.-O. Dauner and G. Müller, Bildung von Cobyrin saure Mittels eines Zelltreien System aus Clostridium Tetanomorphum, *Z. physiol. Chem.*, *356*, 1353 (1975).

38. A. R. Battersby, K. Jones, E. McDonald, J. A. Robinson and H. R. Morris, The Structure and Chemistry of Isobacteriochlorins from *Desulphovibrio gigas*, *Tetrahedron Letts.*, 2213 (1977); A. R. Battersby, E. McDonald, H. R. Morris, M. Thompson, D. C. Williams, V. Ya Bykhovsky, N. Zaitseva and V. Bukin, Biosynthesis of Vitamin B_{12}: Structural Studies on the Corriphyrins from *Propionibacterium shermanii* and the Link with Sirohydrochlorin, *Tetrahedron Letts.*, 2217 (1977).

39. A. R. Battersby, E. McDonald, M.Thompson and V. Ya Bykhovsky, Proof of A-B structure for Sirohydrochlorin by its Specific Incorporation into Cobyrinic Acid, *J. Chem. Soc., Chem. Commun.*, 150 (1978).

40. A. R. Battersby, G. W. J. Matcham, E. McDonald, R. Neier, M. Thompson, W.-D. Woggon, V. Ya Bykovsky and H. R. Morris, Structure of the Trimethylisobacteriochlorin from *Propionibacterium shermanii*, *J. Chem. Soc., Chem. Commun.*, 185 (1979); N. G. Lewis, R. Neier, G. W. J. Matcham, E. McDonald and A. R. Battersby, Vitamin B_{12}: Experiments on Loss of C-20 from the Precursor Macrocycle, *J. Chem. Soc., Chem. Commun.*, 541 (1979).
41. A. R. Battersby, K. Frobel, F. Hammerschmidt and C. Jones, Isolation of 15,23-Dihydrosirohydrochlorin, a Biosynthetic Intermediate: Structural Studies and Incorporation Experiments, *J. Chem. Soc., Chem. Commun.*, 455 (1982).
42. L. Mombelli, C. Nussbaumer, H. Weber, G. Müller and D. Arigoni, Biosynthesis of vitamin B_{12}: Mode of incorporation of factor III into cobyrinic acid, *Proc. Natl. Acad. Sci. U.S.A.*, *78*, 11(1981).
43. A. R. Battersby, M. J. Bushell, C. Jones, N. G. Lewis and A. Pfenninger, Identity of Fragment Extruded During Ring Contraction to the Corrin Macrocycle, *Proc. Natl. Acad. Sci. U.S.A.*, *78*, 13 (1981).
44. M. Imfeld, D. Arigoni, R. Deeg and G. Müller, in "Vitamin B_{12}", Factor I ex *Clostridium tetanomorphum*: Proof of Structure and Relationship to Vitamin B_{12} Biosynthesis, eds. B. J. Zagalak and W. Friedrich, de Gruyter, Berlin, p. 315 (1979).
45. A. R. Battersby, R. D. Brunt, F. J. Leeper and I. Grgurina,Vitamin B_{12}: Synthesis of (±)-[5-^{13}C]Faktor-1 Ester: Determination of the Oxidation State of Precorrin-1, *J. Chem. Soc., Chem. Commun.*, 428 (1989).
46. H. C. Uzar and A. R. Battersby,Vitamin B_{12}: Pulse Labelling to Locate the Fourth Methylation Site, *J. Chem. Soc., Chem. Commun.*, 1204 (1982).
47. H. C. Uzar and A. R. Battersby, Vitamin B_{12}: Order of the Later *C*-Methylation Steps, *J. Chem. Soc., Chem. Commun.*, 585 (1985); A. R. Battersby, H. C. Uzar, T. A. Carpenter and F. J. Leeper, Development of a Pulse Labelling Method to Determine the *C*-Methylation Sequence for Vitamin B_{12}, *J. Chem. Soc., Perkin Trans. 1*, 1689 (1987).
48. (*a*) A. I. Scott, N. E. Mackenzie, P. J. Santander, P. E. Fagerness, G. Müller, E. Schneider, R. Sedlmeier and G. Wörner, Biosynthesis of Vitamin B_{12}: Timing of the Methylation steps between Uro'gen III and Cobyrinic Acid, *Bioorg. Chem.*, *12*, 356(1984); (*b*) A. I. Scott, H. J. Williams, N. J. Stolowich, P. Karuso and M. D. Gonzalez, Temporal Resolution of the Methylation Sequence of Vitamin B_{12} Biosynthesis, *J. Amer. Chem. Soc.*, *111*, 1897 (1989).
49. A. R. Battersby, F. Blanche, S. Handa, D. Thibaut, C. L. Gibson and F. J. Leeper, Vitamin B_{12}: When is the 12β-Methyl Group of the Vitamin Generated by Acetate Decarboxylation?, *J. Chem. Soc., Chem. Commun.*, 1117 (1988).
50. S. P. D. Turner, M. H. Block, Z.-C. Sheng, S. C. Zimmerman, and A. R. Battersby, Synthetic Studies on Vitamin B_{12}. Synthesis of (±)-Faktor-1 Octamethyl Ester, *J. Chem. Soc. Perkin Trans. 1*, 1577 (1988).
51. M. H. Block, S. C. Zimmerman, G. B. Henderson, S. P. D. Turner, S. W. Westwood , F. J. Leeper and A. R. Battersby, Synthesis of Sirohydrochlorin and of its Octamethyl Ester, *J. Chem. Soc., Chem. Commun.*, 1061 (1985).
52. W. G. Whittingham, M. K. Ellis, P. Guerry, G. B. Henderson, B. Müller, D. A. Taylor, F. J. Leeper and A. R. Battersby, Syntheses Relevant to Vitamin B_{12} Biosynthesis: Synthesis of the 2,7,20-Trimethylisobacteriochlorin, *J. Chem. Soc., Chem. Commun.*, 1116 (1989).
53. B. Müller, A. N. Collins, M. K. Ellis, W. G. Whittingham, F. J. Leeper and A. R. Battersby, Syntheses Relevant to Vitamin B_{12} Biosynthesis: Synthesis of the 2,7,12,20-Tetramethylisobacteriochlorin, *J. Chem. Soc., Chem. Commun.*, 1119 (1989).

54. P. J. Harrison, Z.-C. Sheng, C. J. R. Fookes, and A. R. Battersby, Development of the Photochemical Route to Isobacteriochlorins, *J. Chem. Soc. Perkin Trans. 1*, 1667 (1987).
55. K. Tomioka, Y.-S. Cho, F. Sato and K. Koga, Highly Stereoselective Construction of Chiral Quaternary Carbon: Asymmetric Synthesis of β,,β-Disubstituted γ-Butyrolactones, *Chem. Lett.*, 1621 (1981).
56. *Inter alia*, C. Leumann, K. Hilpert, J. Schreiber and A. Eschenmoser, Chemistry of Pyrrocorphins: *C*-Methylations at the Periphery of Pyrrocorphins and Related Corphinoid Ligand Systems, *J. Chem. Soc., Chem. Commun.*, 1404 (1983).
57. C. L. Gibson and A. R. Battersby, Biosynthesis of Vitamin B_{12}: Synthesis and Peripheral *C*-Methylation of Pyrrocorphins Carrying Ester Side Chains, *J. Chem. Soc., Chem. Commun.*, 590 (1989).
58. C. L. Gibson and A. R. Battersby, Biosynthesis of Vitamin B_{12}: Regio-control in Peripheral *C*-Methylation of 20-Methylpyrrocorphins carrying Ester Side Chains, *J. Chem. Soc., Chem. Commun.*, 1223 (1989).
59. C. L. Gibson, F. Blanche and A. R. Battersby, unpublished work, Cambridge, 1988.

ON THE METHYLATION PROCESS AND COBALT INSERTION IN COBYRINIC ACID BIOSYNTHESIS

G. Müller, K. Hlineny, E. Savvidis, F. Zipfel, J. Schmiedl and E. Schneider

Institut für Organische Chemie, Biochemie and Isotopenforschung
Universität Stuttgart, Azenbergstrasse 18, D-7000 Stuttgart 1
Federal Republic of Germany

Following the demonstration by Bernhauer and co-workers that vitamin B_{12} is formed from cobyrinic acid, which acts as the late tetrapyrrolic precursor[1], our studies have been concerned with the problem of how cobyrinic acid is biosynthesized using the following approaches: (1) Seeking and trapping of intermediates from radioactively labeled precursors. (2) Investigations on the utilization of the trapped and labeled compounds for cobyrinic acid biosynthesis. (3) Structural elucidation of the trapped intermediates. (4) Verification of single reactions of established steps with the appropriate purified enzymes. The biological systems (i.e. intact cells, cell-free extracts and the partially purified enzyme systems thereof which can synthesize cobyrinic acid) necessary to perform suitable experiments on this theme were prepared from the corrinoid producers *Propionibacterium shermanii* and *Clostridium tetanomorphum*.

After the establishment of the intermediacy and overall conversion of uroporphyrinogen III to cobyrinic acid[2-4] it was shown that C-methylation rather than the acetate decarboxylation at C_{12} is the next step on the cobyrinic acid pathway.[5] This, as well as the first part of the methylation sequence, was clearly demonstrated by the detection and biochemical conversion of the aromatized mono-, di- and trimethylated intermediates to cobyrinic acid. These so-called factors I, II and III (corresponding to methylation at C_2, C_7 and C_{20}, respectively) were trapped from cobyrinic acid forming systems by cobalt free incubations and with the special aid of the [^3H/^{14}C]- double labeling technique using ^3H-labeled Met or SAM + ^{14}C-labeled ALA (or vice versa).[5,7,16b] Meanwhile it became generally accepted and established (in part)[5c,8-10] that the true intermediates are the **reduced** forms - the so-called precorrins-1, -2 and -3[11,12] - and that no net change of oxidation level is associated with the overall transformation of uroporphyrinogen III to cobyrinic acid, a

Chemical Aspects of Enzyme Biotechnology
Edited by T.O. Baldwin *et al.*, Plenum Press, New York, 1990

result in accord with the extensive model studies by Eschenmoser and co-workers[13-15] as well as the outcome of studies on the C_{20}- elimination where it was clearly shown that C_{20} and its attached methyl group are released as a 2 carbon fragment which was trapped as acetic acid[16-18] (Scheme 1).

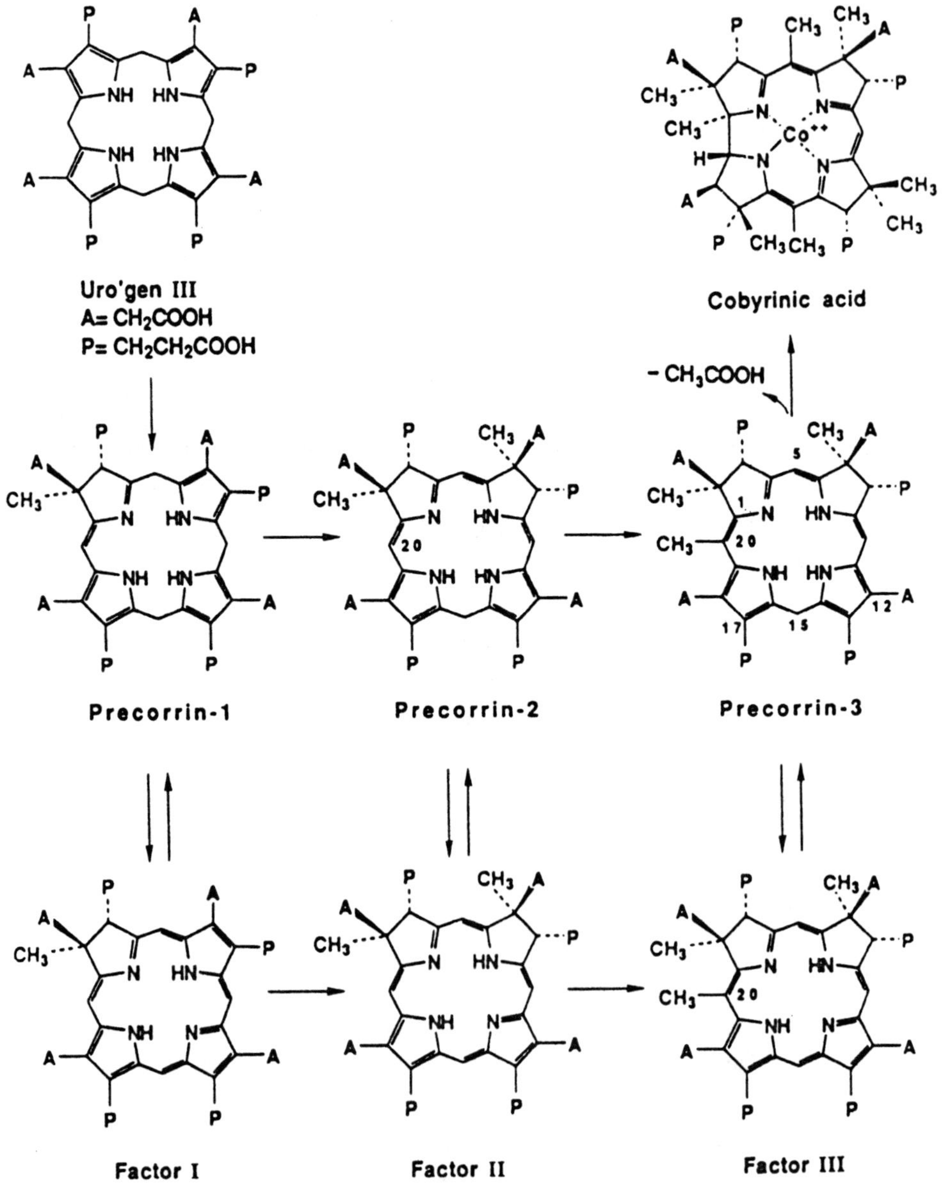

Scheme 1: Intermediates and aromatized isolates of cobyrinic acid pathway

The SAM dependent methylase from *P. shermanii* which acts on uroporphyrinogen III (SAM : Uroporphyrinogen III - methyltransferase) was enriched 70-fold. This preparation was used for the first examinations[5c,19,20] which led to the following findings: (1) Only one enzyme is responsible for C_2 - and C_7 methylation - the product is not factor II but a reduced form of it (now called precorrin-2[11,12]). (2) The protein acts on a reduced form of factor I (NaHg-reduction) forming precorrin-2 - isolated as factor II - but not on factor II or a reduced form of this. (3) It also realizes two methylations on uroporphyrinogen I as well as on C_{12} - decarboxylated uroporphyrinogen III leading to the corresponding isobacteriochlorins but does not act on coproporphyrinogens.

Since it was expected that C_{12}-acetate decarboxylation is the next step after C_{20}-methylation C_{12}-decarboxylated factor II was examined extensively, especially its chromatographic behaviour in comparison with the isobacteriochlorins factors II and III. That knowledge should have helped in the detection of the corresponding trimethylated and decarboxylated isobacteriochlorin. However using cobalt-free (but cobyrinic acid forming) systems from *P. shermanii* and *C. tetanomorphum* and the well established $^{3}H/^{14}C$-technique we failed to trap this hypothetical intermediate. Only recently it has been clearly demonstrated that C_{12} - acetate-decarboxylation does not directly follow C_{20} - methylation on the pathway[20-22] (see below).

These summarized results go back to the late seventies and early eighties at a time when the studies for elucidation of the pathway of corrin biosynthesis had been so promising. Thereafter much unsuccessful effort has been spent in the search for further intermediates. Even the realization of enzymatic reactions for the same purpose has not been possible because of the missing substrates. Meanwhile hypothetical and available substrates were checked for their intermediacy.

In the late seventies hydrogenobyrinic acid - a, c - diamide from *Rhodopseudomonas spheroides* became available with the help of a procedure described by Koppenhagen.[23] Our first experiments with this cobalt-free cobyrinate were concerned with examinations on enzymatic cobalt insertion into this diamide using $[^{57}Co^{2+}]$ and the cobyrinic acid forming systems from *P. shermanii* and *C. tetanomorphum*.[24] The failure to demonstrate the expected enzymatic cobalt incorporation led to the transformation of hydrogenobyrinic acid - a, c- diamid into hydrogenobyrinic acid-heptamethylester.[25] A chemical conversion turned out well by methanolysis using BF_3 / MeOH. The extensively analysed product (Uv/Vis, Fluorescence, MS, ^{1}H-NMR, CD, Chemical cobalt insertion) was hydrolysed then and used for enzymatic studies.

Unexpectedly the cobyrinic acid synthesizing cell extracts of *P. shermanii* did not even catalyze a cobalt-insertion into hydrogenobyrinic acid. But unlike the a, c - diamide the hydrogenobyrinic acid by itself inhibited the formation of cobyrinic acid from the well known substrates.[25]

In 1983 Arigoni and co-worker[26] described the preparation of norcorrinoids which are demethylated at positions 5 and 15 in cobyrinic acid and the dicyanocomplex of 5, 15 - bisnorcobyrinic acid which was tested for intermediacy.[27] Because of the negative result of this experiment we have prepared the cobaltmethyl-, cobalt-5'-deoxyadenosyl- and diaquo-analogues of the bis-norcobyrinate but none of these proved to be a biochemical precursor of cobyrinic acid in the *P. shermanii* system.[28]

^{13}C - pulse labeling experiments have helped then to elucidate the final methylation sequence on the way from precorrin-3 to cobyrinic acid[11,12,18,29] In these experiments the cobyrinic acid synthesizing system of *P. shermanii* containing Co^{2+} + substrate (e.g. factor III) received a pulse of [$^{12}CH_3$]SAM, followed at carefully varied time intervals by a second pulse of [$^{13}CH_3$]SAM (or vice versa). At the end of the defined overall incubation time enriched cobyrinic acid was isolated, converted to cobester[6] and the latter whose ^{13}C-NMR spectrum had been completely assigned previously[5b,18].was analysed by ^{13}C-NMR.

Following a large number of incubations for the purpose of optimization, highly reproducible conditions were attained and gave rise to a spectrum which reveals a clear distinction between the peak heights of the five methyl groups at $C_{17} < C_{12} < C_1 < C_5 < C_{15}$ (Fig. 1) when compared with the spectrum obtained by [$^{13}CH_3$]SAM addition at the very outset. Thus the overall methylation sequence can be described as $C_2 > C_7 > C_{20} > C_{17} > C_{12} > C_1 > C_5 > C_{15}$.[21] It has to be emphasized that with the *P. shermanii* systems the ^{13}C-pulsations which lead to the differentiations of the ^{13}C-methyl group signals only succeed when cobalt is present before pulsation.

This outcome has again brought the question into focus whether C_{12}-acetate decarboxylation happens before C_{17}-methlyation of factor III (and also the factor II analogue) with the cobyrinic acid synthesizing cell-free system of *P. shermanii.* For these experiments [5,15 - $^{14}C_2$] C_{12} - decarboxylated uroporphyrinogen III was converted with cell-free extracts of *P. shermanii* to the corresponding decarboxylated analogons of factors II and III (Scheme 2). Table 1 demonstrates the comparative incorporation experiments with the corrin-synthesizing system of *P. shermanii.* It is noteworthy that reisolated and decarboxylated [^{14}C] factor III which was incubated together with ^3H - labeled factor III (Expt. 4) did not show any ^3H - radioactivity. This is further evidence that the cobyrinic acid forming system does not catalyze the decarboxylation of precorrin-3 and that C_{12}-decarboxylated precorrin-3 does not appear on the pathway. So C_{12} - acetate decarboxylation happens at some stage after the fourth methylation (at C_{17}) but - because of mechanistic reasons - before the fifth methylation (at C_{12}).[21,22]

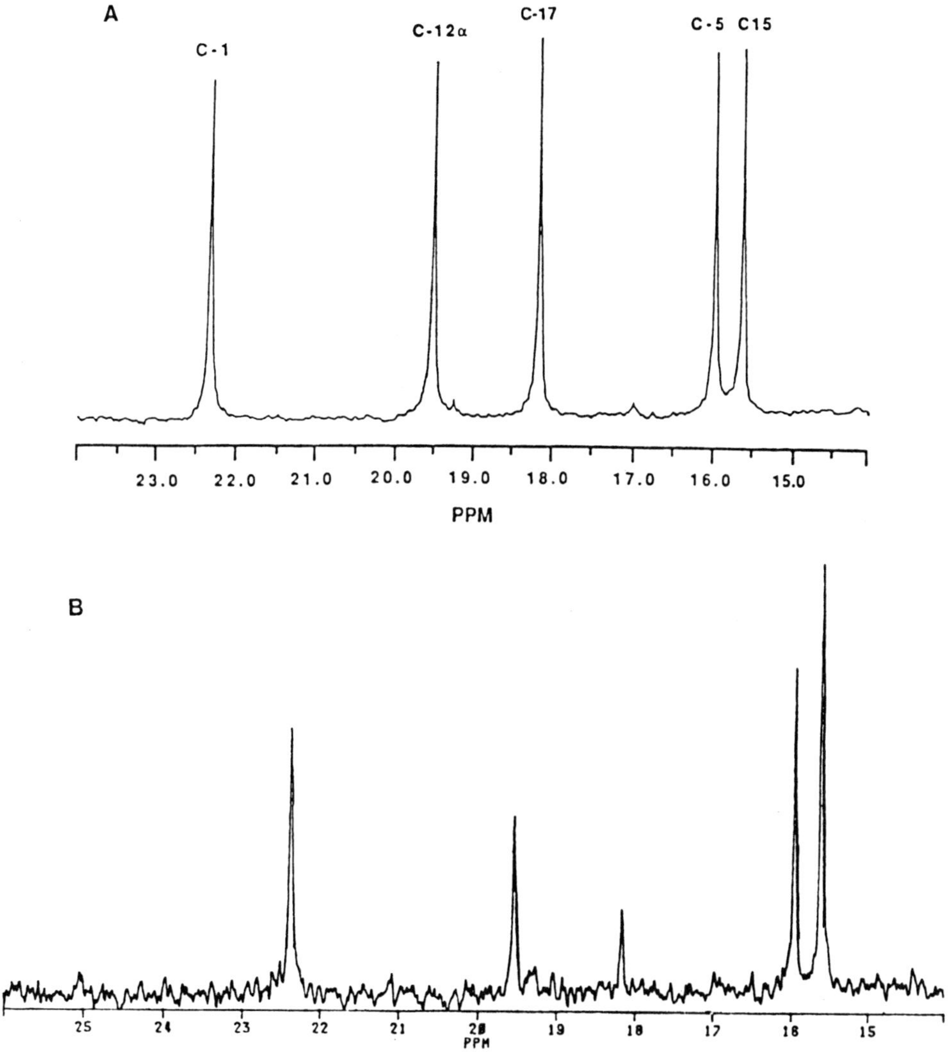

Figure 1 125 MHz ^{13}C - NMR spectra of cobester in 0.25 ml KCN - saturated benzene - d_6. Conversion of precorrin-2 into cobyrinic acid by cell extract of *P. shermanii.* A: Co^{2+} + $[^{13}CH_3]$SAM - addition immediately, 6 h, (100 μg). B: Co^{2+} immediately $[^{13}CH_3]$SAM after 4 h for 1,5 h (60 μg).

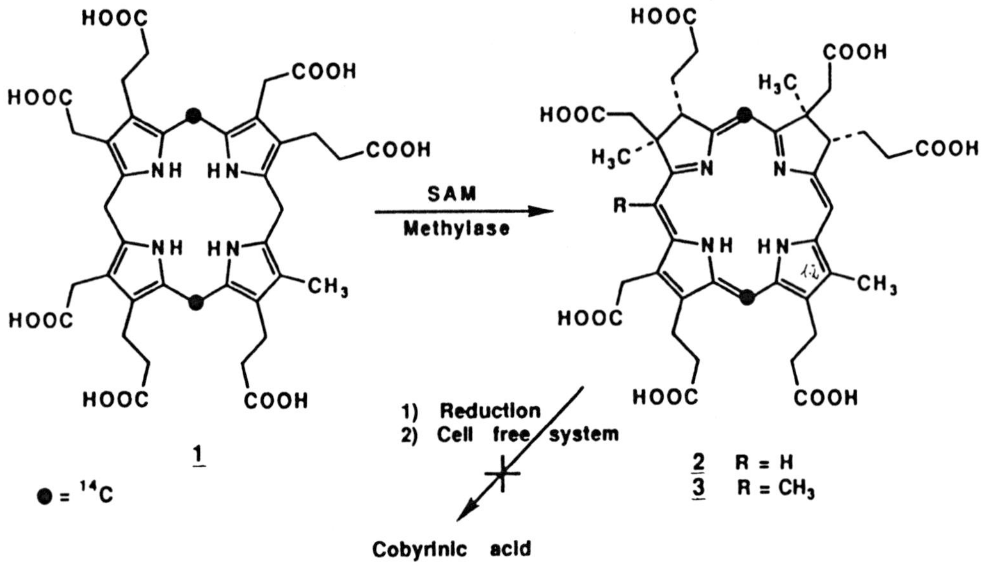

Scheme 2. Preparation[a] and conversion[b] of [5,15-$^{14}C_2$] C_{12}-acetate-decarboxylated factor III to cobyrinic acid. [a] Cell-free extract of 100 g cells + 12 µmol 1 + 40 µmol SAM, 30 °C, 30 h, yield: 900 nmol decarboxylated F II **2**, 330 nmol decarboxylated factor III **3**.

Table 1 Conversion of Factors II and III and their Ring C-Decarboxylated Analogons to Cobyrinic Acid by Cell Extracts of P. shermanii[h].

Expte.	substrates	nmol	dpm/nmol	cobyrinic acid[a] nmol[b]	dpm/nmol	nmol[c]
1	[^{14}C] factor II[d]	400	4 880	86	3 020	53
2	[^{14}C] decarboxy-factor II[e]	400	15 268	39	10	0
3	[^{14}C] decarboxy-factor III[e]	300	15 268	35	10	0
4	[^{3}H] factor III[f]	460	8 650	85	1 036	15
	+ [^{14}C] decarboxy-factor III[e,g]	516	15 268		10	o

[a] isolated and estimated as cobester, [b] cobyrinic acid is formed in all incubations from endogenous substrates, [c] based on radiochemical yield and specific radioactivity of substrate, [d] biosynthesized from [4-^{14}C] ALA, [e] biosynthesized from [5,15-$^{14}C_2$] C_{12}-decarboxyuroporphyrinogen III, [f] biosynthesized with [methyl-^{3}H] SAM, [g] reisolated [^{14}C] decarboxyfactor III did not contain any ^{3}H-radioactivity, [h] from 100 g cells in each case + 20 µmol SAM + 5 µmol Co^{2+}.

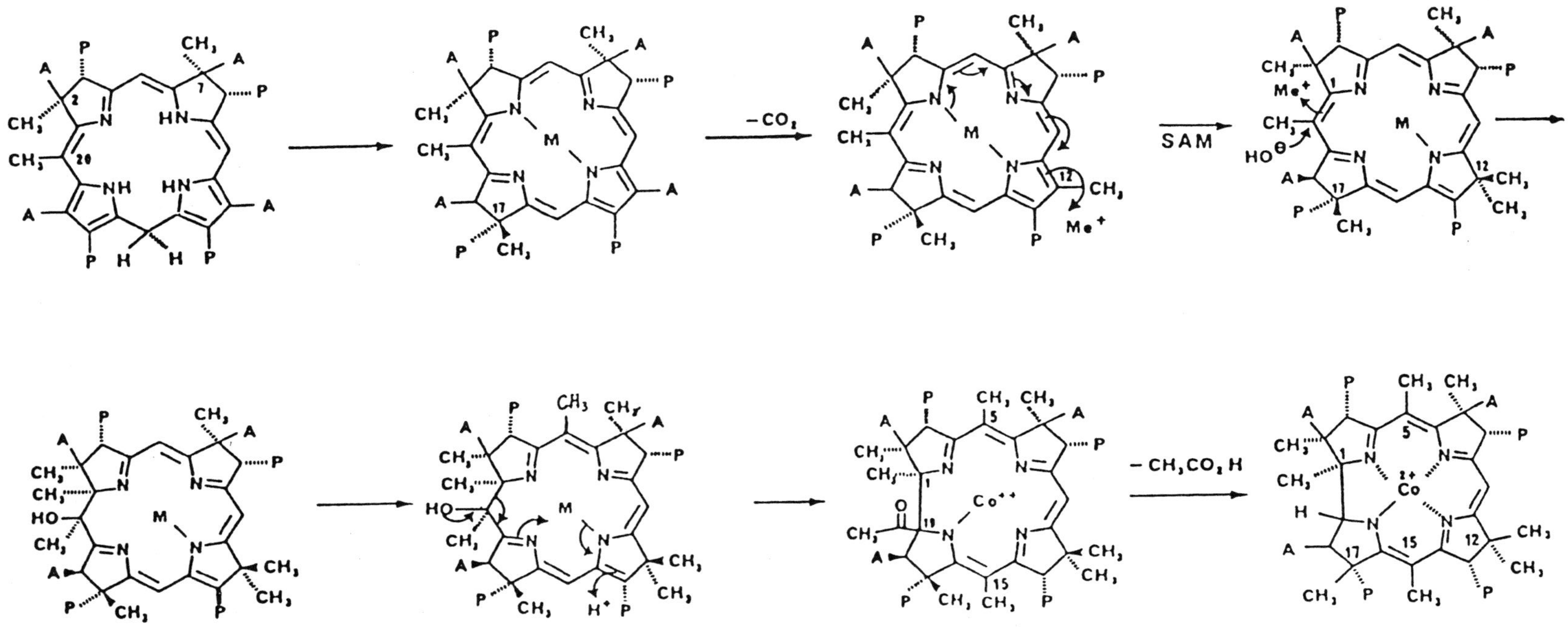

Cobyrinsäure

Scheme 3. Hypothetical steps on the way precorrin-3 → cobyrinic acid schedule research-planning

(M: unknown ion, cobalt or hydrogen)

On the basis of these and the foregoing results the scheduled structures of scheme 3 that are in accord with Eschenmoser's *in vitro* analogy[13-15] have been proposed as the targets of the search for the missing intermediates of corrin biosynthesis. The well established $^{3}H/^{14}C$ - double labeling technique should help again to find the hypothetical intermediates. Our aim was now the detection and isolation of pyrrocorphin - and corphin systems of the pathway, although the isolation of the C_{19}-acetyl - cobyrinate had been our previous aim.

Although hydrogenobyrinic acid and 5,15 -Bis-norcobyrinate are not involved in corrin biosynthesis we failed to isolate this hypothetical - C_{19} - acetylcobyrinate even from huge amounts of cobalt cultivated cells (1 kg) or cobyrinic acid synthesizing systems which have been fed with radioactive substrates. This was not expected since cobalt is essential for trapping the C_{20}-fragment.[17] It is remarkable that these experiments usually led to the isolation of 18, 19-didehydrocobyrinic acid-hexamethylester-C-amide.

Then we tried to stabilize and trap metal-free intermediates from cell-free systems and cell suspensions of *P. shermanii* by different cations such as Cu^{2+}, Ni^{2+}, Mn^{2+} and Zn^{2+}. Quite a few unknown tetrapyrrolic compounds have been detected and characterized. Most of these have not been of biosynthetic interest particularly because of the number of introduced methyl groups. Since C_2 - and C_7 - methylation is catalyzed by only one enzyme, it is remarkable that Cu^{2+} yields considerable amounts of copper complexed factor I but also factor I. Substantial amounts of Ni-II-factor II as well as another dimethylated chromophore containing nickel have been isolated from the corresponding preparations. Coproporphyrin appears complexed with manganese and cobalt. The application of Zn^{2+} then led to the first tetramethylated corphinates of natural origin[30,31] and of course these were expected to be complexed intermediates of the corrin-pathway.

Extensive chromatographic examinations with the methylesters have demonstrated the existence of two general classes of compounds and led to the separation of at least 5 compounds, so-called factors S_0 - S_4. All 5 compounds showed identical $^{3}H/^{14}C$ - ratios and mass spectra obtained from methyl - and ethylesters and the deuteriomethylated (via [CD_3] Met) series confirmed that they were isomeric, thus demonstrating the presence of at least 5 tetramethylated, hexahydroporphyrinic, octacarboxylic zinc chloride complexes. The two different chromophore classes show slightly different UV/Vis-spectra (Figure 2) but differ strongly in fluorescence. It has not been possible to convert one class of chromophore into the other. But the single chromophores seem to isomerize or tautomerise into one another within one general class.

It has been possible to demetalize (MeOH / HCl) and remetalize all the factors S within the mentioned isomerisation. The ^{3}H / ^{14}C - double labeled factors S as well as their metal-free ligands and also the cobalt complexes thereof have been used for incorporation experiments with cell-free extracts of *P. shermanii* in order to check their intermediacy on the cobyrinic acid pathway but these resulted in insignificant incorporation rates in all cases.

All NMR work has been carried out in the laboratory of Prof. A. I. Scott, Texas A&M University.

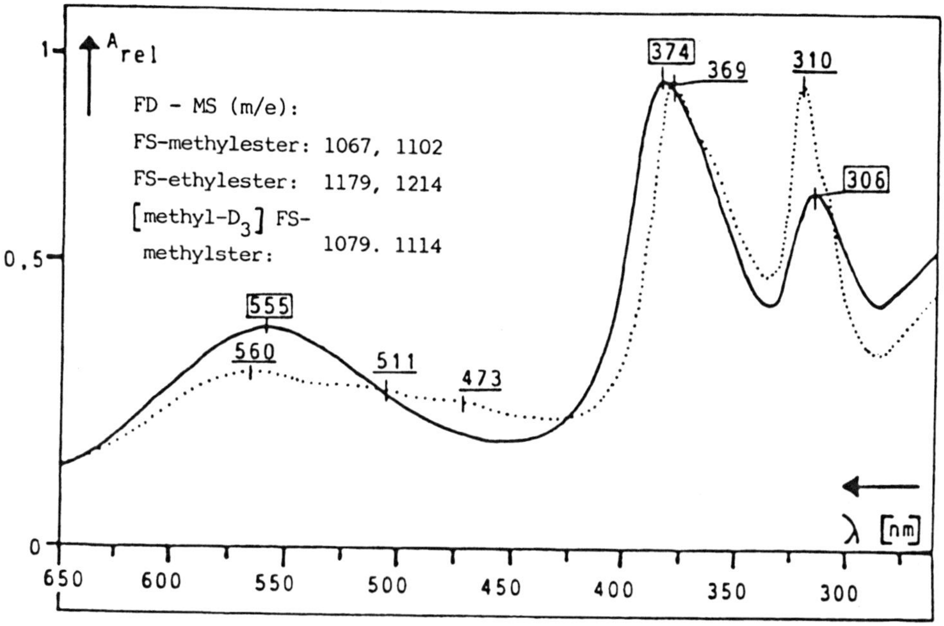

Figure 2 UV/Vis - spectrua of factors —— S_3 and · · · · S_1 -methylester in $CH_2 Cl_2$

^{13}C-NMR analyses with one compound of each class (S_3 and S_1) prepared from [5 - ^{13}C] ALA have clearly shown then that both compounds are derived from uroporphyrinogen I. Later the uro'gen I origin has been demonstrated for all 5 compounds enzymatically with the help of uro'gen I as substrate. The further structural elucidation, i.e. the deduction of the positions of the introduced methyl groups have been done with specimens (as methylesters) that were double-labeled with corresponding incubations and by the aid of $[^{13}CH_3]$ SAM as well as various samples of ALA with ^{13}C - labels at C_3, C_4 and C_5 respectively. Altogether a tremendous[30,31] work has led to the structures shown in scheme 4. In fact factor S_1 proved to be a typical corphinate that bears a methyl group at each β-carbon containing an acetic acid side chain. In factor S_3 only three of four methyl groups are situated at the β-positions of acetic acid sidechains but the fourth is introduced by an α-alkylation which is another mode of alkylation that happens in cobyrinic acid biosynthesis but different from that in the corrin system - this α-methylic group is adjacent to a meso-methylene carbon.

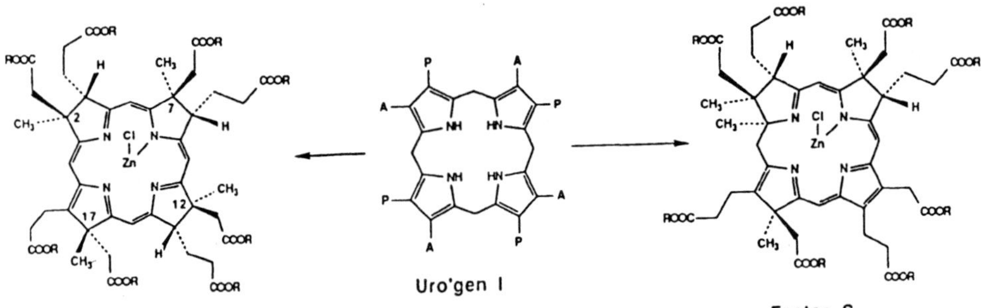

Factor S_1 Uro'gen I Factor S_3

Scheme 4 Conversion of uro'gen I into factor S_1 and S_3 by cell-free extracts of *P. shermanii* + SAM + Zn^{2+}, formula demonstrate one possible structural isomer of each isolate.[30,31]

At the moment the role of factors S is unknown but indeed these corphinoids are detectable in low amounts even in *P. shermanii* cells that have not been handled with additional substrates. In any case the presence of zinc is essential for their formation. The conversion of uro'gen I derived factor II to factors S by its synthesizing systems of *P. shermanii* has so far been unsuccessful. It remains open whether these methylations leading to factors S are non specific and realized by methylases responsible for B_{12} biosynthesis.

The timing of cobalt insertion remained the point in question during the above studies. The first experiment on the cobalt problem had already been done as early as 1976.[32] The ^{57}Co - complex of ^{14}C - labeled factor II was fed to a cobyrinic acid forming cell extract of *C. tetanomorphum* and the $^{57}Co/^{14}C$ - ratio of the isolated cobester fitted well enough with the ratio of introduced Co - factor II. But almost at the same time of this experiment, the trimethylated and **metal-free** factor III was detected and shortly thereafter its intermediacy demonstrated.

Recently we tried to answer the cobalt question with the help of ^{13}C - pulse experiments by using a combined Co^{2+}/[$^{13}CH_3$]SAM - pulse. During one series of incubations of exactly the same kind (i.e. cobyrinic acid synthesizing system and tetrapyrrolic substrate) each incubation received Co^{2+} + [$^{13}CH_3$]SAM simultaneously but the individual pulsations were done in a sequence of different time intervals (every hour for 9 hours). The substrates were either ALA, factor II, factor III or endogenous tetrapyrroles in each of 4 different series of experiments. Cobyrinic acid was isolated as cobester and the latter analysed by NMR as before. Surprisingly - in all 4 cases - the ^{13}C signals did not show any differentiations within one particular series of incubations i.e. all NMR-spectra showed the same ratio of signal intensity for the $^{13}CH_3$ derived methyl groups. These results indicated that cobalt insertion happens either after C_{20}- but before C_{17}-methylation **or** when the methylation process is already complete.

Then we learned that a partly purified enzyme system from *P. shermanii* which is usually used for factor II - preparation yields notable amounts of Co - factor III when incubated in presence of cobalt ion whereas cobalt-free incubations yield considerable amounts of factor II, but only insignificant amounts of factor III. The question then arose whether this unexpected result, surveyed in Table 2, is evidence for early cobalt incorporation. Strong reinforcement of this idea then came from studies concerned with the isobacteriochlorin content of freshly cultivated cells of *P. shermanii*. Interestingly enough, Co - factor III was the predominant isolate from those cells which were cultivated ***in presence of cobalt whereas the cobalt-free cells yielded much more factor II than factor III*** (Table 3); all of the tetrapyrrolic compounds being isolated and estimated as their methyl esters.

Table 2 Isobacteriochlorins and Cobyrinic Acid from ALA + SAM and Enzyme Preparations of 100 g Wet Cells of P. shermanii [a].

Expte.	time of incubation (h)	Co^{2+}	isolated methylesters [nmol] of				
			F II	F III	Co-F II	Co-F III	Cobyrinic acid
1	32	-	2 300	38	-	-	-
2	4	+	122	-	5	5	22
3	8	+	194	-	5	10	37
4	20	+	90	-	5	207	40
5	32	+	42	-	5	625	48

[a] added substrates: ALA: 100 µmoles, SAM: 40 µmoles, Co^{2+}: 10 µmoles and co-factors.

Table 3 Isobacteriochlorins and Corrins Isolated from 100 g Wet Cells of Propionibacterium shermanii [a].

cultivation of cells [b]	F II	F III	Co-F II	Co-F III	Cobyrinic acid (Cba)	Cba-Mono-amid	Cba-Di-amid
	isolated methylesters [nmol] of						
A	85	18					
B	<5	<5	<5	150	5	17	43

[a] Isolated as methylesters; [b] A: cobalt-free B: in presence of Co^{2+}.

Next the cobyrinic acid forming systems prepared from cobalt containing and cobalt-free cultures were compared and studied more closely with the purpose of gaining insight into cobyrinic acid synthesis as well as the nature and mode of formation of precursors. Since cobyrinic acid synthesizing cell-free extracts of *P. shermanii* still contain endogenous substrates for the biosynthesis, the cobalt-free cell extracts were partly enriched (by several steps) with the object of obtaining an enzyme preparation (A) which synthesizes cobyrinic acid ***only with additional substrates*** (Table 4, Expts. 1-3). When the same purification steps are applied to cell extracts prepared from cobalt containing cultivations of *P. shermanii* the enzyme preparation (B) still synthesizes cobyrinic acid **but without extra substrates** (Expts. 4-7). From these findings it was concluded that system (B) contains cobalt containing intermediates, perhaps protein-bound. To support this concept, a cobalt-free cell extract (as in A) was preincubated for 3 hours (i) with precorrin-3 (as factor III) and (ii) with precorrin-3 + Co^{2+} under exactly the same conditions (System C, Expts. 8 and 9). These preincubated extracts were then subjected the purification steps developed as described above. Thereafter (i) was incubated with Co^{2+} and (ii) without Co^{2+}. As expected cobyrinic acid was formed only by the latter incubation (Expt. 9). The enriched enzyme preparation which presumably contained the cobalt chelates of the pathway was next incubated with [$^{13}CH_3$]SAM and the isolated cobester was analysed by NMR. Unexpectedly intensity ratios of the ^{13}C signals was exactly the same as when precorrin-2 or precorrin-3 + Co^{2+} is converted to cobyrinic acid. These and the previous results indicated (see Scheme 1) that a cobalt chelate of precorrin-3 or of its C_{12}-decarboxylated analogue does appear on the pathway.

Next the radioactively labeled cobalt complexes of factors II and III were prepared either by chemical cobalt insertion or - in case of cobalt factor III - with the help of cobyrinic acid forming systems from *P. shermanii*. These were analysed and compared by UV/Vis, MS, TLC and ^3H/^{14}C - LSC. The ratioactive substrates were [methyl-^3H] Met or SAM and [4-^{14}C} ALA.

Table 4 Cobyrinic Acid Formation with Various Enzyme Preparations (system) prepared from Differently Cultivated Cells of P. shermanii [a].

Expte.	system [b]	substrates/ nmol	preincubation of cell extract with substrate / nmol	Co^{2+}	isolated cobester / nmol
1	A	-	-	+	3
2	A	ALA / 10 000	-	+	160
3	A	F III / 1 000	-	+	150
4	B	-	-	-	80
5	B	-	-	+	94
6	B	ALA / 10 000	-	-	90
7	B	F II / 400	-	+	170
8	C	-	F III / 1 000	+	3
9	C	-	F III / 1 000 + Co^{2+} / 8 000	-	80

[a] Prepared of 100 g wet cells in each case. [b] A: enzyme system prepared from cobalt-free cell cultures. B: prepared as A but from cells cultivated in presence of Co^{2+}. C: cell-free extract as in A has been preincubated (3h) with substrate before further enrichment.

Although it is believed that the entire biosynthesis is processed at the hexahydroporphyrinc oxidation level we were curious to know if it is possible to channel Co-III-factor III into the process. The corresponding incorporation experiments were carried out under comparative (Table 5) and competitive conditions (Table 6) with the cobyrinic acid forming cell extracts of *P. shermanii* or with the enriched enzyme preparations thereof (which lack necessary substrates) and the radioactively labeled factors II and III and the cobalt complexes thereof.

Table 5 Conversion of $[^3H/^{14}C]$- Double Labeled Isobacteriochlorins [a] to Cobyrinic Acid [b] by Cell-Free Extracts of P. shermanii [c].

Expte.	added substrates	nmol	dpm/nmol ^3H	dpm/nmol ^{14}C	$^3H/^{14}C$ -ratio	cobyrinic acid [b] dpm ^3H	dpm ^{14}C	$^3H/^{14}C$ -ratio	nmol [d]
1	F II	1100	4134	1295	3.2	74 900	25 450	2.9	18
2	Co-FIII	1100	5458	1217	4.5	92 000	24 750	3.7	20
3	F II	446	1215	245	4.96	16 560	3 360	4.9	13.5
4	Co-FIII	392	1285	750	1.7	12 000	10 160	1.18	14

[a] Isobacteriochlorins have been synthesized from $[4\text{-}^{14}C]$ ALA + $[methyl\text{-}^3H]$ SAM by enzyme-preparations of P. shermanii, isolated as methylesters and hydrolysed (1M piperidine) before use, [b] isolated and estimated as cobester, [c] from 30 g wet cells in each case, [d] based on reversed isotope dilution and specific radioactivity of substrate.

The incorporation of factors II and III as well as the positive incorporations with cobalt factor III came up to our expectations (Tables 5 and 6). Because of the well known steps shown in Scheme 1, the positive incorporation rates obtained with cobalt factor II were unexpected (Table 6). Experiment 4 of Table 6 illustrates the bioconversion of both cobalt factor II as well as cobalt factor III to cobyrinic acid with the substrate-free enzyme preparation of *P. shermanii*. But when [methyl-^{3}H]factors II and III + [^{14}C] cobalt factor II were fed to the cobyrinic acid forming systems (Expts. 5 and 6) a significant percentage (10 - 18%) of the isolated cobesters was derived from a cobalt-free isobacteriochlorin. So at some stage - either before or after the reduction of cobalt factor II by the cell-free system - some cobalt was released and reused for biosynthesis.

Table 6 Conversion of Radioactively Labeled Isobacteriochlorins [a] to Cobyrinic Acid by Substrate-Free Enzyme Preparation of P.shermanii [b,c].

Expte.	substrates	nmol	dpm/nmol	isolated cobester nmol [d]	isolated cobester dpm	isolated cobester nmol [e]
1	[methyl-^{3}H] F III	500	6633	47.4	208800	47.2
2	[methyl-^{3}H] F III	500	6633	46	175000	39.5
	+ [^{14}C] F II	500	2740		23100	8.4
3	[methyl-^{3}H] Co-FIII	430	3175	40	90000	42.5
	+ [^{14}C] F II	430	2563		330	0.13
4	[methyl-^{3}H] Co-FIII	500	6633	38	90600	20.5
	+ [^{14}C] Co-F II	500	2740		59340	21.6
5	[^{14}C] Co-F II	430	1306	27.5	36822	28
	+ [methyl-^{3}H] F III	430	2094		4702	3.3
6	[^{14}C] Co-F II	640	2110	70	129995	62
	+ [methyl-^{3}H] F II	640	1410		18030	12.7

[a] radioactive substrates were synthesized with cell suspensions of P.shermanii from [4-^{14}C]ALA and / or [methyl-^{3}H]Met, cobalt insertion was carried out chemically. [b] prepared from 80 g wet cells in each case, [c] time of incubation 16 h, [d] estimated spectrophotometrically, after purification to constant specific radioactivities, [e] estimated with help of specific radioactiv of substrates.

Finally a careful analysis of the isolates from an incubation containing [^{14}C]Co-factor II + [methyl - ^{3}H] SAM unambiguously demonstrated the third methylation on a dimethylated cobalt complex as well as the cobyrinic acid formation (Table 7).

Table 7 Conversion of [^{14}C] Co-F II [a] into Co-F III and Cobyrinic Acid by Enzyme Preparation of P.shermanii [b] and [^{3}H-methyl] SAM [c].

Isolates [d]	nmol	dpm / nmol ^{3}H	dpm / nmol ^{14}C	^{3}H / ^{14}C-ratio
Co-F II	90	180	990	0.18
Co-F III	135	5 980	987	6.00
Cobyrinic acid	35	33 500	986	34.00

[a] [^{14}C] Co-F II: 1 µmol, 1 000 dpm/nmol, prepared of [4-^{14}C] ALA. [b] prepared of 100 g wet cells, 8 h incubated. [c] [methyl-^{3}H] SAM: 16 µmol, 6 800 dpm/nmol. [d] isolated as methylesters.

From these results it is concluded that in the *P. shermanii* system at least the trimethylated precorrin (precorrin -3) appears as a cobalt complex on the cobyrinic acid pathway and the indications are strong that a dimethylated cobalt complex also appears as a precorrin on the pathway.

It is evident from the above results that a tremendous amount of work remains to be done concerning the mechanism of cobalt insertion as well as the definition of the oxidation level of the substrates and finally the discovery of the remaining intermediates.

ACKNOWLEDGEMENT

We acknowledge Deutsche Forschungsgemeinschaft for financial support for this research and the collaborations between Prof. A. I. Scott (Texas A&M University), Dr. F. Blanche (Rhône-Poulenc-Santé, Centre de recherches de Vitry) and Stuttgart.

REFERENCES

1. K. Bernhauer, F. Wagner, H. Michna, P. Rapp and H. Vogelmann, Biosynthesewege von der Cobyrinsäure zur Cobyrsäure und Cobinamid bei Propionibacterium shermanii, *Hoppe Seylers Z. Physiol. Chem.* **349**: 1297 (1968).

2. A. I. Scott, N. Georgopapadakou, K. S. Ho, S. Klioze, E. Lee, S. L. Lee, G. H. Jemme III, C. A. Townsend and I. A. Armitage, Concerning the Intermediacy of Uro'gen III and of a Heptacarboxylic Uro'gen in Corrinoid Biosynthesis, *J. Am. Chem. Soc.* **97**: 2548 (1975).

3. H.-O. Dauner and G. Müller, Bildung von Cobyrinsäure Mittels eines Zellfreien Systems aus Clostridium Tetanomorphum, *Hoppe Seylers, Z. Physiol. Chem.* **356**: 1353 (1975).

4. A. R. Battersby, M. Ihara, E. McDonald, F. Satoh and D. C. Williams, Derivation of Cobyrinic Acid from Uroporphyrinogen III, *J. Chem. Soc., Chem. Commun.* 436 (1975).

5. (a) A. R. Battersby, Recent Biosynthetic Researches on Vitamin B_{12}, in: "Vitamin B_{12}" eds. B. J. Zagalak and W. Friedrich, de Gruyter, Berlin, p. 217 (1979).
(b) A. I. Scott, Intermediary metabolism of Cobyrinic Acid Biosynthesis, ibid p.248.
(c) G. Müller, R. Deeg, K. D. Gneuss, G. Gunzer and H.-P. Kriemler, On the Methylation Process of Cobyrinic Acid Biosynthesis, ibid p.279.
(d) V. Ya. Bykhovsky, Biogenesis of Tetrapyrrole Compounds and Regulations, ibid p.293.
(e) M. Imfield, D. Arigoni, R. Deeg and G. Müller, Faktor I ex Clostridium tetanomorphum; Proof of Structure and Relationship to Vitamin B_{12} Biosynthesis, ibid p.316.

6. (a) R. Deeg, H.-P. Kriemler, K.-H. Bergmann and G.Müller, Neuartige, methylierte Hydroporphyrine und deren Bedeutung bei der Cobyrinsäure-Bildung, *Hoppe Seylers Z. Physiol. Chem.* **358**: 339 (1977).
(b) K. H. Bergmann, R. Deeg, K.-D. Gneuss, H.-P. Kriemler and G. Müller, Gewinnung von Zwischenprodukten der Cobyrinsäure-Biosynthese mit Zellsuspensionen von Propionibacterium shermanii, ibid 358: 1315 (1977).

7. G. Müller, K. D. Gneuss, H.-P. Kriemler, A. J. Irwin and A. I. Scott, Structure of Factor III, A Trimethylisobacteriochlorin Intermediate in the Biosynthesis of Vitamin B_{12}, *Tetrahedran Suppl.* **37**: 81 (1981).

8. A. R. Battersby, K. Frobel, F. Hammerschmidt and C. Jones, Isolation of 15, 23-Dihydrosirohydrochlorin, a Biosynthetic Intermediate, *J. Chem. Soc., Chem. Commun.* 455 (1982).

9. M. J. Warren, N. J. Stolowich, P. J. Santander, C. A. Roessner, B. A. Sowa and A. I. Scott, Enzymatic Synthesis of Dihydrosirohydrochlorin (precorrin-2) and of a Novel Pyrrocorphin by Uroporhyrinogen III-Methylase, *FEBS Lett.*, **261**: 76 (1990).

10. F. Blanche, L. Debussche, D. Thibaut, J. Crouzet, B. Cameron, Purification and Characterization of S-Adenosyl-L-Methionine: Uroporphyrinogen III-Methyltransferase from Pseudomonas Denitrificans, *J. Bacteriol.* 171: (1989).

11. H. C. Uzar, A. R. Battersby, T. A. Carpenter and F. J. Leeper, Development of a Pulse Labelling Method to Determine the C-Methylation Sequence for Vitamin B_{12}, *J. Chem. Soc. Perkin Trans. 1*, 1689 (1987).

12. A. I. Scott, H. J. Williams, N. J. Stolowich, P. Karuso, M. D. Gonzalez, G. Müller, K. Hlineny, E. Savvidis, E. Schneider, U. Traub-Eberhard and G. Wirth, Temporal Resolution of the Methylation Sequence of Vitamin B_{12} Biosynthesis, *J. Am. Chem. Soc.* **111**: 1898 (1989).

13. V. Rasetti A. Pfalz, C. Kratky and A. Eschenmoser, Ring Contraction of Hydroporphinoid to Corrinoid Complexes, *Proc. Natl. Acad. Sci., USA*, **78**: 16 (1981).

14. V. Rasetti, K. Hilpert, A. Fäßler, A. Pfalz and A. Eschenmoser, Dihydrocorphinol Corrin Ringkontraktion: Eine potentiell biomimetische Bildungsweise der Corrinstruktur, *Angew. Chem.* **93**: 1108 (1981).

15. A. Eschenmoser, Vitamin B_{12}: Experimente zur Frage nach dem Ursprung seiner molekularen Struktur, *Angew. Chem.* **100**: 5 (1988).

16. (a) C. Nussbaumer, M. Imfeld, G. Wörner, G. Müller and D. Arigoni, Biosynthesis of Vitamin B_{12}, Mode of incorporation of factor III into cobyrinic acid, *Proc. Natl. Acad. Sci. USA*, **78**: 9 (1981).
(b) L. Mombelli, C. Nussbaumer, H. Weber, G. Müller and D. Arigoni, Nature of the volatile fragment generated during formation of the corrinring system, ibid p.11.
(c) A. R. Battersby, M. J. Bushell, C. Jones, N. G. Lewis and P. Pfenninger, Identity of fragment extruded during ring contraction to the corrin macrocycle, ibid p.13.

17. J. Friedle, Zur C-20-Eliminierung bei der Cobyrinsäure-Biosynthese, Dissertation, Universität Stuttgart (1982).

18. A. I. Scott, N. E. Mackenzie, P. J. Santander, P. E. Fagerness, G. Müller, E. Schneider, R. Sedlmeier and G. Wörner, Biosynthesis of Vitamin B_{12}: Timing of the Methylation Steps between Uro'gen III and Cobyrinic Acid, *Biorg. Chem.* **12**: 356 (1984).

19. G. Gunzer, Enzymatische Studien für die Aufgliederung des Gesamtvorgangs der mikrobiellen Cobyrinsäure-Bildung in Einzelreaktionen, Dissertation, Universität Stuttgart (1978).

20. H. Weber, Chemie und Biochemie Methylierter Isobacteriochlorine, Dissertation, Universität Stuttgart (1982).

21. A. I. Scott, H. J. Williams, N. J. Stolowich, P. Karuso, M. D. Gonzalez, G. Müller, K. Hlineny, E. Savvidis, E. Schneider, U. Traub-Eberhard and G. Wirth, Temporal Resolution of the Methylation Sequence of Vitamin B_{12} Biosynthesis, *J. Am. Chem. Soc.* **111**: 1897 (1989).

22. F. Blanche, S. Manela, D. Thibaut, C. L. Gibson, F. J. Leeper and A. R. Battersby, When is the 12 β-Methyl Group Generated by Acetate Decarboxylation, *J. Chem. Soc., Chem. Commun.*, 1117 (1988).

23. B. Dresow, G. Schlingmann, L. Ernst and V. B. Koppenhagen, Extracellular Metal-free Corrinoids from Rhodopseudomonas spheroides, *J. Biol. Chem.* **255**: 7637 (1980).

24. J. Friedle, Enzymatische Studien zur Vitamin B_{12}-Bildung, Diplomarbeit, Universität Stuttgart (1979).

25. T. E. Podschun and G. Müller, Hydrogenobyrinsäure und Vitamin B_{12}, *Angew. Chem.* **97**: 63 (1985).

26. C. Nussbaumer and D. Arigoni, Einfacher Zugang zu 5-Nor-und 5,15-Bisnorcobester, *Angew. Chem.* **95**: 746 (1983).

27. C. Nussbaumer, Zur chemischen Reaktivität und zur Biosynthese der Cobyrinsäure, Dissertation, ETH Zürich, Nr. 7623 (1984).

28. J. Bier, Untersuchungen zur Auffindung neuer Substrate und Enzyme der Cobyrinsäure-Biosynthese, Dissertation, Universität Stuttgart (1986).

29. H. C. Uzar and A. R. Battersby, Vitamin B_{12}: Order of the Later C-Methylation Steps, *J. Chem. Soc., Chem. Commun.*, 585 (1985).

30. G. Müller, J. Schmiedel, E. Schneider, R. Sedlmeier, G. Wörner, A. I. Scott, H. J. Williams, P. E. Fagerness, N. E. Mackenzie and H. P. Kriemler, Structure of Factor S_3, a Metabolite Derived from Uroporphyrinogen I, *J. Am. Chem. Soc.***108**: 7875 (1986).

31. G. Müller, J. Schmiedl, L. Savvidis, G. Wirth, A. I. Scott, P. J. Santander, H. J.

Williams and N. J. Stolowich, Factor S_1, a Natural Corphin from Propionibacterium shermanii, *J. Am. Chem. Soc.* **109**: 6902 (1987).

32. R. Deeg, Zur Vitamin B_{12}-Biosynthese, Dissertation, Universität Stuttgart (1978).

33. For review: A. I. Scott: in Chapter 9 of this volume; see also A. I. Scott *Accounts Chem. Research* (1990) in press.

BIOCHEMICAL AND GENETIC STUDIES ON VITAMIN B_{12} SYNTHESIS IN Pseudomonas dentrificans

J. Crouzet, F. Blanche, B. Cameron, D. Thibaut and L. Debussche

Department of Biotechnology and Department of Analytical Chemistry, Rhône-Poulenc Santé, Centre de Recherche de Vitry Alfortville, 13 Quai Jules Guesde BP 14, 94403 Vitry sur Seine Cédex, FRANCE

Intoduction

Vitamin B_{12} is produced industrially by Pseudomonas denitrificans or by bacteria belonging to Propionibacterium species (i. e. Propionibacterium shermanii or Propionibacterium freudenreichii). Industrial strains were obtained after long lasting and intense programs aimed at improving the fermentation caracteristics and the productivity of natural isolates (Florent, 1986). These programs consist of numerous mutagenesis steps, each of them introducing a property resulting in improved characteristics of the strain for industrial purpose (i. e. higher productivity, higher growth rate in a defined medium, growth to a higher biomass, etc). At each mutagenesis step the clone of interest is selected randomly either (i) for its property to produce higher level of vitamin B_{12} or (ii) for its resistance to an antibiotic for instance or (iii) for a phenotype related to an improved production of vitamin B_{12} (Florent, 1986). These selection programs have allowed various companies to improve the production of the starting bacteria. For instance the Pseudomonas denitrificans wild type isolate, strain MB580, from which the industrial strain used by Rhône-Poulenc Santé derives produces 2.5 mg/l of cobalamin. In the same conditions, the industrial strain presently used produces much more (Florent, 1986).

For any set of fermentation conditions, the production of vitamin B_{12} of a strain is set both by its growth habit in the fermentor and by interactions of a series of genetically determined physiological parameters and a set of environmental factors. The genetically controlled parameters include among others : the level of enzymes in the biosynthetic pathway, the intracellular supply of precursors and maybe the export of the final product. Recombinant DNA techniques allow to act on the genetically controlled parameters. Genetic engineering has been successful to

Chemical Aspects of Enzyme Biotechnology
Edited by T.O. Baldwin *et al.*, Plenum Press, New York, 1990

improve microorganisms metabolites productivity. For instance, cephalosporin C production has been increased at the industrial scale in Cephalosporium acremonium by recombinant DNA techniques (Skatrud et al., 1989); similar approaches have been used for improving the production of amino acids by Escherichia coli or by Corynebacteria (Miwa et al., 1983; Ozaki et al., 1985; Sano et al., 1987). It is therefore expected that recombinant DNA techniques should also improve the production of vitamin B_{12} of industrial strains.

Cloning *Pseudomonas denitrificans* genes involved in cobalamin synthesis

We have reported the cloning of *Pseudomonas denitrificans* genes involved in cobalamin synthesis (*cob* genes) (Cameron et al., 1989). Those genes are from strain SC510 which is closely related to the industrial strain used by Rhône-Poulenc Santé. At least 14 *cob* genes have been cloned (see Figure 1); 12 of them are implicated in the conversion of uroporphyrinogen III (uro'gen III) to cobinamide and two are implicated in the conversion of cobinamide to coenzyme B_{12}, see Figure 1. These genes have been identified by heterologous complementation of *Pseudomonas putida* and *Agrobacterium tumefaciens* Cob mutants. In *P. putida* the Cob mutants were identified as being unable to use ethanolamine as a source of nitrogen in the absence of added cobalamin (it has been reported in other organisms (Babior, 1982) that the deamination of ethanolamine requires coenzyme B_{12} as a cofactor); in *A. tumefaciens* the Cob mutants were simply screened for their reduced cobalamin synthesis. The mutants were complemented for the restoration of the lost phenotype, by a genomic library from *P. denitrificans*. The cloned genes are grouped in 4 genomic regions and we do not know whether those Cob *loci* are clustered or located at distant positions on the chromosome of *P. denitrificans*. Clustering of Cob genes has already been reported in *B. megaterium* and *S. typhimurium* (Wolf and Brey, 1986; Jeter et al., 1984; Jeter and Roth 1987). In *B. megaterium*, all the Cob mutations found are linked by cotransduction (Wolf and Brey; 1986, Brey et al., 1986); in *S. typhimurium*, except for the *cysG* gene, the *cobA* (Escalante and Roth, 1990) and the *cobD* loci (Grabau and Roth, 1989), all the identified genes are located in the same region on the chromosome.
The aim of our study is to understand the genetic and the biochemical organization of the cobalamin pathway in order to achieve improvement of industrial cobalamin producing strains.

Biochemical studies on the cobalamin pathway.

Purification and characterization of *S*-Adenosyl-L-Methionine:Uroporphyrinogen III methyltransferase from *Pseudomonas denitrificans*

S-Adenosyl-L-Methionine:Uroporphyrinogen III methyltransferase (SUMT), the enzyme which catalyzes *C*-methylations of uroporphyrinogen III at C-2 and C-7 to give

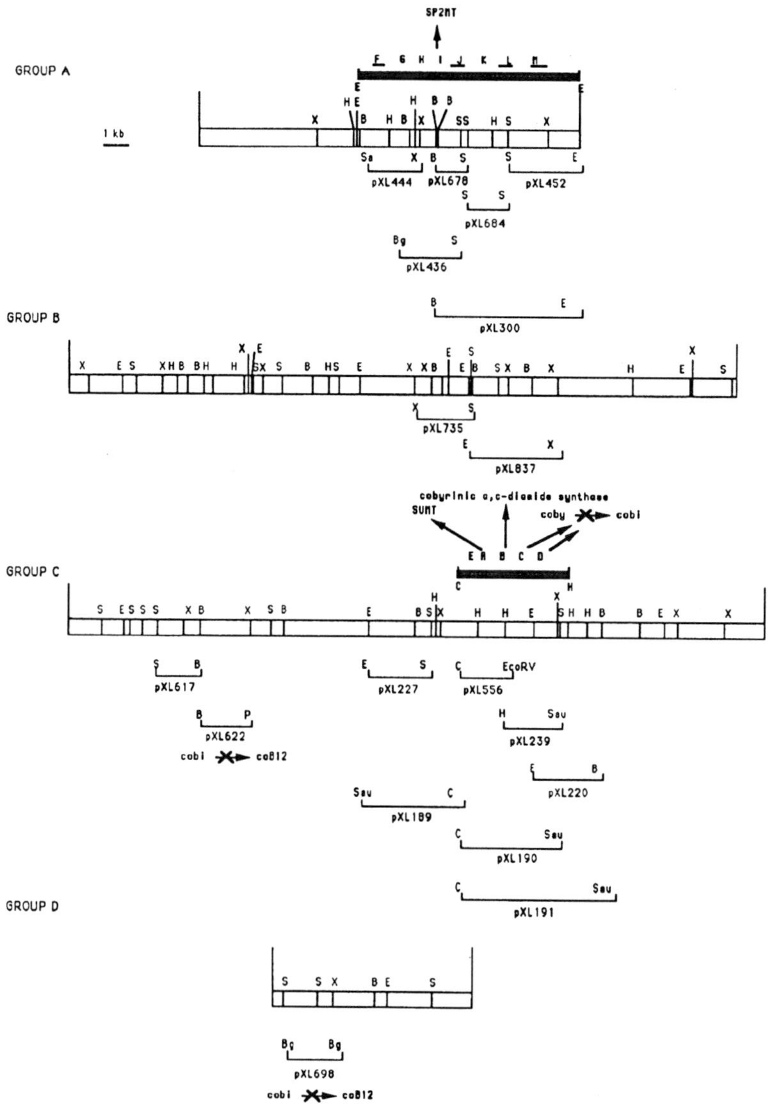

Figure 1. Physical map of the four loci carrying *Pseudomonas denitrificans* *cob* genes. Inserts of the 14 previously identified subclones are presented (pXL444, pXL678, pXL452, pXL684, pXL436, pXL735, pXL837, pXL617, pXL622, pXL227, pXL556, pXL239, pXL220 and pXL698), each of them carrying at least one *cob* genes; in addition inserts of plasmids pXL300, pXL189, pXL190, pXL191 are shown, it has been postulated that they contain additional *cob* genes (Cameron et al., 1989). The 13 sequenced *cob* genes are shown as well as the enzymatic activities catalyzed by the encoded proteins; genes *cobF*, *cobJ*, *cobL* and *cobM* which are proposed to code for potential methyltransferases are underlined. As previously described (Cameron et al., 1989), plasmids pXL622 and pXL698 carry genes involved in the conversion of cobinamide into coenzyme B_{12}. B, *Bam*HI; Bg, *Bgl*II; C, *Cla*I; E, *Eco*RI; H, *Hind*III; P, *Pst*I; S, *Sst*I; Sa, *Sal*I, Sau, *Sau*3AI; X, *Xho*I; coB12, coenzyme B_{12}; coby, cobyric acid; cobi, cobinamide.

precorrin-2 was purified about 150 fold from extracts of a recombinant strain of *Pseudomonas denitrificans* (Blanche et al, 1989). Our team has been the first one to report the purification of an enzyme catalyzing such an activity. This enzyme is a key one of the pathway since uro'gen III is at the branch point of two divergent pathways. The purified enzyme is an homodimer of *Mr* 30,000 +/-1,000 estimated by gel electrophoresis under denaturing conditions. The pH optimum is 7.7, the K_m for *S*-adenosyl-L-methionine and uroporphyrinogen III are 6.3 and 1.0 mM respectively, and the turnover number is 38 h^{-1}. The enzyme activity is shown to be completely insensitive to feedback inhibition by cobalamin and corrinoids intermediates tested at physiological conditions (Blanche et al., 1989). At uro'gen III concentrations above 2 mM, SUMT exhibits a substrate inhibition phenomenon. Since SUMT was purified from an industrial strain one cannot exclude that such a regulation mechanism has appeared during one of the mutagenesis steps. In order to answerthis question, SUMT was purified from the wild type *P. denitrificans* and was shown to exhibit the same substrate inhibition phenomenon. Whether this regulation is of physiological significance or not is a question that remains to be answered. One possibility is that substrate inhibition results in favoring hemes synthesis over cobalamin synthesis when uro'gen III accumulates at concentrations higher than 2 mM. For an obligate aerobic bacteria such as *P. denitrificans*, the rate of heme synthesis is certainly higher than of cobalamin synthesis, since cobalamin is a cofactor needed at very low level for bacterial growth (i. e. less than 200 molecules per bacterium). It would be of interest to know if SUMT from other organisms exhibit the same properties.

Purification of *S*-Adenosyl-L-Methionine:precorrin-2 methyltransferase from *Pseudomonas denitrificans*

S-Adenosyl-L-Methionine:precorrin-2 methyltransferase (SP_2MT), which catalyzes the C-20 methylation of precorrin-2 to precorrin-3, was purified to homogeneity from extracts of a recombinant strain of *Pseudomonas denitrificans* (Thibaut et al., submitted for publication). For enzyme purification and characterization, an enzyme assay was developped. The enzyme is an homodimer of *Mr* = 26,000. Evidence was obtained that the chemically reduced form of sirohydrochlorin (dihydro-sirohydrochlorin), is methylated at C-20 to form precorrin-3 by pure SP_2MT. The purification of *P. denitrificans* SP_2MT allows the *in vitro* synthesis of high level of precorrin-3 (Blanche et al., 1988).

Genetic studies on cobalamin biosynthesis

A 5.4 kb fragment from complementation group C and an 8.7 kb fragment from complementation group A, as previouly defined (Cameron et al., 1989) were genetically studied and sequenced. These two fragments *P. denitrificans* were previously shown to carry at least 3 and 5 *cob* genes respectively (Cameron et al., 1989).

Genetic studies and nucleotide sequence of a *P. denitrificans* 5.4 kb fragment

A 5.4 kb fragment from complementation group C was sequenced. This fragment is bounded by a *Cla*I site at its left and by an *Hind*III site at its right end (see Figure 1). It was chosen because (i) when a subfragment is amplified in *Pseudomonas denitrificans* SC510 Rif^r, SUMT activity is increased by a factor of 50 and (ii) it was shown, by complementation analysis, to carry at least three *cob* genes. Five open reading frames (named ORF1 to ORF5), characterized by a high coding probability according to analysis by Staden's program (Staden et McLachlan, 1982; Staden, 1984) were identified on the same strand, see Figure 2 (going 5'→3' from *Cla*I to *Hind*III sites). In four out of five ORFs (OFR2 to ORF5), either the ribosome binding site or the initiation codon overlaps with the termination codon of the preceding ORF, suggesting that these ORFs may be translationally coupled (Normak et al., 1983). ORF1 and ORF2 are separated by an intergenic region of 130 bp.
The 5.4 kb fragment was subjected to a genetic analysis in order to determine if the 5 open reading frames are *cob* genes (Crouzet et al., submitted for publication). The complementation of the different mutants allowed us to identified 5 different genes. Because insertions in each gene are always contained in an unique ORF, we concluded that the 5 identified ORF are *cob* genes, named *cobA* to *cobE*, see Figure 2.

CobA is the structural gene encoding SUMT

Since plasmid pXL190, the insert of which is part of the 5.4 kb fragment, was shown to lead to a 50-fold increase in SUMT activity in strain SC510 Rif^r (Blanche et al., 1989), it was likely that one of the identified *cob* gene was the structural gene for SUMT. NH_2-terminus sequencing of the purified SUMT was performed. The sequence of the first 10 aminoacids was identical to the first ten amino acids of the protein encoded by the *cobA* gene (data not shown). *CobA* is predicted to encode a protein of *Mr* 29, 200 which is in good agreement with the biochemical determination (Blanche et al., 1989). Therefore, *cobA* is the structural gene encoding SUMT. No significant homology was found in protein data bases with *P. denitrificans* SUMT. On the contrary, the protein encoded by the *E. coli cysG* gene (Peakman et al., submitted to publication) shows striking homology to COBA (data not shown). More than 41% of strict homology is observed between the two proteins. This homology takes place on most of COBA sequence but corresponds to the carboxyterminus of the *E. coli* protein. Interestingly, there are three domains of higher homology between COBA and CYSG (data not shown; Crouzet et al., submitted for publication); between these domains the strict homology is higher than 70 %. It is probable that these domains are involved in the active site of the enzyme. The *S. typhimurium cysG* gene is believed to catalyze at least the same reaction as SUMT (Jeter et al., 1984) since *cysG* mutants are blocked in both cobalamin and siroheme synthesis. It has been recently

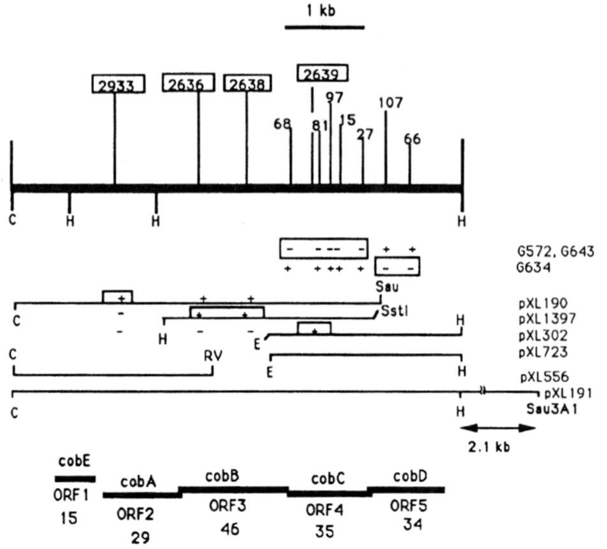

Figure 2 . Genetic analysis of the 5.4 kb fragment from complementation group C (Cameron et al., 1989). The chromosomal Tn*5*Sp^r insertions in strain SBL25 Rif^r (Blanche et al., 1989) whose numbers are boxed are shown as well as the Tn*5* insertions obtained on plasmid pXL723 which are numbered 15, 27, 66, 68, 81, 97 and 107. Plasmid pXL723 has been obtained by subcloning the 2.3 kb *Eco*RI-*Hind*III into pRK290 (Ditta et al., 1980). For the Tn*5* and the Tn*5*Sp^r insertions, a - (respectively +) sign beneath shows that insertion does inactivate (respectively does not) the complementation of mutants G572, G643 or G634. The inserts of plasmids studied for complementation of mutants are shown. The vectors in which the indicated inserts are cloned are all incQ plasmids already described (Cameron et al., 1989). Above each insert + or - signs indicate that the plasmid respectively does or does not complement Cob mutants that are aligned on those signs. The position of each ORF are indicated as well as the name of the corresponding *cob* gene; the molecular weight, in KD, of the protein encoded by each *cob* gene is also indicated.

The insertions that prevent the complementation of G572, G643 and G634 are indicated. For each class, the insertions are mapped in the same open frame. It is therefore concluded that ORF 4 and 5 are two *cob* genes, named *cobC* and *cobD* respectively.

Insertion 2639 is found in ORF4 and is complemented by pXL302 which contains *cobC* and *cobD*. Insertions 2636 and 2638, mapped in ORF3, are not complemented by pXL302, but are complemented by pXL1397 which contains ORF3 and ORF4. ORF3 is called *cobB*. Insertion 2933, mapped in ORF2, is complemented by plasmid pXL190 but not by plasmid pXL1397 that complements mutants for which the Tn*5*Sp^r has been mapped into ORF3 and 4. Therefore this insertion inactivates another *cob* gene, *cobA*, corresponding to ORF 2. It will be described elsewhere that ORF1 is also a *cob* gene named *cobE* (Crouzet et al., submitted for publication).

published that *E. coli* *CYSG* has a SUMT activity (Warren et al., 1990a; Warren et al., 1990b), therefore the homology detected between *P. denitrificans* SUMT and *E. coli* CYSG protein must be specific of SUMT activity. Interestingly, *cobA* is clustered with *cob* genes in *P. denitrificans* contrary to *S. typhimurium* where *cysG* gene is not found near the *cobI* to *cobIII* region (Jeter et al., 1984).

Identification of the structural gene encoding Cobyrinic acid *a,c*-diamide synthase

The purification of *P. denitrificans* Cobyrinic acid *a,c*-diamide synthase will be described elsewhere (Debussche et al., submitted for publication); it is an homodimer *Mr* 2*45, 00. This activity is responsible for the amidation of carboxylic groups at positions *a* and *c* of cobyrinic acid. NH_2 groups are provided by glutamine and one molecule of ATP is hydrolyzed for each amidation. The scheme of the reaction is shown in Figure 3. The amplified enzymatic activity has been purified from strain SC510 Rif^r pXL191 (Debussche et al., submitted for publication). Since plasmid pXL191 carries the 5.4 kb fragment (see Figure 2), it was investigated if the structural gene for Cobyrinic acid *a,c*-diamide synthase was one of the identified *cob* genes. For this purpose the sequence of the NH_2-terminal first 15 aminoacids of the purified Cobyrinic acid *a,c*-diamide synthase was determined and was found to match exactly the aminoterminal sequence of COBB protein except that the methionine encoded by the initiation codon has been removed. *CobB* gene is predicted to encode a protein of molecular weight 45.6 kD which is in good agreement with the size of the purified monomer. It is concluded that *cobB* is the structural gene for Cobyrinic acid *a,c*-diamide synthase.

Biochemical studies on *cobC* and *cobD* mutants

One approach to localize the activities of proteins COBC and COBD, along the cobalamin pathway is to study the intracellular accumulation of biosynthetic intermediates in mutants complemented by *cobC* and *cobD*. As a matter of fact, if an accumulation of intermediates is higher in the mutant than in the parent strain, it is probable that the mutant is blocked in a step which utilizes the accumulated intermediates as precursors or substrates. *Agrobacterium tumefaciens* Cob mutants G634, G643 (Cameron et al., 1989) and a *Pseudomonas putida* Cob G572 mutant (Cameron et al., 1989) were studied for their intracellular content of corrinoids. These Cob mutants are *A. tumefaciens* and *P. putida* strains complemented by *P. denitrificans* *cob* genes; it is likely that these bacteria share the same cobalamin pathway. This would implicate that the Cob mutants are blocked in the step catalyzed by the product of the complementing gene. Studies on these mutants should therefore give indications on the activity of the protein encoded by the complementing gene. Strain G643, carries a mutation in *cobC* (as described elsewhere; Crouzet et al., submitted for publication). This mutant accumulates cobyric acid and cobyrinic acid pentaamide (data not shown, to be published elsewhere). In addition, this mutant is blocked

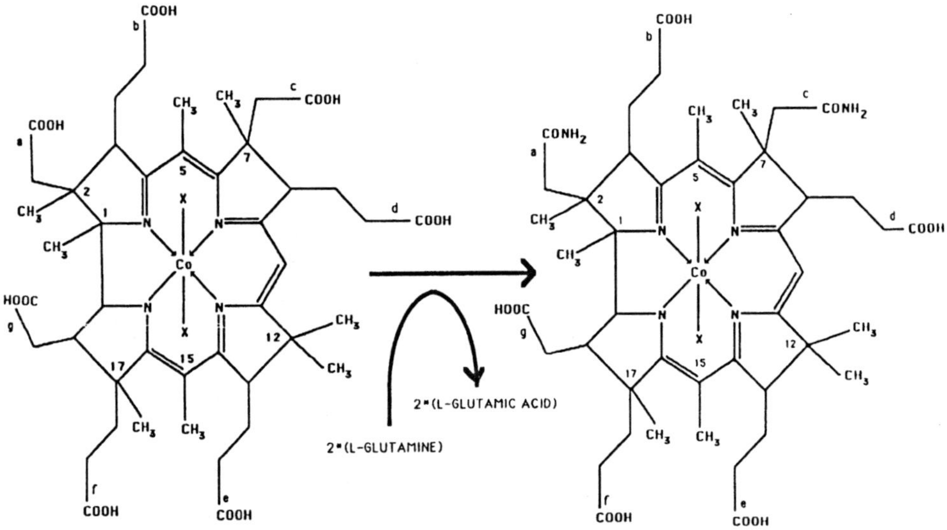

COBYRINIC ACID

COBYRINIC ACID a,c-DIAMIDE

Figure 3 . Reaction catalyzed by Cobyrinic acid _a,c_-diamide synthase.

before cobinamide as previously established (Cameron et al., 1989), and as it is accumulating the intermediate preceding cobinamide, it is likely that protein COBC is involved in the transformation of cobyric acid into cobinamide. Strain G634, an Agrobacterium tumefaciens Cob mutant previously described (Cameron et al., 1989) is blocked before cobinamide and is complemented by cobD. This mutant accumulates cobyric acid and must be also blocked after cobyric acid. CobD gene, as cobC, codes for a protein that is involved in the transformation of cobyric acid into cobinamide.

DNA sequence of the 8.7 kb EcoRI fragment from plasmid pXL151

The 8.7 kb EcoRI DNA fragment, from plasmid pXL151 (Cameron et al., 1989) was sequenced. When amplified, this fragment leads to the complementation of all the Cob mutants classified in complementation group A. Therefore this fragment must contain all the cob genes that complement these mutants. It contains 8753 bp. 8 open reading frames (ORF), named ORF6 to ORF13 are found, as presented in Figure 4. Six ORFs (from 7 to 12) show the characteristics of ORFs that are translationally coupled; the stop codons of ORF7 to ORF11 are either overlapping with the initiation codon of the next ORF or in the case of ORF7, are only two bases apart. On the contrary, a 100 bp intergenic region separates ORF6 from ORF7 and ORF12 is found 69 bp upstream from ORF13. No sequence with a dyad symmetry capable of forming a ρ-independent terminator (Platt, 1986) appears in the intergenic regions between ORF6 and ORF7, ORF12 and ORF13 nor after ORF13. The same GC content was found in this 8.7 kb fragment as in the P. denitrificans 5.4 kb fragment carrying cobA to cobE genes, i. e. 65.4 and 65.7 %, respectively.

Genetic analysis of the 8.7 kb fragment

A mutation analysis of the 8.7 kb EcoRI was carried out in order to localize the cob genes among the 8 identified ORFs. The 8.7 kb fragment was cloned in pRK290 (Ditta et al., 1980) EcoRI site; the resulting plasmid, pXL367, was mutagenized with Tn5 and Tn3lacZ transposons (DeBruinj, 1984; Stachel., 1985). 29 Tn5 insertions and 13 Tn3lacZ insertions were selected in the 8.7 kb fragment and then mapped (see Figure 4). These pXL367 derivatives were introduced by bacterial mating into Agrobacterium tumefaciens Cob mutants already described (Cameron et al., 1989). The mutants (G164, G609, G610, G611, G612, G613, G614, G615, G616, G620 and G638) were chosen because they represent each complementation class previously identified on group A (Cameron et al., 1989), and they are all complemented by plasmid pXL367. Complementation of the Cob mutants by the pXL367 derivatives was examined. The results of this study are presented in Figure 4. The insertions are classified into 9 groups. One group contains all the insertions that still lead to the complementation of all the mutants; the other groups, from 1 to 8 correspond to insertions for which the plasmids do not complement specific

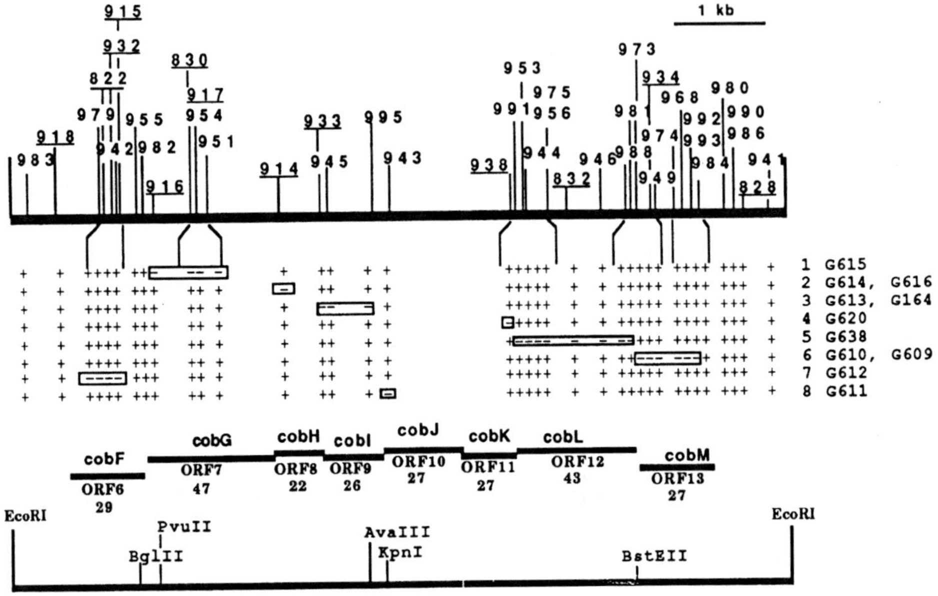

Figure 4 . Genetic anlysis of the 8.7 kb EcoRI fragment. Tn5 and Tn3lacZ insertions were selected into the 8.7 kb EcoRI. They are inserted in plasmid pXL367. The name of each transposon insertion and the site of insertion are indicated. Tn3lacZ insertions are underlined. ORF6 to 13 deduced from DNA sequence are represented at the bottom of the figure with the name of the corrresponding *cob* gene and the molecular weight, in KD, of the encoded protein. + signs, or - respectively, are aligned vertically with insertion site to indicate that the plasmid carrying the corresponding insertion does, or does not, respectively, complement *Agrobacterium tumefaciens* Cob mutant the names of which are aligned horizontally with the symbol; the number of each group of inactivating insertions is indicated with the mutants name. - signs for mutants belonging to the same groups are boxed. The groups and the associated Cob mutants are the following: 1, G615; 2, G614 and G616; 3, G613 and G164; 4, G620; 5, G638; 6, G610 and G609; 7, G612; 8, G611.

mutants; these insertions are inactivating insertions. There are as many inactivating insertions groups as ORFs on the fragment and for each group, transposons are always inserted within a unique ORF where transposons from other groups are never mapped. All the inactivating insertions are within ORFs. Nearly all of the non inactivating insertions are mapped outside ORFs. This study allows the identification of 8 genes, involved in cobalamin synthesis, named cobF to cobM and corresponding to ORF6 to ORF13 respectively (see Figure 4). Based on the organization of the ORFs described previously, it is probable that these genes or most of them are part of the same operon, however it remains to be demonstrated.
The proteins encoded by genes cobF to cobM named COBF to COBM were analyzed with the program of Hopp and Woods (1981) that draws hydrophilicity plots. The hydrophylicity plots of all these COB proteins are typical of cytoplasmic proteins (data not shown). The codon usage of these genes is very similar to the one observed for cobA to cobD genes (data not shown) indicating that all P. denitrificans sequences cob genes share the same codon usage, although one of them (cobE) is slightly different.

CobI is the structural gene encoding SP_2MT

As described elsewhere, the SP_2MT purification has been performed. The enzyme activity has been purified from strain SC510 Rif^r pXL253 since this plasmid allows a significant amplification of this activity. pXL253 has been obtained by cloning, into pKT230 (Bagdasarian et al., 1981) EcoRI site, the sequenced 8.7 kb fragment. It is therefore likely that that this fragment carries SP_2MT structural gene. The purified SP_2MT was subjected to NH_2-terminus sequencing. The first 15 identified amino acids are strictly identical with the aminoterminal sequence of the COBI protein encoded by the cobI gene (data not shown), with the exception that the aminoterminal methionine has been removed. The molecular weight of the protein encoded by cobI is 25,800 which is very close to the estimated size of SP_2MT monomer. It is concluded that cobI is the structural gene for SP_2MT.
A search for homologous proteins in the EMBL bank using the Kanehisa program (1984) did not reveal any significant homology with the COBF to COBM proteins. A computer search for similarity between COBA and COBI proteins was done. 3 regions of homology were found (data not shown, to be published elsewhere). The strict homologies range from 40 to 60 %. Interestingly the 3 regions of homology between COBA and COBI correspond exactly to the COBA regions showing homology with E. coli CYSG protein (data not shown). CYSG catalyzes at least the same reaction as COBA (i. e. the transfer of two methyls from SAM to uro'gen III at C-2 and C-7) as suggested (Warren et al., 1990a; Warren et al., 1990b). Since COBI substrate is slightly different from COBA and CYSG substrates the homologies between these 3 proteins might reflect (i) similar structural domains between SAM binding proteins catalyzing SAM-dependent methyl transfer reaction in the cobalamin pathway or (ii) structural similarities between uro'gen III, precorrin-1 and precorrin-2.

Sequence homology between six COB proteins

Apart from the C-methylations catalyzed by SUMT and SP_2MT, five other methyl transfers occur in the pathway and the corresponding enzymes have not yet been identified. We looked for homology between other COB proteins and COBA, COBI and CYSG. In the 5.4 kb fragment carrying 5 cob genes, none of the four proteins, COBB to COBE, presents homology with COBA or CYSG. Among proteins COBF to COBM, apart for COBI, which has already been studied, COBF, COBJ, COBL and COBM show homologies within the three regions of homologies previously presented (data not shown; Crouzet et al., submitted for publication). For each region, a consensus sequence was raised (data not shown) and compared to proteins in Genbank, using Kanehisa's program (1984). Except for COBJ and COBL, region 1, COBF and COBM, region 2, the scores were higher between the consensus sequence and sequences in the regions of homology than with any protein segment in Genbank. On the contrary, homologies within the third region are not significant. These observations shows that, the four proteins have relevant homologies with COBA and/or COBI within conserved region 1 or 2. In addition, the regions involved in the homologies are located in the same parts of the proteins, amino acids 1 to 45, region 1, 60 to 120, region 2 and 85 to 171, region 3. Since the homologies involve conserved regions between 3 SAM methyltransferases, it is postulated that COBF, COBJ, COBL and COBM are probably SAM methyltransferases, but one cannot exclude that these sequences homologies might also reflect structural similarities between the different substrates. All these COB proteins have similar molecular weight (around 28,000) except COBL, the molecular weight of which is 43,000; protein COBL may have a second domain catalyzing another reaction because it is a longer protein and its carboxy end does not contains any of the previously defined homologous region.

COBF affinity to SAM

One way to test the hypothesis that proteins COBF, COBJ, COBL and COBM might be SAM-dependent methyltransferases is, to purify one of them and show that it exhibits SAM affinity to a comparable extent as COBA or COBI. A purification of COBF was performed. The purified protein has the same molecular weight as the one proposed for the COBF protein. NH_2-terminus sequencing of the purifed preparation was done. It is a mixture of two proteins one being initiated at the proposed ATG initiation codon, the second one being translated from the initiation codon which is 15 bp downstream in the sequence. This result will be discussed elsewhere (manuscript in preparation), however this purified preparation was used for SAM binding assay. Purified COBA protein (Blanche et al., 1989) and BSA were also analyzed in the same conditions. Both COBA and COBF proteins show affinity for SAM (see Figure 5) because a radioactivity peak is eluted out of the column at the same retention time as

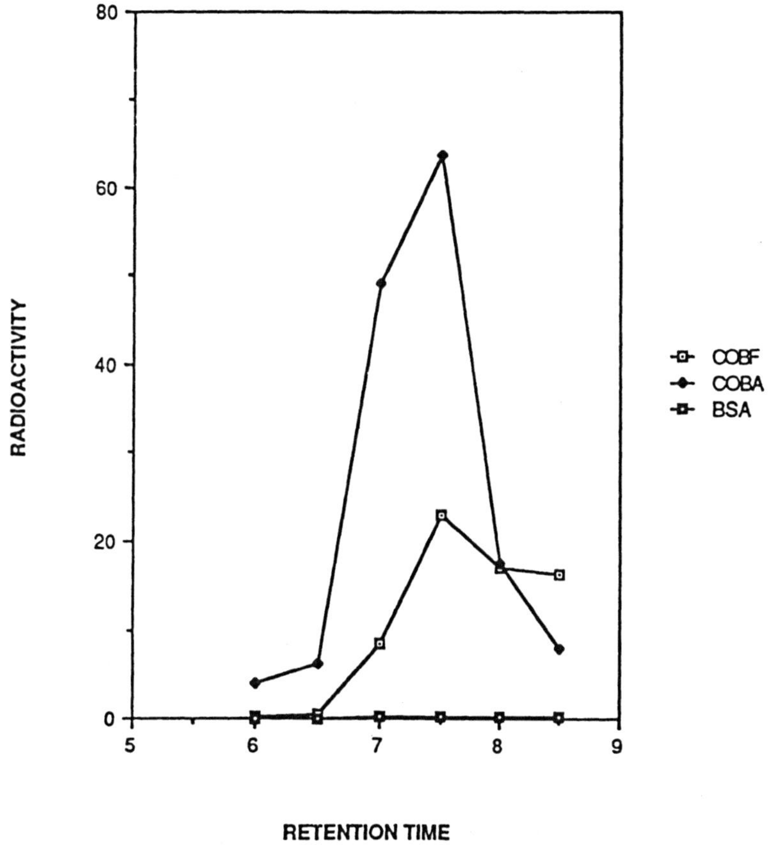

Figure 5. SAM binding affinity of COBA and COBF. The purified proteins were incubated with [methyl-^{3}H]-SAM; after incubation, the samples were injected onto a TSK-125 column (Bio-Rad). Fractions were collected and radioactivity present in each fraction was counted. SAM and protein retention times were determined by on-line monitoring of the absorbance of the eluent at 280 nm. Retention time on the column are indicated in min; radioactivity in fractions representing 0.5 min of elution time is indicated in arbitrary units. Pure BSA has been included in the test. Free SAM is eluted around 10.5 min. Proteins COBA and COBF are eluted around 7.5 min.

the purified proteins alone (7.5 min). This radioactivity peak does not correspond to free SAM, because SAM is eluted out of the column much later (10.5 min.). Contrary to COBA and COBF, BSA does not bind SAM. Therefore it is concluded that COBF, as COBA, is a SAM binding protein confirming the hypothesis based on sequence homology. We propose that COBF, as well as COBJ, COBL and COBM are SAM methyltransferases. It implies that proteins, COBF, COBJ, COBL and COBM play a role in the pathway between precorrin-3 and cobyrinic acid since the methyl groups that are transferred to the macrocycle nucleus, apart from those that are transferred by SUMT and SP_2MT, are introduced between precorrin-3 and cobyrinic acid.

Biochemical study on Cob mutants complemented by the *cobF* to *cobL* genes

Mutants used in the genetic study were analyzed as described (Crouzet et al., submitted for publication) for their intracellular corrinoids content (Blanche et al., submitted for publication). No mutant accumulates any corrinoid (from cobyrinic acid to coenzyme B_{12}) above 1 mg/l threshold, which is significant since other mutants already studied in similar conditions, do accumulate corrinoids (Crouzet et al., submitted for publication). The absence of cobyrinic acid and downstream intermediates in the mutants indicate that they are blocked before cobyrinic acid. The observation that these mutants are blocked before cobyrinic acid is consistent for mutant G613 which is a *cobI* mutant since it is complemented by *P. denitrificans cobI* gene. Based on the fact that *cobI* is the structural gene for SP_2MT this mutant is blocked well before cobyrinic acid. This result shows that, except *cobI*, all *cobF* to *cobM* genes code for proteins acting between precorrin-3 and cobyrinic acid.

Conclusions

We have performed genetic and biochemical studies on the coenzyme B_{12} pathway, from uro'gen III to cobalamin. Among the cloned *Pseudomonas denitrificans* genes, 13 of them have been identified by genetic analysis. *cobA*, *cobI* and *cobB* are the structural gene for SUMT, SP_2MT and Cobyrinic acid *a*,*c*-diamide synthase, respectively. Genetic engineering techniques through gene amplification has allowed us to purify at least three different activities involved in the pathway. Biochemical studies on mutants complemented by the other *cob* genes revealed that proteins COBF to COBH and COBJ to COBM play a role in the pathway between precorrin-3 and cobyrinic acid; four of them (COBF, COBJ, COBL and COBM) being other methyltransferases of the pathway. Proteins COBC and COBD should play a role in the transformation of cobyric acid into cobinamide. The function or a role in a specific part of the pathway of protein COBE remains to be found. Interestingly, proteins comparisons have allowed us to propose that some identified COB proteins are methyltransferases. It will be of interest to see if the enzymes catalyzing the amidations show the same properties;

in such a case COBB proteins should have at least one homologous COB protein. Recombinant DNA is shown in this study to be an important tool in studying the pathway of cobalamin synthesis.

The 13 identified cob genes correspond to 9 complementation groups determined previously by complementation (Cameron et al., 1989). Therefore the number of genes involved in cobalamin synthesis is higher than the one determined by the initial analysis. 14 genes were identified by complementation data; with the results presented in this paper, it can be deduced that there are at least 18 cob genes; however this is certainly an underestimation.

We are currently performing the same approach with the other P. denitrificans cob genes and will soon have completed our study; this will allow us to know the exact number of genes involved in coenzyme B_{12} synthesis. Investigations, based on intermediates accumulations in mutants complemented by the corresponding cob genes, will also allow us to locate the product of all the identified genes in different parts of the pathway: (i) between precorrin-3 and cobyrinic acid, (ii) between cobyrinic acid and coenzyme B_{12}. In the latter steps, since the intermediates are known, and since we have developped a system to separate all of them (Blanche et al., submitted for publication), we will precisely affect a catalytic function for each gene, as it has been done for cobC and cobD. Such investigations are in progress. Overexpression of protein involved in the pathway by gene amplification, as described here, will successfully give to the biochemists a tool for finding unknown activities of the pathway. Genetic fusions between cob genes and a reporter gene will allow us to study the regulation of the cob genes in P. denitrificans. Such work is under progress. This is likely to lead to the isolation of mutants that will have lost some regulations and might be overproducers of coenzyme B_{12}.

References

Babior B. M. 1982. Ethanolamine ammonia-lyase. p. 107-144. In D. Dolphin (ed.), B12 , voll. John Willey & Sons, Inc., New-York.

Bagdasarian, M., R. Lurz, B. Rückert, F. C. Franklin, M. M. Bagdasarian, J. Frey, and K. Timmis. 1981. Specific-purpose plasmid cloning vectors. II. Broad host range, high copy number, RSF1010-derived vectors, and a host vector system for gene cloning in Pseudomonas. Gene, 16, 237-247.

Blanche, F., D. Thibaut, M. Couder and J. C. Muller. Identification of corrinoids precursors of cobalamin from Pseudomonas denitrificans by high-performance liquid chromatography. Submitted to publication.

Blanche, F., L. Debussche, D. Thibaut, J. Crouzet and B. Cameron. 1989. Purification and Characterisation of S-Adenosyl-L-Methionine:Uroporphyrinogen III methyltransferase from Pseudomonas denitrificans. J. Bacteriol., 171, 4222-4231.

Blanche, F., S. Handa, D. Thibaut, C. L. Gibson, F. J. Leeper and A. R. Battersby. 1988. Biosynthesis of vitamin B_{12}: when is the 12β-methyl group of the vitamin generated by acetate decarboxylation ? J. Chem. Soc., Chem. Commun., 1988, 1117-1119.
Brey, R. N., Banner C. D. B., Wolf J. B., 1986. Cloning of Multiple Genes Involved with Cobalamin (Vitamin B12) Biosynthesis in Bacillus megaterium. J. Bacteriol., 167, 623-630.
Cameron, B., K. Briggs, S; Pridmore, G. Brefort and J. Crouzet, 1989. Cloning and analysis of genes involved in coenzyme B_{12} biosynthesis in Pseudomonas denitrificans . J. Bacteriol., 171, 547-557.
Crouzet, J., B. Cameron, L. Cauchois, S. Rigault, M.-C. Rouyez, F. Blanche, D. Thibaut and L. Debussche. Genetic and sequence analysis of an 8.7 kb Pseudomonas denitrificans fragment carrying 8 genes involved in the transformation of precorrin-2 into cobyrinic acid. Submitted to publication.
Crouzet, J., L. Cauchois, F. Blanche, L. Debussche, D. Thibaut, M.-C. Rouyez, S. Rigault, J.-F. Mayaux and B. Cameron. Nucleotide sequence of 5.4 kb DNA fragment from Pseudomonas denitrificans containing five cob genes and identification of structural genes encoding SAM:uroporphyrinogen III methyltransferase and Cobyrinic acid a,c-diamide synthase. Submitted to publication.
De Bruijn, F. J. and J. R. Lupski. 1984. The use of transposon Tn5 mutagenesis in the rapid generation of correlated physical and genetic maps of DNA segments cloned into multicopy plasmids-a review. Gene, 27, 131-149.
Debussche, L. , D. Thibaut, B. Cameron, J. Crouzet, F. Blanche. Purification and characterization of cobyrinic acid a,c-diamide synthase from Pseudomonas denitrificans. Submitted to publication.
Ditta G., Stanfield S., Corbin D., Helinski D. R., 1980. Broad host range DNA cloning system for Gram-negative bacteria: Construction of a gene bank of Rhizobium melitoti. Proc. Natl. Acad. Sci. U. S. A., 77, 7347-7351.
Escalante-Semerena, J. C., S.-J. Suh and J. R. Roth. 1990. cobA function is required for both de novo cobalamin biosynthesis and assimilation of exogenous corrinoids in Salmonella typhimurium. J. Bacteriol., 172, 273-280.
Florent, J. 1986. Vitamins. p115-158. In H.-J. Rehm and G. Reed (ed.), Biotechnology, vol.4, VCH Verlagsgesellschaft mbH, Weinheim.
Grabau, C. and J. Roth. 1989. A vitamin B_{12} mutant blocked in aminopropanol synthesis. Abstr. Annu. Meet. Soc. Microbiol., p. 180.
Hopp, T. P. and K. R. Woods. 1981. Prediction of protein antigenic determinants from amino acids sequences. Proc. Natl. Acad. Sci. USA, 78, 3824-3828.
Jeter, R. M., Olivera B. M., Roth J. R., 1984. Salmonella typhimurium synthetises cobalamin (vitamin B12) de novo under anaerobic growth conditions. J. Bacteriol., 159:206-213.
Jeter, R. M. and J. R. Roth. 1987. Cobalamin (Vitamin B12) Biosynthetic Genes of Salmonella typhimurium. J. Bacteriol. 169, 3189-3198.
Kanehisa, M. 1984. Use of statistical criteria for screening

potential homologies in nucleic acids sequences. Nucleic Acids Res., 12, 203-215.

Macdonald, H. and J. Cole. 1985. Molecular cloning and fuctional analysis of the cysG and nirB genes of E. coli K12, Two closely-linked genes required for NADH-dependant reductase activity. Mol. en. Genet. 200, 328-334.

Miwa, K., T. tsuchida, O. Kurahashi, S. Nakamori and K. Sano. 1983. Construction of L-threonine overproducing strains of Escherichia coli K-12 using recombinant DNA techniques. Agric. Biol. Chem., 47, 2329-2334.

Normark, S., S. Bergtröm, T. Edlund, T. Grundström, B. Jaurin, F. Lindberg and O. Olsson. 1983. Overlapping genes. Ann. Rev. Genet., 17, 499-525.

Ozaki, A., R. Katsumata, T. Oka and A. Furuya. 1985. Cloning of the genes concerned in phenylalanine biosynthesis in Corynebacterium glutamicum and its application to breeding of a phenylalanine-producing strain. Agric. Biol. Chem., 49, 2925-2930.

Peakman, T., J. Crouzet, J.-F. Mayaux, S. Busby, S. Mohan, R. Nicholson and J. Cole. Nucleotide sequence, organisation and structural analysis of the products of the nirB to cysG region of the E. coli chromosome. Submitted to publication.

Platt T. 1986. Transcription termination and the regulation of gene expression. Ann. Rev. Biochem., 55, 339-372.

Sano, K., K. Ito, K. Miwa and S. Nakamori. 1987. Amplification of the phosphoenolpyruvate carboxylase gene of Brevibacterium lactofermentum to improve amino acid production. Agric. Biol. Chem., 51:597-599.

Skatrud, P. L., A. J. Tietz, T. D. Ingolia, C. A. Cantwell, D. L. Fisher, J. L. Chapman and S. Queener. 1989. Use of recombinant DNA to improve production of cephalosporin C by Cephalosporium acremonium. Bio/Technology, 7, 477-485.

Stachel, S. E., G. An, C. Flores and E. W. Nester. 1985. A Tn3lacZ transposon for the random generation of β-galactosidase gene fusions: application to the analysis of gene expression in Agrobacterium. Embo J., 4, 891-898.

Staden R. and A. D. McLachlan. 1982. Codon preference and its use in identifying protein coding regions in long DNA sequences. Nucleic Acid Res., 10, 141-156.

Staden R. 1984. Measurements of the effects that coding for a protein has on a DNA sequence and their use for finding genes. Nucl. Acid Res., 12, 551-567.

Thibaut, D., M. Couder, J. Crouzet, L. Debussche, B. Cameron and F. Blanche. Assay and purification of S-adenosyl-L-methionine:precorrin-2 methyltransferase from Pseudomonas denitrificans. Submitted to publication.

Warren, M., C. A. Roessner, P. J. Santander and A. I. Scott. 1990a. The Escherichia coli cysG encodes S-adenosylmethionine-dependant uroporphyrinogen III methylase. Biochem. J., 265, 725-729.

Warren, M., N. J. Stolowich, P. J. Santander C. A. Roessner, B. A. Sowa and A. I. Scott. 1990b. Enzymatic synthesis of dihydrosirohydrochlorin (precorrin-2) and a novel pyrrocorphin by uroporphyrinogen III methylase. FEBS Lett., 261, 76-80.

Wolf, J. B., and R. N. Brey. 1986. Isolation and genetic characterisation of Bacillus megaterium Cobalamin biosynthesis-deficient mutants. J. Bacteriol., 166, 51-58.

GENETIC APPROACHES TO THE SYNTHESIS AND PHYSIOLOGICAL SIGNIFICANCE OF B12 IN SALMONELLA TYPHIMURIUM

John R. Roth, Charlotte Grabau and Thomas G. Doak

Department of Biology
University of Utah
Salt Lake City, Utah 84112

ABSTRACT

In the bacterium Salmonella typhimurium the biosynthetic genes for cobalamin are located in three separated regions of the genetic map. Most synthetic genes (approximately 20 of 25) are located in a single large operon whose expression is stimulated by cAMP and by a reduced cell interior; this operon is repressed by adenosyl-B12. Nutritional requirements of mutants suggest that an early intermediate in corrinoid biosynthesis (prior to cobyric acid in the pathway) is adenosylated and that adenosyl cobinamide is an obligatory precursor to completion of cobalamin synthesis (addition of dimethyl benzimidazole). Only one gene has been associated with the ability to perform adenosylation in the de novoe pathway; this gene is one of several alternatives needed to adenosylate exogenouse corrinoids. One gene has been associated with ability to synthesize aminopropanol.

Salmonella uses about 1% of its genome to encode the synthesis and transport of B12. While this large genetic investment suggests an important physiological role for the cofactor, the four known B12-dependent functions do not appear to be essential for growth of wild type Salmonella under the laboratory conditions tested. It is particularly puzzling that Salmonella only synthesizes B12 when growing anaerobically.

INTRODUCTION

Several years ago, we discovered that the enteric bacterium Salmonella typhimurium is able to synthesize cobalamin de novo (1). It is surprising that this enormous pathway was not discovered earlier in the 30-year history of intensive genetic investigation of Salmonella (2). The explanation of this delay lies in the fact that most genetic work has been done on aerobically-grown cultures and Salmonella synthesizes B12 only during anaerobic growth. The finding of B12 synthesis in Salmonella opened two areas of investigation. First, since Salmonella has a well-developed genetic system, an opportunity was provided to apply genetic methods to the study of B12 synthesis and its regulation. Second, since synthesis of B12 is seen only under anaerobic conditions, questions are raised regarding the physiological importance of B12 to Salmonella and more generally to all organisms. Since B12 is not uniformly used by all life forms and is only produced anaerobicaly by Salmonella, one can ask what aspects of metabolism or life style have selected for continued use of B12 by some life forms while other groups have abandoned use of the cofactor? Below we will review our observations regarding the synthesis of cobalamin in Salmonella and speculate about the general physiological significance of this cofactor.

SYNTHESIS OF COBALAMIN IN SALMONELLA

Mutations affecting B12 synthetic enzymes

Isolation of mutants The starting material for any genetic analysis is mutants. Mutants lacking cobalamin synthetic enzymes were identified by taking advantage of Salmonella's B12-dependent methionine synthetic enzyme (metH). In strains lacking the B12-independent alternative enzyme (metE), methionine synthesis is dependent on the metH enzyme and therefore on availability of B12. Anaerobically, metE mutants make B12, activate their MetH enzyme and synthesize methionine. Derivatives of metE mutants that cannot make B12 grow on minimal medium only if either methionine or B12 is provided exogenously. Such mutants are designated cob for their defect in cobalamin synthesis.

Three loci on the genetic map Mutations affecting cobalamin synthesis map at three widely separated locations in the Salmonella chromosome. (See Figure 1.) The main gene cluster lies near the his locus at minute 42 of the genetic map (1,3). A single gene, cobA, lies near the trp locus at minute 34 (4). The third region, close to the nadD gene at minute 14, seems to include at least two genes (cobD and E) (Charlotte Grabau, unpublished). These regions are designated on the map in Figure 1.

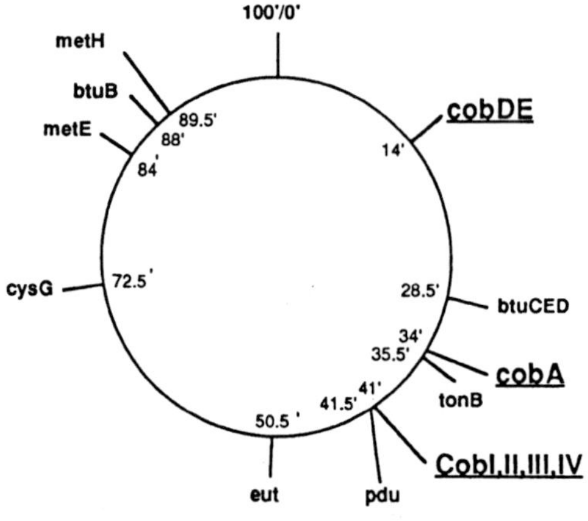

Figure 1 Genetic map of Salmonella typhimurium

Included in Figure 1 are genes for B12 transport (btu) which have been characterized in *E. coli* by Kadner, Bradbeer and coworkers (5-6); the existence of analogous genes in Salmonella has been confirmed (Tom Doak and Charlotte Grabau, unpublished).

While all known *cob* mutations map in one of these regions, we can not exclude the existence of B12 synthetic genes mapping in additional regions. Below we will discuss each of the known regions and what is known about its encoded functions.

The main operon

Genetic Map. The general structure of the main *cob* gene cluster (2) is presented in Figure 2 . As is indicated in the map, there are three general regions, designated I, II, and III. Mutations can be placed into these groups by determining

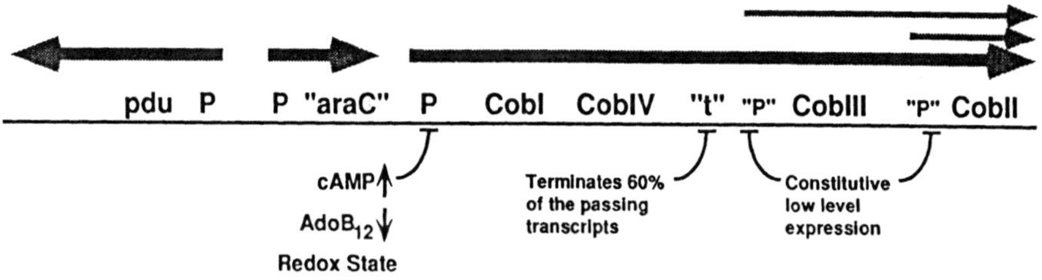

Figure 2 Genetic map of the main Cob operon including promoters, control sites and the adjacent *pdu* operon.

their nutritional requirements. Mutations in region I cause a failure to make B12 (scored as *metH* activity) that is circumvented by addition of cobinamide (Cbi); thus, this region appears to encode enzymes involved in elaboration of the corrinoid ring. Mutations in region II cause a defect that can be remedied if dimethylbenzimidazole (DMB) is provided; these mutations appear to affect synthetic enzymes for this constituent of B12. Mutants for region III fail to make B12 even when both Cbi and DMB are given; they appear to be unable to complete the reactions involved in modifying and joining these intermediates to form B12. Within the CobI region is a subset of mutations that cause a B12 synthetic defect that can be remedied by addition of cobalt at high concentration; it seems likely that these mutations (CobIV) affect a high affinity transport system for cobalt. This possibility is supported by sequence data for this region which predicts protein sequences similar to those of known transport systems.

Synthetic functions (aerobic and anaerobic). The functions outlined above were defined by mutant phenotypes determined for anaerobic cells. In the presence of oxygen, even wild type cells are unable to make B12; the failure to make B12 aerobically is due to a failure of the CobI pathway, completion of the corrinoid ring. The other Cob functions (II and III) are active under both aerobic and anaerobic conditions. Thus wild type Salmonella behaves, under aerobic conditions, like a CobI mutant in that it is unable to make B12 de novo, but can do so if cobinamide is provided. The failure to make B12 aerobically is not due to a lack of expression of the relevant genes; no B12 is made aerobically even when regulatory mutations provide a high level of expression of the CobI genes. Thus the failure to make B12 is likely to be due to oxygen sensitivity of early synthetic intermediates or enzymes involved in early steps of the pathway.

DNA sequence of the main operon. The first 17kb of the main operon have been cloned (D. Andersson and J. Escalante, unpublished results) and sequenced in collaboration with George Church, Steve Kieffer-Higgins and Mark Rubenfeld. This sequence, which is still being refined, includes the CobI, and III regions but very little of the CobII region. The missing CobII region has been difficult to clone, but preliminary results (Ping Chen, unpublished) suggest that lambda vectors including this region have now been identified. Reliable assessment of reading frames in the cob operon must await final details of sequence determination, but inspection of the current sequence data suggests the existence of 13 genes in the CobI region (including CobIV) and 3 genes in the CobIII region. Complementation studies of the CobIII and CobII regions suggest five functions in CobIII and four in CobII (Jorge Escalante and coworkers, unpublished results).

Just to the left (counterclockwise) of the CobI operon promoter is a sequence which shares homology with the positive regulatory protein of the arabinose operon (araC). It is not yet clear whether this protein is involved in control of either the cob or pdu operons. Analysis of the CobI sequence also reveals three open reading frames that are homologous to componenets of a class of tranport systems. These genes are likely to be involved in cobalt transport.

Number of functions involved in providing B12

From the above discussion, we believe that the main cob gene cluster includes about 20 genes. The cobA region, discussed below, seems to include only one gene (4a). The cobD region includes two genes (cobD and E) based on mutant phenotypes (Charlotte Grabau, unpublished), however no complementation tests have been done and more genes are certainly possible. Based on the work of Bradbeer and Kadner (5 and 6; for review see 7), transport of B12 requires the function of at least four genes that act uniquely on B12. In summary, available data suggest that at least 25 genes are dedicated to the transport and synthesis of B12 in Salmonella. Allowing for the possibility that a few genes remain to be discovered, Salmonella seems to use nearly 1% of its 3000 genes to provide B12. Below we will discuss the significance of this genetic investment.

Adenosylation of corrinoids

Evidence that the *cobA* protein plays a role in adenosylation (aerobic studies) Aerobic cultures of wild type Salmonella (which do not produce B12) can utilize ethanolamine as a carbon and nitrogen source if some form of exogenous corrinoid is provided. Several corrinoids (Me-B12, CN-B12 or Ado-B12) permit aerobic growth on ethanolamine. Since the cofactor of ethanolamine ammonia lyase (eutBC) is adenosyl cobalamin (Ado-B12), we conclude that these corrinoids provide (or can be converted to) internal Ado-B12 in the presence of oxygen. (See Figure 3 for a summary of the inferred pathways, which are discussed below.)

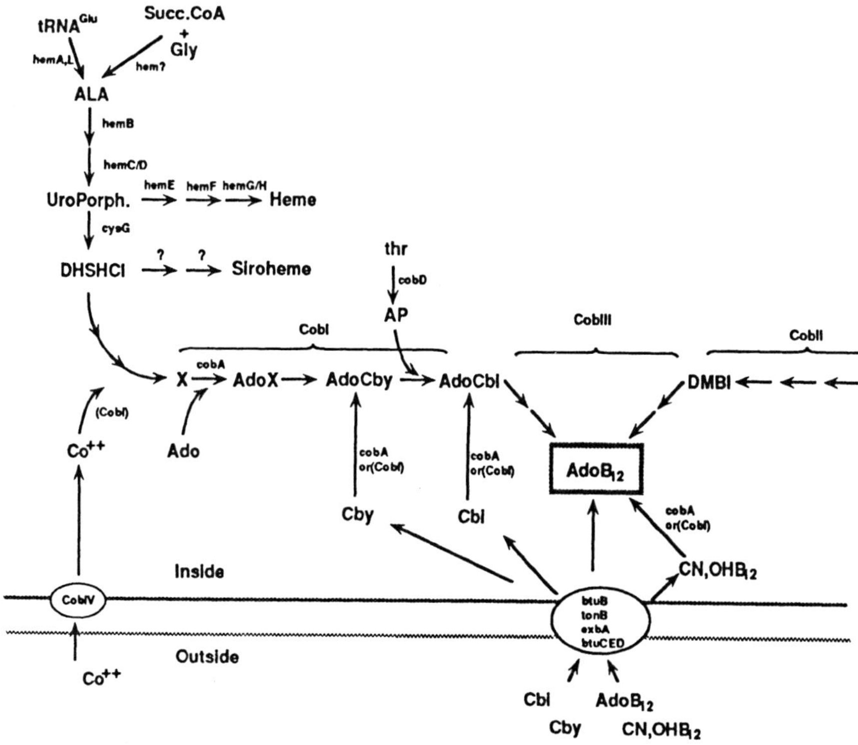

Figure 3 Outline of the cobalamin synthetic pathway with aspects revealed by genetic studies.

When cobA mutants are tested aerobically, they can use Ado-B12 but cannot use CN-B12 or Me-B12 to permit growth on ethanolamine. The requirement of these mutants for an adenosylated corrinoid to support use of ethanolamine, suggests that the mutant defect must lie in formation of adenosylated corrinoids (either cobalt reduction or adenosyl transfer) (4a).

Ado-Cby and Ado-Cbi (not Cby and Cbi) are biosynthetic intermediates. Aerobic cultures of cobA mutants can use either CN-B12 or Ado-B12 to activate the metH enzyme. This is consistent with the cobA defect in adenosylation described above, since the metH enzyme (unlike ethanolmine lyase) can use non-adenosylated forms of B12 as a cofactor. Surprisingly, aerobic cultures of cobA mutants cannot use CN-Cbi or CN-Cby as a precursor of corrinoids for activation of MetH enzyme, but can use Ado-Cbi or Ado-Cby. Since the metH enzyme does not require adenosylated B12, we conclude that cobA mutants require adenosylated precursors in order to complete the addition of DMB and formation of cobalamin. We conclude that formation of cobalamin requires corrinoid adenosylation at a step prior to addition of DMB and that cobA mutants can not convert either CN-Cby or CN-Cbi to their adenosylated forms or to any form of B12 under aerobic conditions.

Adenosylation in the de novo pathway requires the cobA gene Under anaerobic conditions, wild type Salmonella can produce Ado-B12 de novo and, based on the reasoning presented above, must be able to adenosylate some early precursor in the CobI pathway; this adenosylation requires cobA function. We conclude this because cobA mutants fail (anaerobically) to make B12 for either the metH or the eutBC functions. However cobA mutants can use Ado-Cby or Ado-Cbi as precursors of B12 under both aerobic and anaerobic conditions.. (Anaerobically they can also use non-adenosylated forms as described below.) This is evidence that the cobA gene product is required for an early step in the de novo CobI pathway (leading to Ado-Cbi formation) as well as for aerobic assimilation of exogenous corrinoids (described above).

Alternative assimilatory functions to CobA are activated under anaerobic conditions With or without oxygen, wild type cells can take up both CN-Cby and CN-Cbi and use them as precursors of cobalamin (scored by MetH activity). Under aerobic conditions cobA mutants fail to use these precursors, but can assimilate either Ado-Cby or Ado-Cbi. In contrast, under anaerobic conditions, cobA mutants can use the non-adenosylated precursors. We interprete this to mean that the cobA function is involved in adenosylation and that genes encoding alternatives to the cobA function are turned on under anaerobic conditions. These alternative routes can provide only for adenosylation of exogenous corrinoids; they do not allow de novo synthesis of B12 (see above).

We have isolated two classes of mutants that lack the anaerobic alternative adenosylation functions. This suggests that at least two proteins may contribute to this alternative. These functions are cobB, and cobX. The cobB gene maps within the CobIII region of the main gene cluster (Escalante, unpublished). The cobX gene lies just outside the main cluster and may be within the operon encoding propandiol utilization (pdu), described below (Tom Doak, unpublished).

Biochemical interpretation of the functions inferred above is not clear. Based on previous enzymology (20a, 20b), one would expect three functions to be required to achieve adenosylation of corrinoids, two reductases and an adenosyltransferase. However, only one gene (cobA) has been identified which is required for aerobic adenosylation of assimilated of corrinoids and anaerobic adenosylation in the de novo pathway. This could mean that the cobA enzyme alone performs all adenosylation functions or that the cobA protein is one of several functions required for adenosylation of corrinoids and the other types have escaped detection. Detection of some classes of mutants could be difficult if multiple enzymes exist, each sufficient to provide the function in vivo. The possible multiplicity of functions able to reduce cobalt or adenosylate corrinoids may be further complicated by the existence of adenosylation functions encoded within the ethanolamine (eut) operon; the eutA gene may contribute to Ado-B12 formation (13). Construction of strains with multiple defects should make it possible to sort out these possibilities.

Synthesis and addition of aminopropanol

It is thought that aminopropanol (AP; derived from threonine) is added to Cby to form Cbi; AP forms part of the sidechain that ultimately will connect DMB to the corrinoid ring. Two classes of mutants appear defective in converting Cby to Cbi. (Based on the reasoning presented above, we expect that the actual reaction is Ado-Cby to Ado-Cbi.) Both mutant classes were detected as being defective in B12 synthesis anaerobically but able to make B12 (either aerobically or anaerobically) if provided with Cbi. Neither class is helped by addition of Cby, suggesting a defect in synthesis or addition of aminopropanol. One class of mutations maps within the CobI region of the main gene cluster. The other class, cobD, maps at minute 14. The cobD mutants are unique in that they can make B12 if aminopropanol is provided, suggesting that they are defective in synthesis of this compound and that aminopropanol *per se* can serve as a precusor of the side chain. The enzymatic conversion of theonine to aminopropanol has not been demonstrated, but, based on the existence of only one mutant class, would appear to be achieved by a single protein. Our results suggest that aminopropanol is not made by a previously suggested (20a, 20b) sequence of known reactions: threonine -> b-keto-a-amino butyrate -> amino acetone -> aminopropanol. The cobD mutants still possess threonine dehydrogenase activity and are able to use threonine as a source of serine (presumably through b-keto-a-amino butyrate). Conversely, mutants unable to use threonine as a serine source retain the ability to make B12. The, as yet undemonstrated, direct decarboxylation of threonine seems more likely. If threonine reacts directly with a B12 intermediate in forming the sidechain, these reaction must be able to use aminopropanol as an alternative substrate.

Synthesis of dimethylbenzimidazol

Enzymes for synthesis of dimethylbenzimidazole are encoded in the CobII region of the main operon. Results of Jorge Escalante and coworkers at the University of Wisconsin suggest a four-enzyme pathway with dimethyl analine as precursor; the source of this precursor has not been identified. The CobII

Table 1. Expression of main cobalamin operon during fermentation and respiration.

	β-galactosidase activity during growth on:			
	Glucose		Glycerol	
Genotype	+ O_2	- O_2	+ O_2	+ Fumarate
CobI::lac	4	20	10	430
CobII::lac	20	40	40	280
CobIII::lac	30	30	30	240

genes have recently been cloned (Ping Chen and Roth, unpublished results), and their base sequence is being determined. Formation of DMB seems to limit the production of B12 under anaerobic growth conditions; addition of endogenous DMB greatly increases cellular B12 levels in wild type cells (Andersson and Doak, unpublished results).

Regulation of B12 synthesis

Operon structure and internal promoters. Our current view of the structure of the main operon is described in Figure 2. It is not yet clear whether the Cob operon and the adjacent genes for propanediol utilization (8) are regulated in some concerted way. The main *cob* operon appears to include a single large transcription unit which is regulated as described below (Tom Doak, unpublished). About 60% of the transcripts from the main (regulated) promoter appear to terminate before the CobIII region. Several internal promoters exist which provide for expression of the CobIII and CobII genes, even when transcription from the main promoter is blocked. These secondary promoters are weak, providing only about 5% of the level of the maximally-induced main promoter. There is no evidence for control of the internal promoters.

The above conclusions are based on use of operon fusions which place the *lac* operon of *E. coli* under the control of promoters in the main *cob* operon. The expression of the *lacZ* gene is assayed under various conditions and in strains carrying a variety of mutations in addition to the operon fusion.

Regulation of the main promoter by catabolite repression

The main operon appears to be subject to catabolite repression in that transcription levels are lower in cells grown with glucose and can be increased by addition of cAMP (9). Operon expression is low in *crp* mutants, which lack the catabolite activator protein. Similarly, operon expression is low in *cya* mutants which are unable to produce cAMP; in the latter mutants, operon expression is increased by addition of cAMP. The catabolite repression system (Crp protein/ cAMP) does not seem to be involved in the redox control described below since normal redox control is seen in *cya*, *crp** double

Table 2. Effect of electron transport inhibition on expression of the main cobalamin operon in the presence of oxygen.

Treatment	β-galactosidase activity of a Cob::lac fusion: Before	After	Induction ratio
Cyanide added	5	98	20
Ubiquinone removed	5	31	6
Heme removed	5	28	6

mutants. These strains cannot synthesize cAMP; they make a form of Crp regulatory protein that activates target genes regardless of the cAMP level (9).

Regulation of the main promoter by redox level

In initial experiments, production of B12 was only seen in cells grown in the absence of oxygen, during fermentation or anaerobic respiration (Table 1). This prompted assays of operon

regulation in response to the presence of oxygen; these assays showed maximal operon expression under conditions of anaerobic respiration with fumarate as an electron acceptor and glycerol as a carbon/energy source (9). During fermentation of glucose, B12 is produced but operon expression is low.

Subsequent investigation of this regulatory behavior demonstrated that the operon is induced in response to a reducing cell interior; oxygen *per se* does not appear to be involved in signaling operon repression (11). The main evidence for this conclusion is that expression of the operon is induced, even in the presence of oxygen, if electron transport is inhibited (Table 2). This has been done by adding cyanide, withholding ubiquinone from an ubiquinone auxotroph or heme from a heme auxotroph. Further evidence is the observation that the level of Cob operon expression is inversely related to the midpoint potential of the electron acceptor used during respiration (11); this suggests that the Cob operon is induced by the "accumulation" of internal reducing agents. The exact nature of the signal compound is not clear.

Mutations (including deletions) in the promoter region of the main operon allow expression of the main Cob operon in the presence of oxygen (10). These mutants retain ability to regulate the operon in response to cAMP and B12, suggesting that they have not fused the operon to a foreign promoter. Many mutations unlinked to the main operon affect the redox regulation; however, due to the variety of metabolic alterations that could affect internal redox conditions, we suspect that very few of these mutations have a direct effect on components of the regulatory machinery. However, true regulatory types may ultimately be found among these mutants.

Table 3. Evidence that Ado-B12 is the effector for endproduct control of main operon transcription

Genotype	β-galactosidase activity during growth with: no addition	plus CN-Cbi	plus CN-B12	plus Ado-B12
cob-24::lac cobA$^+$	364	116	25	27
cob-24::lac cobA::Cam	373	343	287	54

Regulation of the main promoter by corrinoids

Transcription of the main operon is repressed by addition of exogenous CN-B12 or Ado-B12 (1, 3). The level of endogenously synthesized B12 are only sufficient to fractionally repress cob transcription. Externally supplied B12 can accumulate to higher intercellular levels and cause a more complete repression. Recent work suggests that the actual effector for repression is Ado-B12 (Tom Doak, unpublished); in strains with a cobA mutation, the main operon is repressed by addition of Ado-B12, but not by CN-B12. Results for aerobically grown cells are in Table 3; under anaerobic conditions the cobA alternative functions are expressed and provide for repression by CN-B12, presumably by allowing synthesis of Ado-B12 (Tom Doak, unpublished).

PHYSIOLOGICAL SIGNIFICANCE OF B12

Salmonella's investment in B12 and its returns

As described above, Salmonella dedicates almost 1% of its genome to the synthesis and transport of B12. One might imagine that this large genetic investment is justified by some essential role of B12 in metabolism. From observation of the known functions of B12 in Salmonella (discussed below), the importance of B12 to Salmonella is not obvious. It is especially puzzling that B12 is only synthesized anaerobically, since none of the known B12-dependent enzymes has a metabolic role that is known to be particularly important in the absence of oxygen. It could be argued that the inherent oxygen-sensitivity of early intermediates in corrinoid ring formation makes it futile to attempt synthesize B12 aerobically. While this might explain the failure to make B12 aerobically, other bacteria are known to synthesize B12 in the presence of oxygen so it must be possible for cells to protect these biosynthetic intermediates. We suspect that the anaerobic production of B12 by Salmonella reflects a metabolic importance of B12 that is realized only under anaerobic conditions. If the chief value of B12 were realized only under anaerobic conditions, there would be no selective pressure for evolving mechanisms for protection of oxygen-sensitive intermediates or biosynthetic enzymes. Later we will speculate on a possible basis for this anaerobic significance.

Known B12-dependent functions in Salmonella

The four known B12-dependent functions are listed below. We will review briefly the known physiological importance of each function with emphasis on experiments we have done. Unless additional B12-dependent enzymes are discovered, the physiological importance of B12 to Salmonella must lie in one of these functions.

Synthesis of methionine (the *metH* and *metE* enzymes) Two enzymes enzymes (encoded by the metH and metE genes) are independently able to catalyse the last step of methionine synthesis (homocysteine methyl transferase). The metE enzyme does not require B12 and is responsible for methionine synthesis when no B12 is available. Under anaerobic conditions, or when B12 is supplied exogenously to aerobic cells, the metE gene is repressed and the alternative B12-dependent enzyme is produced. Because of this, mutants defective for the metE enzyme are isolated aerobically as auxotrophs whose nutritional requirement can be satisfied by either methionine or by B12 (which activates the alternative metH enzyme).

The metE enzyme has a 100-fold lower turnover number than the B12-dependent alternative. This would appear to make the B12-independent route costly to the cell, since during synthesis of methionine via the metE route this enzyme is present as several percent of total cell protein. It could be argued that the importance of B12 to Salmonella lies in supporting a more efficient means of methionine synthesis under anaerobic conditions when energy might limit growth. We have compared the growth rate of Salmonella strains using the two alternative routes of methionine synthesis in aerobic glucose-limited chemostats (Dan Andersson, unpublished). With B12 supplied exogenously, strains using the metH route showed no growth advantage over those using the inefficient, B12-independent metE enzyme. Thus, if the significance of B12 to Salmonella lies in support of methionine synthesis, this value was not apparent under the conditions tested.

Utilization of ethanolamine Ethanolamine can provide a source of energy, carbon and nitrogen. Utilization of ethanolamine requires the presence of Ado-B12 as cofactor for the B12-dependent enzyme, ethanol ammonia lysase. We have investigated ethanolamine utilization by Salmonella (12, 13). We have only been able to demonstrate utilization of ethanolamine as carbon and energy souce for cells growing aerobically. Slow growth is achieved when ethanolamine is used as a nitrogen source anaerobically. These results suggest that ethanolamine is most valuable to Salmonella under aerobic conditions (when B12 is not synthesized and when use of ethanolamine requires exogenous B12). Based on these simple observations, the ethanolamine pathway does not seem a likely candidate for explaning the physiological value of the B12 biosynthetic genes. Induction of the ethanolamine operon under aerobic conditions requires provision of both ethanolamine and B12 exogenously (12,13); under anaerobic conditions, B12 is synthesized and ethanolamine alone is sufficient for induction (David Roof, unpublished). The existence of this regulatory mechanism is consistent with observations (described above) that B12 is not made by aerobically and thus is an "undependable" cofactor for

Salmonella. Except for B12-dependent enzymes [ethanolamine lyase, metH and propanediol dehydratase (described below)], we know of no enzyme whose production is regulated by the presence of its cofactor.

Utilization of propanediol Although Salmonella can not ferment propanediol, it does synthesize the B12-dependent enzyme propanediol dehydratase and can use propanediol as a carbon and energy source under aerobic conditions (8). Although Salmonella, using endogenous B12, degrades propanediol anaerobically (14), this degradation does not provide a carbon or energy source. Therefore, as is true for ethanolamine, the value of propanediol seems greater under aerobic conditions (when no B12 is produced). Catabolism of these substrates aerobically requires an exogenous B12 source.

Contrary to the above rationalization, the genetic location of the genes for propanediol utilization suggests that we should look for anaerobic conditions under which propanediol utilization is important to Salmonella. The genes for propanediol use (pdu) map immediately adjacent to the main cob operon (see Figure 2). Although the two operons are transcribed divergently, we suspect that their regulatory mechanisms are interrelated. If so, it would suggest that an important use of B12 is connected with propanediol utilization.

Synthesis of Q-base The fourth B12-dependent function in *E. coli* (and presumably in Salmonella as well) is in synthesis of the hypermodified tRNA base, queuosine. This base is found adjacent to the anticodon of several tRNAs. It is not clear what role this modified base plays in cell physiology, since mutants that fail to make the base grow well, at least under laboratory conditions (15). B12 is required only for the last step in synthesis of the base; in the absence of B12 the precursor epoxyqueosine is formed but is not converted to queosine. Since tRNA's have been shown to play a role in regulation of genes having attenuation mechanisms (19), it is conceivable that synthesis of B12 under anaerobic conditions could, by altering tRNA structure, lead to differences in gene expression. Thus B12 could mediate gene regulation in response to reducing conditions.

Summary of B12-dependent functions

None of the functions described above seems critical to Salmonella and none seems to have unique value under anaerobic conditions. There is an alternative (metE) for the B12-dependent methionine synthetic enzyme (metH). Both ethanolamine and propanediol, whose degradation requires B12, seem to be most useful under aerobic conditions, when B12 is not made. There is no obvious selective value for the Q-base of tRNAs.

This line of reasoning is, of course, dangerous, since it applies only to large differences in growth detected under particular laboratory conditions. In nature, even minor differences in growth rate might be of critical importance over long periods of time. Since our understanding of bacterial ecology is rather scant, it is easy to imagine that in nature, one or all of these four functions are very important. It is also possible that these functions are not critical to the growth of cells, but are important for some aspect of interaction with the host organisms inhabited by these bacteria.

One could argue that the above list of B12-dependent functions does not include some undiscovered critical function that is of obvious importance. While the existence of additional B12-dependent enzymes is possible, their discovery would not provide an easy answer to the question of significance. Mutants which are unable to synthesize B12, including deletion mutants that lack the entire large operon, are able to grow well both aerobically and anaerobically. They can ferment a variety of carbon sources and can use many more by anaerobic respiration if an alternative electron acceptor (NO_3, DMSO, TMAO, fumarate) is provided. The only physiological defect we observe for cob mutations in otherwise wild type bacteria is failure to use ethanolamine as a nitrogen source under anaerobic conditions; even in wild type cells, growth on ethanolamine under these conditions is very poor. [Salmonella and *E. coli* possess an anaerobic alternative to the oxygen-dependent nucleotide reductase, but the anaerobic form does not appear to require B12. (16 and David Roof, unpublished)]

General speculation on the role of B12 in nature

Based on our experience with B12 in Salmonella, we would like to speculate about the general physiological significance of this cofactor. Cobalamin is "undependable" in Salmonella, being synthesized only under anaerobic conditions. Similarly in nature B12 seems to be unevenly distributed, present in some organisms but not others. Could these clues suggest a generalization?

An old vitamin in a new world It has been argued that B12, despite its structural complexity, formed prebiotically on this planet and was used by primitive life forms when the envirionment was still anaerobic and reducing (17, 18). These early life forms then developed the ability to synthesize B12. Since these early days, plants and oxygen have appeared. We suggest that the chief metabolic value of B12 is still realized under the reducing conditions and this consideration might be used to rationalize the current distribution of B12 usage in nature and the patterns of gene regulation that control its synthesis in organisms such as Salmonella.

Production of electron sinks Most of the known B12-dependent enzymes in bacteria support fermentation or the ability to utilize small molecules as a carbon and energy source in the absence of oxygen or alternative electron acceptors. In many of these cases, the B12-catalyzed rearrangement generates (e. g., by a deamination or dehydration) a reducible product. Two prime examples are ethanolamine lyase and propanediol dehydratase, both of which generate aldehydes that could be reduced and serve as electron sinks for elimination of excess reducing power. Salmonella appears to have lost the simple fermentations (glycerol, propanediol) but may still use B12 as a means of providing electron sinks (acetaldehyde and propionaldehyde) under reducing conditions. It seems consistent with this idea that a reducing cell interior (not oxygen *per se*) signals induction of B12 production in Salmonella. We would like to consider the possibility that the predominant function of B12 in nature is to support metabolism which allows life in reducing environments.

Distribution of B12 reflects need for reducible metabolites

The above view of B12 fits with the general observations in bacteria. Most anaerobes have maintained B12 production and the cofactor plays an important role in supporting their anaerobic life style. For these organisms the conditions of life remain much as they were during the "heyday" of B12, the primitive reducing world. Salmonella, a facultative anaerobe, has maintained the ability to synthesize B12, but uses this ability only when growing anaerobically. This aspect of Salmonella's behavior fits with the general pattern even though we have not yet identified conditions for which B12 improves Salmonella's ability to grow anaerobically. While Salmonella lacks the ability to perform the classic B12-dependent fermentations (e. g., glycerol, propanediol) we predict that it still uses the products of B12 reactions as sites to get rid of excess reducing power. To demonstrate the value of these electron sinks, we will need to find mixtures of substrates which can support anaerobic fermentation of Salmonella only if B12 synthesis is in place.

If we extend this idea to plants, their exptreme exposure to oxygen would obviate the need for reducible metabolites and might explain their loss of B12 usage. We suggest that predecessors of modern animals may have become dependent on B12 for catalysis of reactions not immediately associated with the need for reducible metabolites (e.g. methyl transfer). Perhaps because of this they have remained dependent on B12 even after the arrival of oxygen and respiration as a source of energy.

Literature Cited

1. Jeter, R. M., B. M. Olivera, and J. R. Roth 1984. *Salmonella typhimuriuim* synthesizes cobalamin (vitamin B_{12}) *de novo* under anaerobic growth conditions. J. Bacteriol. 170: 2078-2082.

2. Sanderson, K. E., and J. R. Roth 1988. The linkage map of Salmonella typhimurium; Edition VII. Microbiological Reviews 52: 485-532.

3. Jeter, R. M., and J. R. Roth. 1987. Cobalamin (vitamin B_{12}) biosynthetic genes of *Salmonella typhimurium*. J. Bacteriol 169: 3189-3198.

4a. Escalante-Semerena, J. C., S.-J. Suh and J. R. Roth. 1990 The *cobA* function is required for both de novo cobalamin biosynthesis and assimilation of exogenous corrinoids in *Salmonella typhimurium*. J. Bacteriol. 172: 273-280.

4b Lundrigan, M. D., and R. J. Kadner. 1989. Altered cobalamin metabolism in *Escherichia coli btuR* mutants affects *btuB* gene regulation. J. Bacteriol. 171: 154-161.

5. Reynolds, R., G. Mottur and C. Bradbeer 1980 Transport of Vitamin B12 in *Escherichia coli* J. Biol. Chem. 255: 4313-4319

6. Bassford Jr., P., and R. J. Kadner 1977 Genetic Analysis of components involved in Vitamin B12 uptake in Escherichia coli. J. Bacteriol. 132: 796-805.

7 Sennett, K., L. Rosenberg and I. Mellman. 1981. Transmembrane transport of cobalamin in prokaryotic and eukaryotic cells. Annu. Rev. Biochem. 50: 1053-1086

8 Jeter, R. M. 1990. Cobalamin dependent 1,2-propanediol utilization by Salmonella typhimurium. submitted to J. Gen. Microbiol.

9. Escalante-Semerena, J. C., and J. R. Roth. 1987. Regulation of cobalamin biosynthetic operons in Salmonella typhimurium. J. Bacteriol. 169: 2251-2258.

10. Andersson, D. and J. R. Roth. 1989. Mutations affecting regulation of the cobinamide biosynthetic genes in Salmonella typhimurium. J. Bacteriol. 171: 6726-6733.

11 Andersson, D. and J. R. Roth. 1989. Redox regulation of the cobinamide biosynthetic genes in Salmonella typhimurium. J. Bacteriol 171: 6734-6739.

12. Roof, D. M., and J. R. Roth. 1988. Ethanolamine utilization in Salmonella typhimurium. J. Bacteriol. 170: 3855-3863.

13 Roof, D. M., and J. R. Roth. 1988. Functions required for vitamin B_{12}-dependent ethanolamine utilization in Salmonella typhimurium. J. Bacteriol. 171:3316 - 3323.

14. Obradors, N., J. Badía, L. Baldomà and J. Aguilar. 1988. Anaerobic metabolism of the L-rhamnose fermentation product 1,2-propanediol in Salmonella typhimurium. J. Bacteriol. 170:2159-2162.

15. Frey, B., J. McCloskey, W. Kersten, and H. Kersten. 1988. New function of vitamin B_{12}: cobamide-dependent reduction of epoxyqueuosine to queuosine in tRNAs of Escherichia coli and Salmonella typhimurium. J. Bacteriol. 170: 2078-2082.

16 Fontecave, M., R. Eliasson, and R. Reichart. 1989 Oxygen-sensitive ribonucleotide triphosphate reductase is present in anaerobic Escherichia coli. Proc.Natl.Acad.Sci (US) 86: 2147-2151.

17 Eschenmoser, A. 1988. Vitamin B_{12}: Experiments concerning the origin of its molecular structure. Angew. Chem. Int. Ed. Engl. 27 5-39

18. Georgopapadakou, N. H. and A. I. Scott 1977. On B_{12} biosynthesis and evolution. J. Theor. Biol. 69: 381-384

19. Johnston, H., W. Barnes, F. Chumley, L. Bossi and J. Roth. 1980. A model for regulation of the histidine operon of Salmonella typhimurium. Proc. Natl Acad Sci USA 77: 508 -512.

20a Vitols, E., G. A. Walker, and F. M. Huennekens. 1966. Enzymatic conversion of vitamin B12 to a cobamide coenzyme, aaaaa05,6 dimethylabenzimidazole) deoxyadenosyl cobamide (adenosyl-B12) J. Biol. Chem. 241:1455

20b Walker, G. A., S. Murphy, and F. M. Huennekens. 1969. Enzymatic conversion of vitamin B12 to adenosyl-B12: evidence for the existence of two separate reducing systems. Arch. Biochem. Biophys.

21 Neuberger, A. and G. H. Tait. 1960. The enzymic conversion of threonine to aminoacetone. Biochim. Biophys. Acta 41: 164-165.

MECHANISTIC AND EVOLUTIONARY ASPECTS OF VITAMIN B_{12} BIOSYNTHESIS

A. Ian Scott

Center for Biological NMR, Department of Chemistry
Texas A&M University, College Station, Texas 77843-3255 USA

INTRODUCTION

For the past 20 years our laboratory has been engaged in the elucidation of the vitamin B_{12} biosynthetic pathway and a previous Account[1] described progress made in the first ten years of this endeavor. By that time (1977), it had been established that 5-aminolevulinic acid (ALA), porphobilinogen (PBG) uro'gen III and three intermediates, Factors I-III, now known to be transformed as the precorrins 1-3, were sequentially formed on the way to the corrin nucleus as summarized in Scheme 1. The ensuing decade has witnessed an exponential increase in the rate of the acquisition of pure enzymes using recombinant DNA techniques and these methods together with the exploration of new NMR pulse sequences have not only cast light on the mechanisms of the B_{12} synthetic enzymes but have given us considerable optimism with regard to the discovery of the remaining intermediates and of the enzymes which produce them.

This lecture will focus on three topics drawn from the "early" and "late" segments of the pathway. The first of these deals with mechanistic and structural proposals for PBG deaminase, the assembly enzyme for tetrapyrrole biosynthesis which, working together with uro'gen III synthase, is responsible for the synthesis of the unsymmetrical type-III macrocycle. The second topic encompasses the sequence of C-methylations connecting uro'gen III with the precorrins, leading finally to cobyrinic acid and the properties of the purified methyl transferase responsible for the first two methyl group insertions into uro'gen III. Thirdly, the discovery of a new class of corphinoids, which reflect a degree of non-specificity of the B_{12} synthesizing enzymes and the consequences of this finding for the biochemical evolution of the B_{12} structure will be discussed. The complex problems inherent in tetrapyrrole and corrin biosynthesis have occupied the attention of several laboratories over the last 20 years. Many of the major groups working on B_{12} biosynthetic and genetic research have contributed to this symposium and complementary studies by Professors A. R. Battersby (Cambridge) G. Müller (Stuttgart), P. M. Jordan (London), which have provided valuable and stimulating results in the area of heme and corrin biosynthesis, will be found elsewhere in this volume.

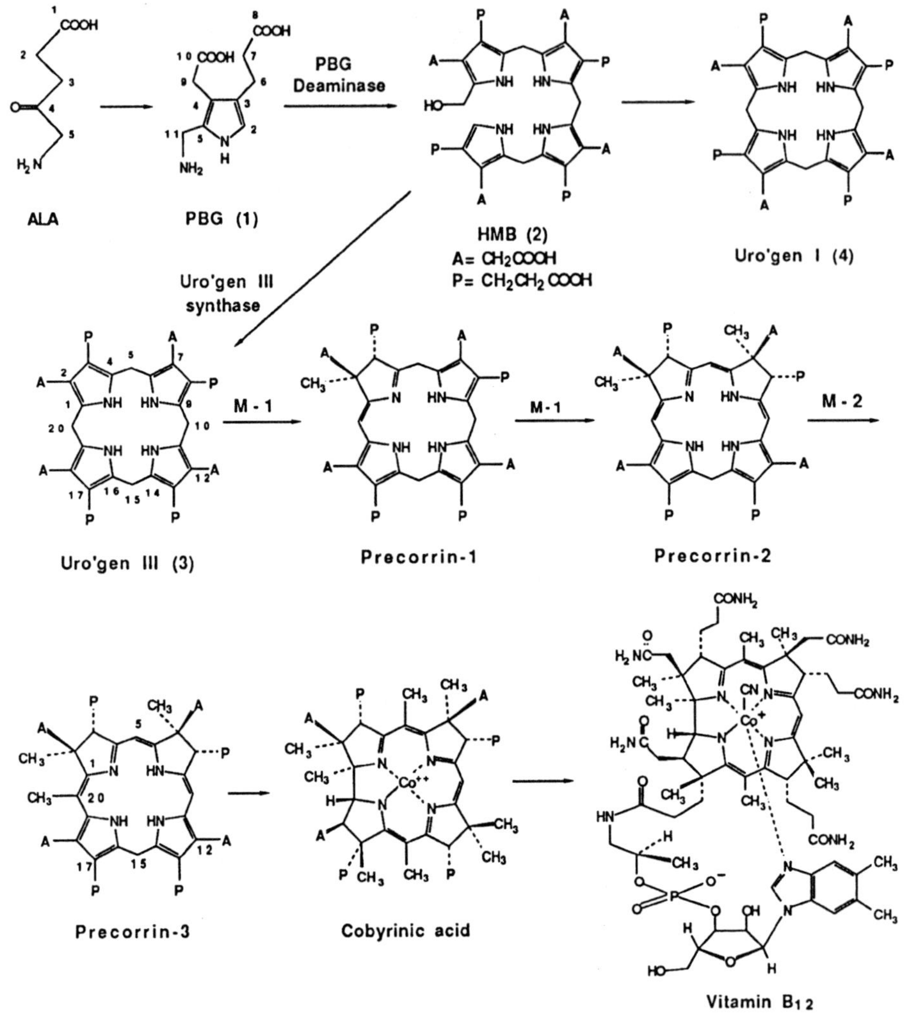

Scheme 1

The Enzymes of Tetrapyrrole Synthesis: PBG Deaminase and Uro'gen III Synthase.

PBG deaminase (EC4.3.1.8) catalyzes the tetramerization of PBG (**1**) to preuro'gen (hydroxymethylbilane, HMB; **2**)[2,3] which is cyclized with rearrangement to the unsymmetrical uro'gen III (**3**) by uro'gen III synthase[3,4,5] (EC4.2.1.75) (Scheme 1). In the absence of the latter enzyme, preuro'gen (**2**) cyclizes to uro'gen I (**4**), which, as discussed below, turns out to be a substrate for the methylases of the vitamin B_{12} pathway. As discussed by Professors Jordan and Battersby in this volume, the sequencing of the *hemC* gene and its expression allowed a detailed study of the mechanism. Based on Jordan's sequence for *hemC* we used genetic engineering to construct a plasmid pBG 101 containing the *Escherichia coli hemC* gene[6,7] for deaminase and described the overproduction of deaminase in 1987. *E. coli* (TBI) transformed with this plasmid produces deaminase at levels greater than 200 times those of the wild strain[7] thereby allowing access to substantial quantities of enzyme for detailed study of the catalytic mechanism.

Previous work with deaminase [8-11] had established that a covalent bond is formed between substrate and enzyme, thus allowing isolation of covalent complexes containing up to 3 PBG units (ES_1-ES_3). Application of ^3H-NMR spectroscopy to the mono PBG adduct (ES-1) revealed a rather broad ^3H chemical shift indicative of covalent bond formation with a cysteine thiol group at the active site.[2] However, with adequate supplies of pure enzyme available in 1987 from the cloning of *hemC*[6] we were able to show that a novel cofactor, derived from PBG during the biosynthesis of deaminase, is covalently attached to one of the four cysteine residues of the enzyme in the form of a dipyrromethane which, in turn, becomes the site of attachment of the succeeding four moles of substrate during the catalytic cycle. Thus, at pH < 4, deaminase (**5**) rapidly develops a chromophore (λmax 485 nm) diagnostic of a pyrromethene (**6**), whilst reaction with Ehrlich's reagent generates a chromophore typical of a dipyrromethane (λmax 560 nm) changing to 490 nm after 5-10 min. The latter chromophoric interchange was identical with that of the Ehrlich reaction of

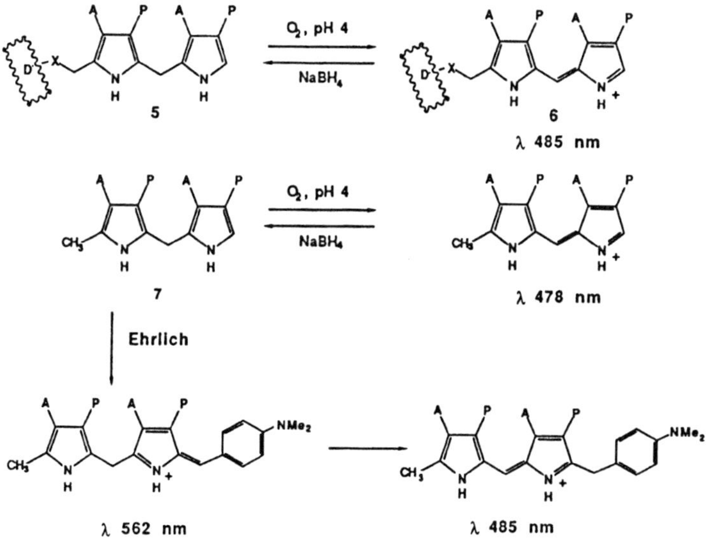

Scheme 2

the synthetic model pyrromethane (**7**) and can be ascribed to the isomerization shown (Scheme 2) for the model system (**7**). Incubation of *E. coli* strain SASX41B (transformed with plasmid pBG 101: *hemA*$^-$ requiring ALA for growth) with 5-^{13}C-ALA afforded highly enriched enzyme for NMR studies. At pH8, the enriched carbons of the dipyrromethane (py-CH_2-py) are clearly recognized at 24.0 ppm (py-CH_2py), 26.7 ppm (py-CH_2X), 118.3 ppm (α-free pyrrole) and 129.7 ppm (α-substituted pyrrole) (Fig. 1A). The signals are sharpened at pH12 (Fig. 1B) and the CH_2-resonance is shifted to δ29.7 in the unfolded enzyme. Comparison with synthetic models reveals that a shift of 26.7 ppm is in the range expected for an α-thiomethyl pyrrole (py-CH_2 SR). Confirmation of the dipyrromethane (rather than

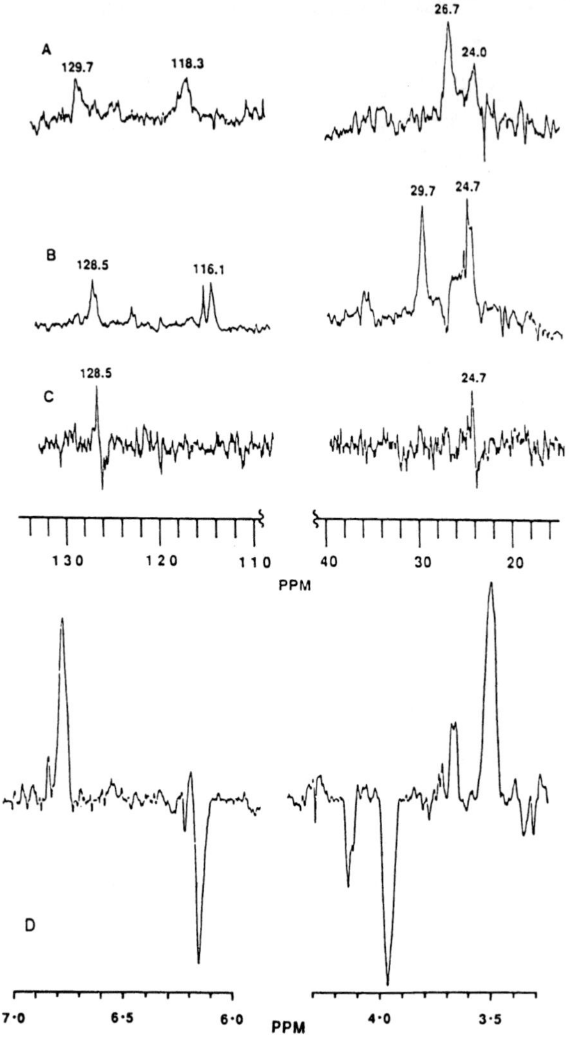

Figure 1

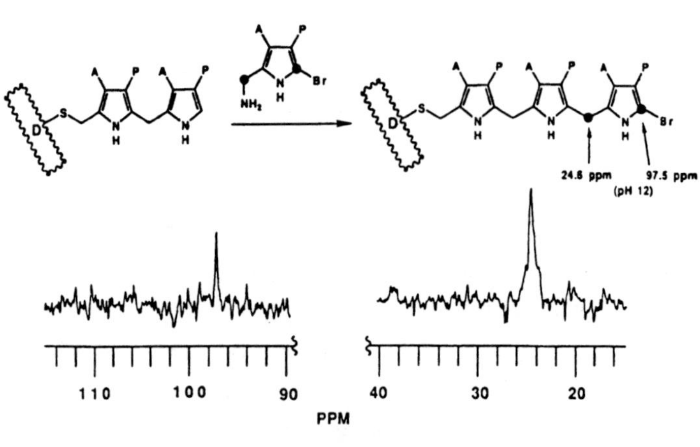

Figure 2

oligo pyrromethane) came from the ^{13}C INADEQUATE spectrum taken at pH12 (Fig. 1C) which reveals the expected coupling only between py-CH_2-py (δ 24.7) and the adjacent substituted pyrrole carbon (δ 128.5 ppm). When the enriched deaminase was studied by INVERSE INEPT spectroscopy, each of the 5 protons attached to ^{13}C-nuclei were observed as shown in Fig. 1D. A specimen of deaminase was covalently inhibited with the suicide inhibitor [2,11-$^{13}C_2$]-2-bromo PBG (8, Fig. 2) to give a CMR spectrum (pH12) consistent only with structure 9. The site of covalent attachment of substrate (and inhibitor) is therefore the free α-pyrrole carbon at the terminus of the dipyrromethane in the native enzyme, leading to the structural and mechanistic proposal for deaminase shown in Scheme 3.

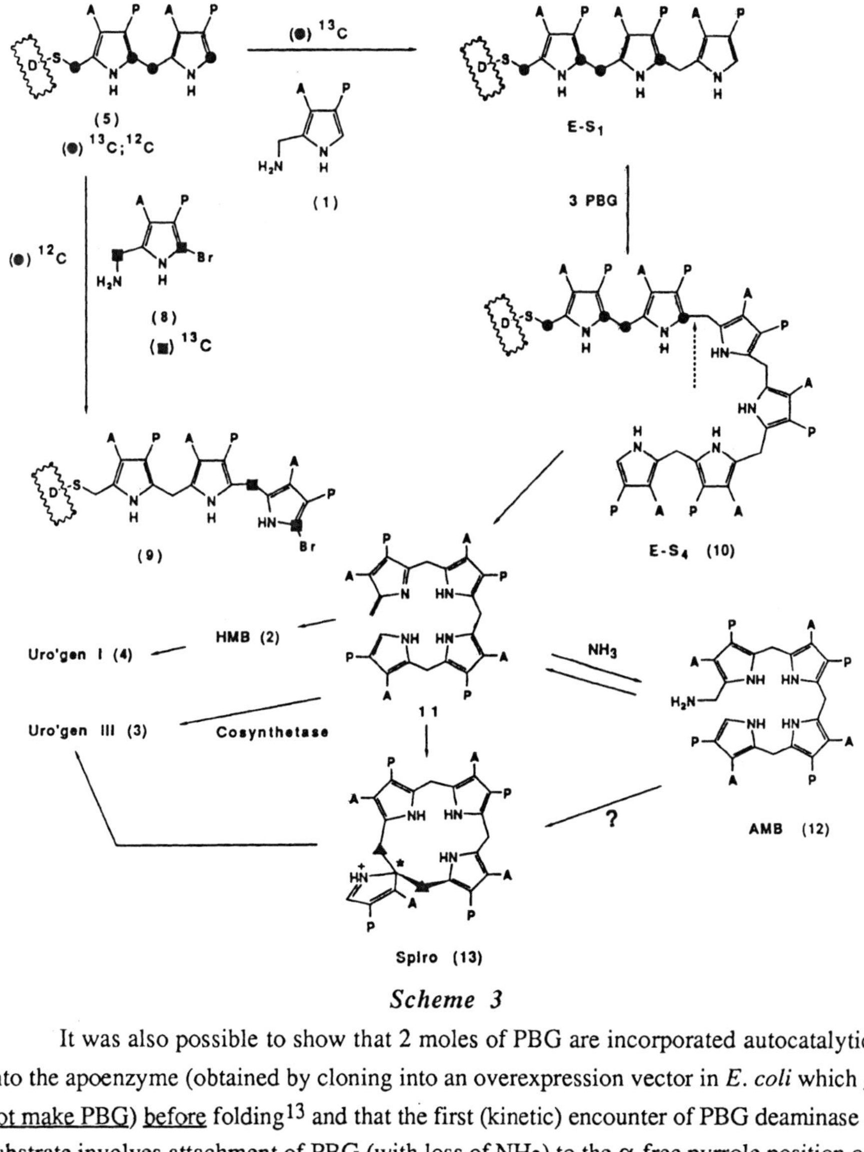

Scheme 3

It was also possible to show that 2 moles of PBG are incorporated autocatalytically into the apoenzyme (obtained by cloning into an overexpression vector in *E. coli* which does not make PBG) before folding[13] and that the first (kinetic) encounter of PBG deaminase with substrate involves attachment of PBG (with loss of NH_3) to the α-free pyrrole position of the dipyrromethane to form the ES_1 complex[14,15] (Scheme 3). The process is repeated until the "tetra PBG" (ES_4) adduct (**10**) is formed. At this juncture site-specific cleavage of the hexapyrrole chain (at →) releases the azafulvene bilane (**11**) which either becomes the substrate of uro'gen III synthase, or in the absence of the latter enzyme, is stereospecifically hydrated[3] to HMB (**2**) at pH12, or is cyclized chemically to uro'gen I (**4**) at pH ≤ 8. It is quite remarkable that within the short period from mid-1987[14,15,16a,17a] to January 1988 three groups cloned and expressed deaminase and by spectroscopic methods defined the dipyrromethane active site cofactor in a series of independent [14,16a,17a] and collaborative[15] publications. The ^{13}C-labeling defines the number of PBG units (two) attached in a head-to-tail motif to the native enzyme at pH8 and reveals the identity of the nucleophilic group (Cys-SH) which anchors the dipyrromethane (and hence the growing oligopyrrolic chain) to the enzyme. Site specific mutagenesis[7] and chemical cleavage[16,17] were employed to determine that Cys-242 is the point of attachment of the cofactor. Thus, replacement[7] of cysteine with serine at residues 99 and 242 (respectively) gave fully active and inactive specimens of the enzyme respectively.

The use of the α-carbon of a dipyrromethane unit as the nucleophilic group responsible for oligomerization of 4 moles of PBG (with loss of NH_3 at each successive encounter with an α-free pyrrole) is not only remarkable for the exquisite specificity and control involved, but is, as far as we know, a process unique in the annals of enzymology in that a substrate is used not only once, but twice in the genesis of the active site cofactor! Even more remarkable is the fact that the apoenzyme is automatically transformed to the active holoenzyme by addition of two substrate PBG units without the intervention of a second enzyme. Crystallization and X-ray diffraction studies of both the native enzyme and several of its genetically altered versions are now in progress.

Uro'gen III Synthase - The Ring D Switch

We now turn briefly to the rearranging enzyme Uro'gen III synthase. The early ideas of Bogorad[18] involving the aminomethyl bilane (AMB;**12**) as substrate were helpful in finally tracking down the elusive species which, after synthesis by PBG deaminase, becomes the substrate for Uro'gen III synthase. This was shown to be HMB (**2**) and at this stage (1978) the lack of activity of AMB (**12**) as substrate relegated the latter bilane to an interesting artefact produced quantitatively by addition of ammonia to deaminase incubations (Scheme 3). However with the acquisition of substantial quantities of pure Uro'gen III synthase obtained by cloning the genes *hemC* and *D* together[19] and overexpression in *E. coli*, the substrate specificity of the synthase (often called cosynthetase) has been reinvestigated. Ever since its conception[20] by Mathewson and Corwin in 1961, the spiro compound (**13**) (Scheme 3) has been a favorite construct with organic chemists, since both its genesis through α-pyrrolic reactivity and its fragmentation - recombination rationalize the intramolecular formation of Uro'gen III from the linear bilane, preuro'gen (HMB;**2**). A careful search for the spiro-compound (**13**) was conducted at subzero temperatures in cryosolvent (-24° C; ethylene glycol/buffer) using various ^{13}C-isotopomers of HMB as substrate.[19] Although the synthase reaction could be slowed down to 20 hours (rather than

20 sec.) no signals corresponding to the quaternary carbon (*δ ~80) ppm or to the α-pyrrolic methylene groups (▲; δ 35-40 ppm) could be observed. During these studies, however, it was found that AMB (**12**) served as a slow but productive substrate for Uro'gen III synthase at high concentrations (1 mmole in enzyme and substrate concentrations) and that if care is not taken to remove the ammonia liberated from PBG by the action of deaminase, not only is the enzymatic formation of AMB observed[21,22] but in presence of Uro'gen III synthase, the product is again Uro'gen III. This raises the interesting question of whether, under certain physiological conditions, the true substrate for Uro'gen III synthase is in fact AMB formed at the locus of deaminase by the ammonia released from PBG. Regardless of which version of the bilane system (HMB, AMB) is used in the experiment, the lack of observation of any intervening species free of enzyme strongly suggests that the intermediate is enzyme-bound and therefore difficult to detect. Present research is directed towards the solution of this problem by more sensitive, low temperature spectroscopy (see also Battersby in this volume).

However we recently proposed[19] a novel alternative to the "spiro intermediate" (**13**) hypothesis for the mechanism of ring D inversion mediated by uro'gen III synthase which uses a self assembly concept involving lactone formation as portrayed in Scheme 4. We suggested that generation of the azafulvene (**12**) as before is followed not by carbon-carbon

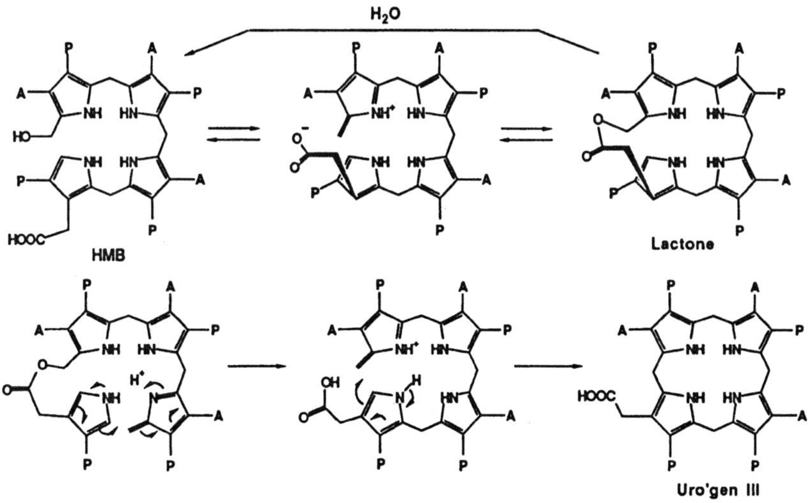

Scheme 4

bond formation (→ **13**) but by closure of the macrocyle *via* the lactone (Scheme 4). The regio-specificity for this step is reflected by the failure[5] of synthetic bilanes lacking the acetic acid side chain at position 17 in ring D to undergo enzyme-catalyzed rearrangement to the type-III system together with the observation that a "switched" bilane carrying a propionate at position 17 (and an acetate at 18) does indeed give a certain amount of (rearranged) uro'gen I (but also Uro'gen III) enzymatically,[5,22] indicating that lactone formation could also be achieved (but less efficiently) by a propionate group at the 17-position. The subsequent chemistry is quite similar to the fragmentation-recombination postulated for the spiro system except that a "twisted lactone" becomes the pivotal intermediate species and the sequence proceeds as shown in Scheme 4. Although this novel hypothesis is rather difficult to test experimentally, specific chemical traps for the lactones combined with the use of ^{18}O labeling are being used to confirm or refute the possible role of such macrocyclic lactones as the basis for the mechanism of Uro'gen III synthase.

Temporal Resolution of the Methylation Sequence

The bioconversion of Uro'gen III to cobyrinic acid is summarized in Scheme 5 where it is postulated that the oxidation level of the various intermediates is maintained at the same level as that of Uro'gen III. The lability of the reduced isobacteriochlorins, the fact that they are normally isolated in the oxidized form, and the requirement for chemical reduction of Factor I (but not Factors II and III) before incorporation into corrin in cell free systems lent credence to this idea but rigorous proof for the *in vivo* oxidation state of Factors I and II has only recently been obtained as discussed below.[23,24] In order to distinguish between the oxidation levels of the isolated chlorins, isobacteriochlorins, corphins, etc.. and to avoid possible confusion in using the term "Factor" (which has other connotations in B_{12} biochemistry e.g. intrinsic Factor/Factor III) the term precorrin-**n** has been suggested[24] for the actual structures of the biosynthesized intermediates after Uro'gen III where **n** denotes the number of SAM derived methyl groups. Thus, although the names Factors I-III for the isolated species will probably survive(for historical reasons) all of the true intermediates probably have the same oxidation as Uro'gen III *viz.* tetrahydro Factor I (precorrin-**1**), dihydro Factor II (precorrin-**2**), and dihydro Factor III (precorrin-**3**) (Scheme 3).

In spite of intensive search, no new intermediates containing four or more methyl groups (up to a possible total of eight) have been isolated, but the biochemical conversion of Factor III to cobyrinic acid must involve the following (see Scheme 5), not

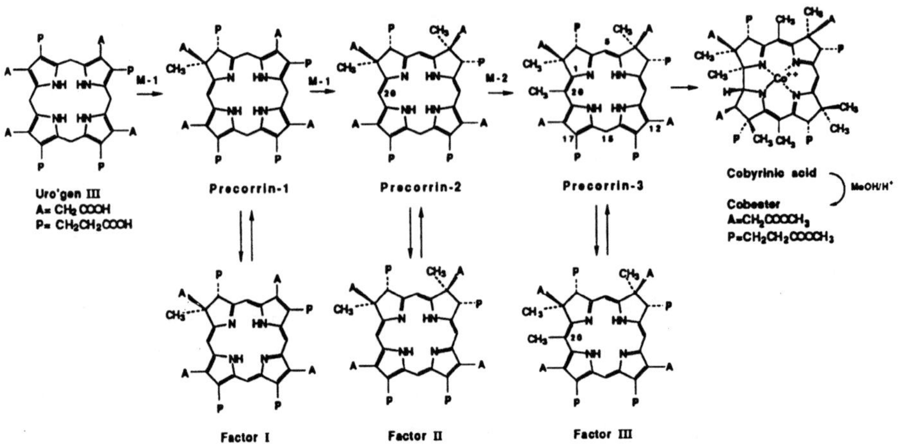

Scheme 5

necessarily in the order indicated: (1) The successive addition of five methyls derived from S-adenosyl methionine (SAM) to *reduced* Factor III (precorrin-**3**). (2) The contraction of the permethylated macrocycle to corrin. (3) The extrusion of C-20 and its attached methyl group leading to the isolation of acetic acid.[25-28] (4) Decarboxylation of the acetic acid side chain in ring C (C-12). (5) Insertion of Co^{+++} after adjustment of oxidation level from Co^{++}. In order to justify the continuation of the search for such intermediates whose inherent lability to oxygen is predictable, ^{13}C pulse labeling methods were applied to the cell-free system which converts Uro'gen III (**3**) to precorrin-**2** and thence to cobyrinic acid, a technique used previously in biochemistry to detect the flux of radio-labels through the intermediates of biosynthetic pathways.

The Pulse Experiments

The methylation sequence can be resolved temporally provided that enzyme-free intermediates accumulate in sufficient pool sizes to affect the intensities of the resultant methyl signals in the CMR spectrum of the target molecule, cobyrinic acid when the cell-free system is challenged with a pulse of $^{12}CH_3$-SAM followed by a second pulse of $^{13}CH_3$-SAM (or vice versa) at carefully chosen intervals in the total incubation time (6-11 hours). By this approach it is possible to "read" the biochemical history of the methylation sequence as reflected in the dilution (or enhancement in the reverse experiment) of $^{13}CH_3$ label in the methionine-derived methyl groups of cobyrinic acid after conversion to cobester, whose ^{13}C-NMR spectra has been assigned.[29,30] Firstly, precorrin-**2** is accumulated in whole cells containing an excess of SAM in the absence of Co^{++}. The cells are then disrupted and Co^{++} added immediately, followed by a pulse of $^{13}CH_3$-SAM (90 atom%) after 4 hours. After a further 1.5 hr. cobyrinic acid is isolated as cobester. The ^{13}C NMR spectrum of this specimen (Fig. 3a) defines the complete methylation sequence, beginning from precorrin-**2**, as C-20 > C-17 > C-12α > C-1 > C-5 > C-15, with a differentiation of 25% (± 5%) in the relative signal intensities for the C_5 and C_{15} methyl groups[31,32] further confirmed by hetero-filtered ^1H spectroscopy of the ^{13}C-enriched sample (Fig. 3b). Methylation at C-20 of precorrin-**2** to give precorrin-**3** is not recorded in the spectrum of cobester since C-20 is lost on the way to cobyrinic acid, together with the attached methyl group, in the form of acetic acid.[25-28] The sequence C-17 > C-12α > C-1 had previously been found in *Clostridium tetanomorphum*[24] and further differentiation between C-5 and C-15 insertion suggests the order C-15 > C-5 in this organism for the last two methylations, i.e. opposite from the *P. shermanii* sequence.

It is now apparent that several discrete methyl transferases are involved in the biosynthesis of cobyrinic acid from uro'gen III, since enzyme-free intermediates must accumulate in order to dilute the ^{13}C label. A rationale for these events is given in Scheme 6, which takes the following facts into account: (a) the methionine derived methyl group at C-20 of precorrin-**3** does not migrate to C-1 and is expelled together with C-20 from a late intermediate (as yet unknown) in the form of acetic acid (b) neither 5,15-norcorrinoids[33] nor descobalto-cobyrinic acid[34] are biochemical precursors of cobyrinic acid (c) regiospecific

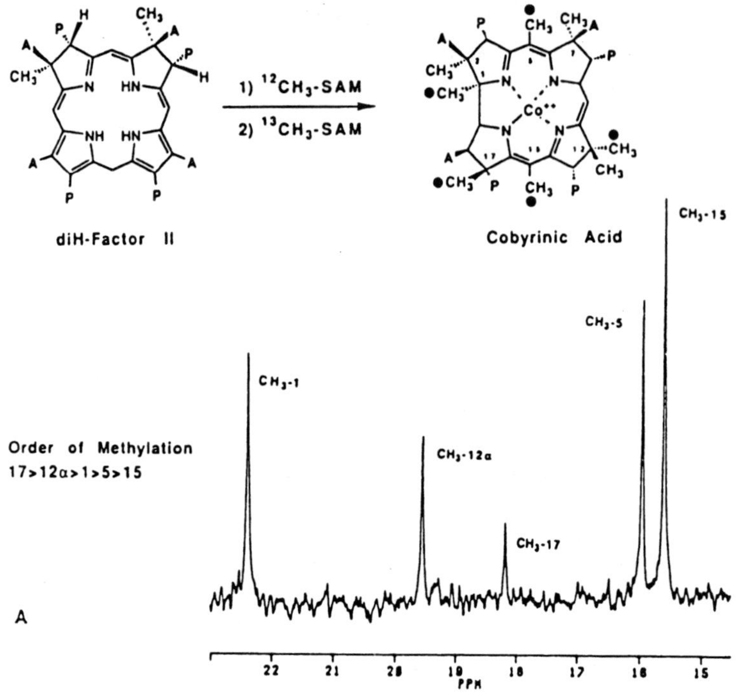

1) $^{12}CH_3$-SAM
2) $^{13}CH_3$-SAM
diH-Factor II
Cobyrinic Acid
Order of Methylation
17>12α>1>5>15
CH3-1
CH3-12α
CH3-17
CH3-5
CH3-15
A
PPM

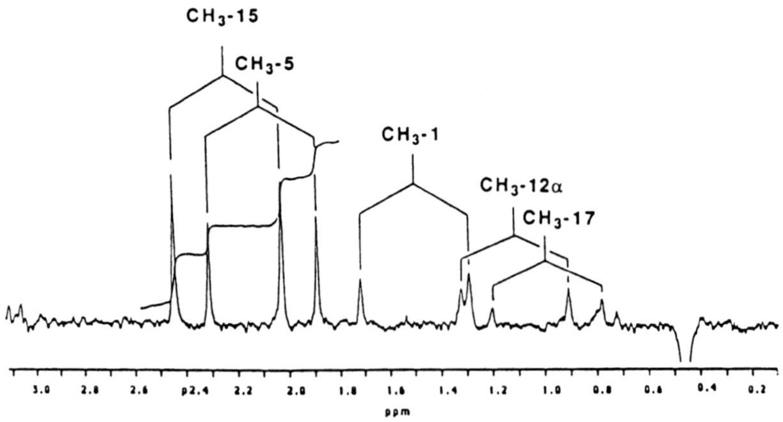

B
CH3-15
CH3-5
CH3-1
CH3-12α
CH3-17
ppm

Figure 3

loss of ^{18}O from [1-^{13}C, 1-$^{18}O_2$]-5-aminolevulinic acid-derived cyanocobalamin from the ring A acetate occurs,[35] in accord with the concept of lactone formation, as portrayed in Scheme 6, where precorrin-**5** is methylated at C-20 followed by C-20 → C-1 migration and lactonization. If the latter mechanism is operative, the C-20 → C-1 migration must be stereospecific, since precorrin-**3** labeled at C-20 with $^{13}CH_3$ is transformed to cobyrinic acid with complete loss of label, a less attractive alternative being direct methylation at C-1.

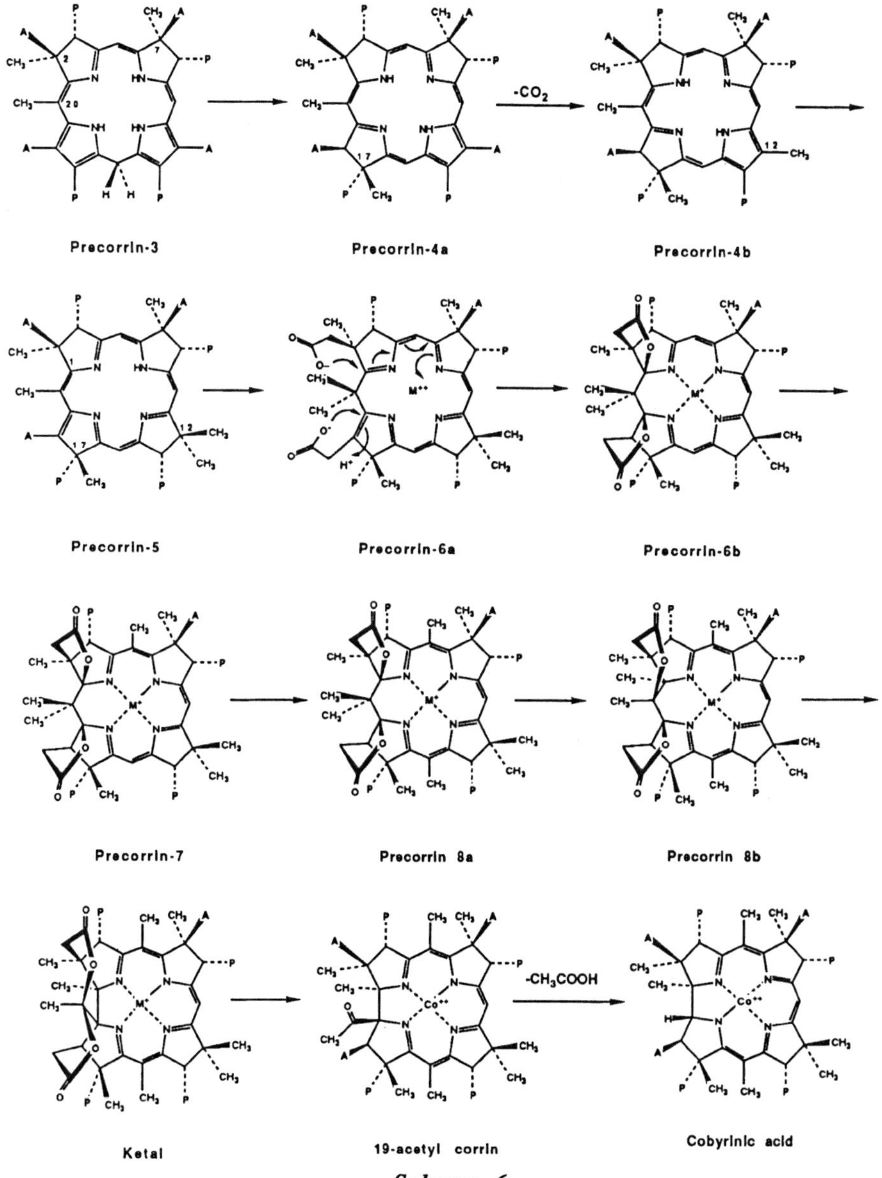

Scheme 6

Timing of the Decarboxylation Step in Ring C

It has also proved possible to define the point in the biosynthetic sequence where ring C-decarboxylation occurs, using synthetic [5,15-$^{14}C_2$]-12-decarboxylated uro'gen III (**14**) as a substrate for the non-specific methylases of *P. shermanii* to prepare the 12-methyl analogs (**15**) and (**16**) of factors II and III respectively (Scheme 7). It was shown by these studies[32] and independently by ^{13}C-labeling[36] that these ring C-decarboxylated analogs are not substrates for the enzymes of corrin biosynthesis, leading to the conclusions that (a) in normal biosynthesis, precorrin-**3** is not the intermediate which is decarboxylated (b) decarboxylation occurs at some stage after the fourth methylation (at C-17) and by mechanistic analogy, before the fifth methylation at C-12. Hence, two pyrrocorphin intermediates *viz.* precorrins **4a, 4b** should intervene between precorrins-**3** and -**5** i.e. precorrin-**3** is C-methylated at C-17 to give precorrin-**4a** followed by decarboxylation (→ **4b**) and subsequent C-methylations at C-12α, C-1, C-5, C-15, in that order as suggested in Scheme 6. Although this sequence differs in the penultimate stages from that reported elsewhere[24], consensus over the timing of decarboxylation in ring C has been reached since identical conclusions have been reported using *Pseudomonas denitrificans*.[36]

The decarboxylation mechanism most plausibly involves the addition of a proton to a ring C exomethylene species. The stereospecificity (retention) of this step has been demonstrated by Kuhn-Roth oxidation of the chirally-labeled C-12ß methyl group of cobester following incorporation of ALA chiral (^2H, ^3H) on the methylene group, which becomes incorporated into the C-12 acetate side chain.[37]

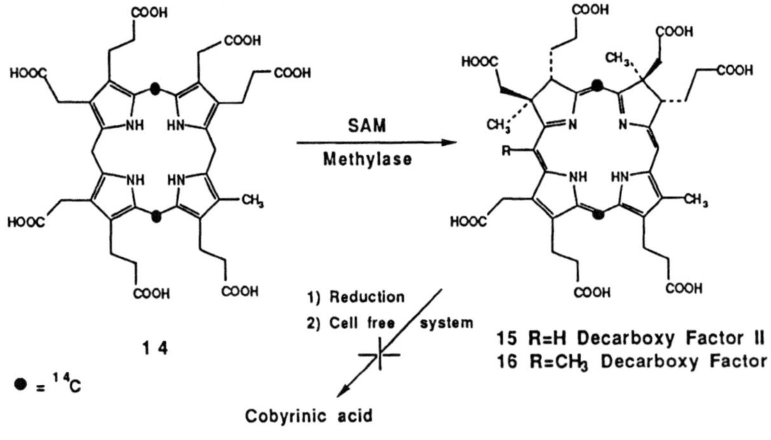

14

● = ^{14}C

15 R=H Decarboxy Factor II
16 R=CH_3 Decarboxy Factor III

Scheme 7

Although factors I-III, when adjusted to the requisite oxidation level (i.e. precorrins 1-3), all serve as excellent substrates for conversion to cobyrinic acid in cell-free systems from *P. shermanii* and *C. tetanomorphum*, direct evidence for their sequential interconversion was obtained only recently. In the presence of $^{13}CH_3$-SAM and the complete cell-free system containing a reducing system (Factor II → precorrin-**2**) a specimen of Factor II is converted to Factor III containing a single ^{13}C-enriched methyl at C-20 (δ 19.2 ppm). The demonstration of this conversion removes any doubt that separate pathways could exist for the biosynthesis of vitamin B_{12} from precorrins-**2** and **3** respectively.[32]

We now return to the construction of a working hypothesis for corrin biosynthesis. The formulations precorrins **6b, 7, 8a, 8b** take into account the idea of lactone formation using rings A and D acetate side chains. The migration C-20 $\rightarrow$ C-1 could be acid or metal ion catalyzed and the resultant C-20 carbonium ion quenched with external hydroxide or by the internal equivalent from the carboxylate anion of the C-2 acetate in ring A (precorrin **8a** $\rightarrow$ **8b**) as suggested by the results of labeling with ^{18}O discussed above.[35] In any event, the resultant dihydrocorphinol-bislactone precorrin-**8b**, is poised to undergo the biochemical counterpart of Eschenmoser ring contraction[38] to the 19-acetyl corrin. Before this happens we suggest that the final methyl groups are added at C-5 (precorrin-**7**), then C-15 (precorrin-**8a**) to take account of the non-incorporation of the 5, 15-norcorrinoids.[33] The resultant precorrin-**8b** (most probably with cobalt in place) then contracts to 19-acetyl corrin which, by loss of acetic acid, leads to cobyrinic acid. Cobalt insertion must precede methylation at C_5 and C_{15} since hydrogenocobyrinic acid does not insert cobalt enzymatically[34]. The valency change $Co^{++} \rightarrow Co^{+++}$ during or after cobalt insertion has so far received no explanation.

The Proton Balance of Vitamin B_{12} Biosynthesis

Armed with the knowledge of the methyl transferase sequences, experiments have been carried out to trace the origin of the methine protons in Vitamin B_{12}. A possibility existed that the proton at C-18 (together with that at C-19) is introduced via reduction of 18, 19 dehydro-cobyrinic acid[33,39] although early experiments using NAD^{3}H and NADP^{3}H showed no incorporation of ^3H radioactivity at C-18 and C-19.[40] On the other hand, adoption of Eschenmoser's *in vitro* model for the dihydrocorphinol $\rightarrow$ acetylcorrin ring contraction and deacetylation[38] as a construct for B_{12} biosynthesis would leave the oxidation level of the pathway unperturbed from uro'gen III to cobyrinic acid. In order to examine the stage at which deacetylation occurs kinetic and equilibrium probes for the protonations at C–18 and C-19 as well as at the other β-positions carrying protons, *viz.* C-3, C-8, and C-13 were developed. Since the loss of acetic acid from precorrin-**8** is biochemically irreversible, position 19 may be susceptible to a large kinetic isotope effect. Thus, by allowing the production of B_{12} to take place in D_2O enriched medium, the early precursors should have reached equilibrium with reference to isotopic replacement at C-3, C-8, C-13, and C-18 (and possibly C-10). When cyanocobalamin (**16a**) was isolated from a fermentation carried out in 50% D_2O in the presence of [4-^{13}C]-ALA the observation of large isotopic shifts on the ^{13}C-enriched positions could be used to assign the position of deuterium incorporation as shown in Fig. 4. Thus, as expected, C-3, C-8 and C-13 showed large α-isotopic shifts. C-17 shows a β-shift in consonance with considerable enrichment at C-18. In contrast, the deuterium enrichment at C-19 although observable was only *ca.* 30% of the content at the other centers while C-10 was devoid of enrichment. The relative ratios of ^2H at C-18 and C-19 (*ca* 3:1) compared with the consistently higher deuterium content at C-3, C-8, and C-13 are in accord with the concept that not only is deacetylation a late step but that it takes place irreversibly under kinetic control, leading to the observed isotope effect.[41] Independent work had shown earlier[42] that in a cell free system, C-18 and C-19 in cobyrinic acid are deuterated in a similar ratio. Since the chemical counterpart [38] of 19-deacylation involves a nickel or cobalt corrin it is tempting to suggest that the insertion of cobalt takes place before

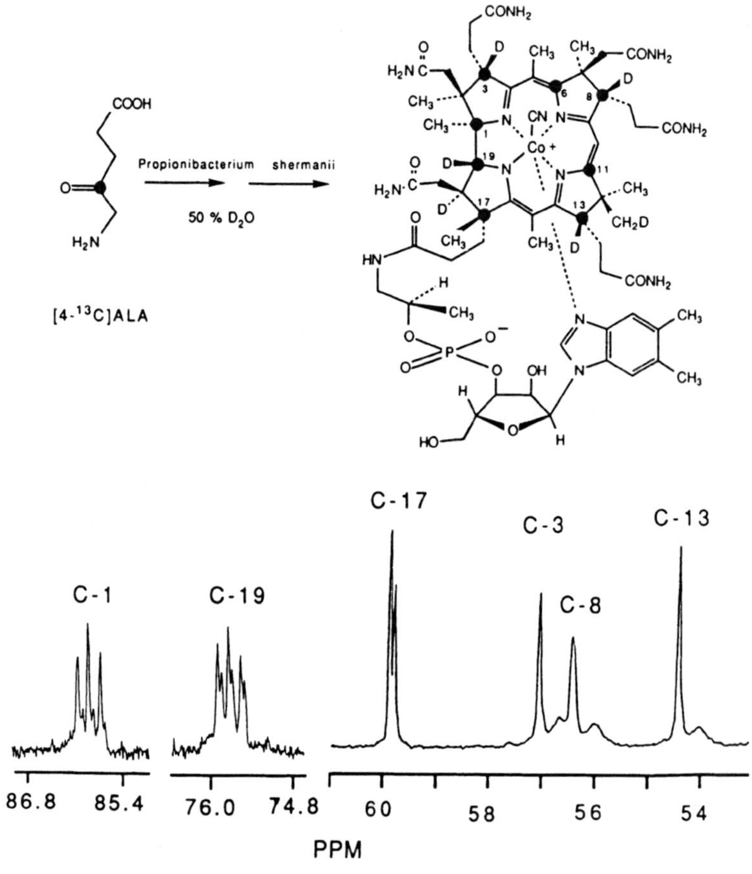

Figure 4

this step. However, as mentioned above, the non-incorporation of cobalt into cobalt-free cobyrinic acid in presence of the corrin synthesizing enzymes[34] is surprising and suggests that although corrin biosynthesis can be achieved in the absence of cobalt in certain bacteria, metal insertion is a late step in cobyrinic acid in most cases.

The possibility exists that a divalent cation e.g. Zn^{++} can mediate in the various methylations and rearrangements leading from the porphinoid to the corrinoid template and can be replaced by Co^{++} at a later stage.

The Methyl Transferases

The first of the methylase enzymes catalyzes the sequential formation of Factors I and II and has been named S-adenosyl methionine Uro'gen III methyl transferase (SUMT). SUMT was first partially purified from *P. shermanii* by G. Müller[43] and recently has been overexpressed in *Pseudomonas denitrificans*.[44] In *E. coli* it was found that the *CysG* gene encodes Uro'gen III methylase (M-1) as part of the synthetic pathway to siroheme, the cofactor for sulfite reductase, and overproduction was achieved by the appropriate genetic engineering.[45] Although, SUMT and M-1 appear to perform the same task, it has beeen found that their substrate specificities differ. Thus, it has been possible to study in detail the reaction catalyzed by M-1 directly using NMR spectroscopy and to provide rigorous proof that the structure of precorrin-**2** is that of the dipyrrocorphin tautomer of dihydro-Factor II (dihydrosirohydrochlorin). Uro'gen III (enriched from [^{13}C-5-ALA] at the positions shown in Scheme 8) was incubated with M-1 and [$^{13}CH_3$]-SAM. The resultant spectrum of the

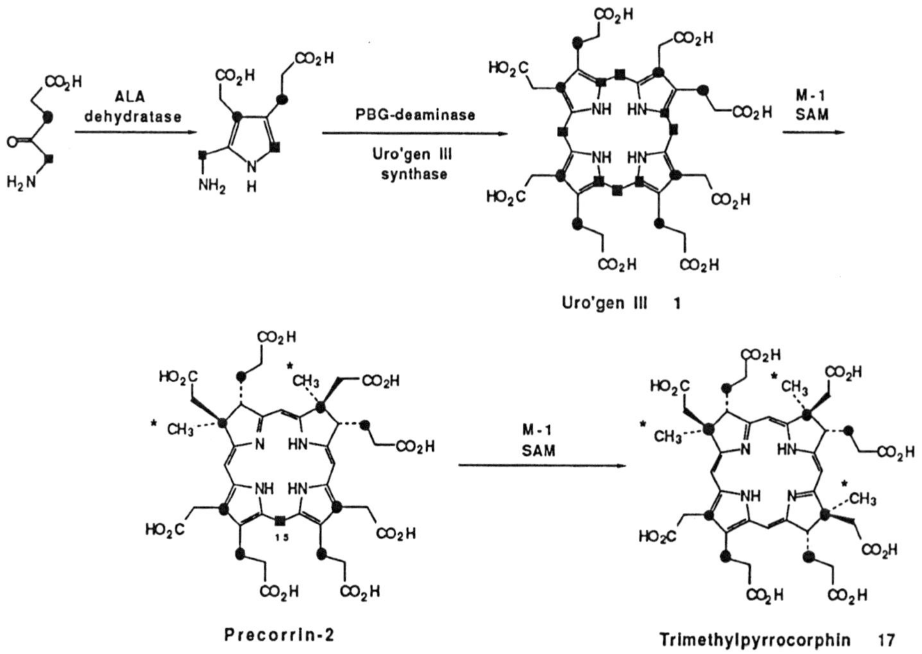

Scheme 8

precorrin-**2** revealed an sp^3 enriched carbon at C-15, thereby locating the reduced center. By using a different set of ^{13}C-labels (• from ^{13}C-3 ALA) and [$^{13}CH_3$]-SAM) the sp^2 carbons at C_{12} and C_{18} were located as well as the sp^3 centers coupled to the pendant ^{13}C-methyl groups at C_2 and C_7. This result confirms an earlier NMR analysis[23] of precorrin-**2** isolated by anaerobic purification of the methyl ester, and shows that no further tautomerism takes place during the latter procedure. The two sets of experiments mutually reinforce the postulate that precorrins-**1**, -**2**, and -**3** all exist as hexahydroporphinoids and recent labeling experiments[46] have provided good evidence that precorrin-**1** is discharged from the methylating enzyme (SUMT) as the species with the structure shown (or an isomer).

However, prolonged incubation (2 hr.) of Uro'gen III with M-1 provided a surprising result for the UV and NMR changed dramatically from that of precorrin-**2** (a dipyrrocorphin) to the chromophore of a pyrrocorphin, hitherto only known only as a synthetic tautomer of hexahydroporphyrin. At first sight, this event seemed to signal a further tautomerism of a dipyrrocorphin to a pyrrocorphin catalyzed by the enzyme but when $^{13}CH_3$-SAM was added to the incubation, it was found that a <u>third</u> methyl group signal

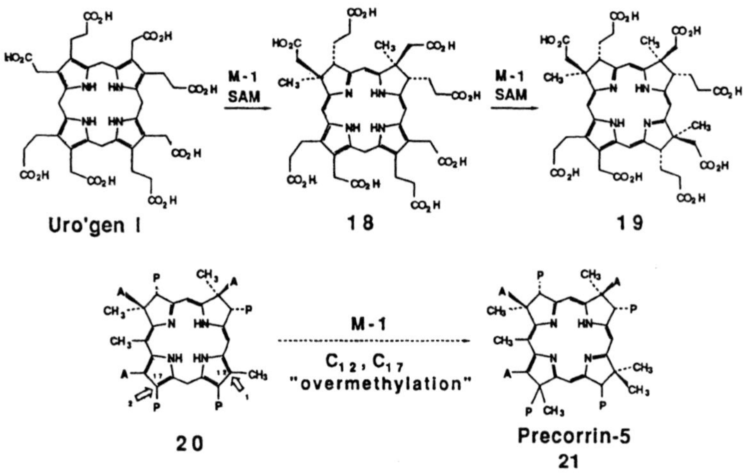

Scheme 9

appeared in the 19-21 ppm region of the NMR spectrum. When Uro'gen III was provided with the ^{13}C labels (•) (as shown in Scheme 8) 3 pairs of doublets appeared in the sp^3 region (δ 50-55 ppm) of the pyrrocorphin product. The necessary pulse labeling experiments together with appropriate FAB-MS data finally led to the structural proposal (**17**)[47,48] for the novel trimethyl pyrrocorphin produced by "overmethylation" of the normal substrate, uro'gen III, in presence of high concentration of enzyme. Thus M-1 has been recruited to insert a ring C methyl and synthesizes the long sought "natural" chromophore corresponding to that of the postulated precorrin-**4** although in this case the regiospecificity is altered from ring D to ring C. This lack of specificity on the part of M-1 was further exploited to synthesize a range of "unatural" isobacteriochlorins and pyrrocorphins based on isomers of Uro'gen III. Thus, Uro'gen I produces 3 methylated products corresponding to precorrin I, precorrin-**2** (**18**), and the type-I pyrrocorphin (**19**) (Scheme 9). Uro'gens II and IV can also serve as substrates for M-1 but not for SUMT! These compounds are reminiscent of a series of tetramethyl type I corphinoids, Factors S_1-S_4 isolated from *P. shermanii*[49,50] which occur as their zinc complexes. When uro'gen I was incubated with SUMT,[51] isolation of Factor II of the type I (Sirohydrochlorin I) family revealed a lack of specificity for this methyltransferase also, although in the latter studies no pyrrocorphins were observed. This may reflect a control mechanism in the *P. denitrificans* enzyme (SUMT) which does not "overmethylate" precorrin-2 as is found for the *E. coli* M-1 from whose physiological function is to manufacture sirohydrochlorin. The fact that *E. coli* does not synthesize B_{12} could reflect an evolutionary process in which the C-methylation machinery has been retained, but is only required to insert the C-2 and C-7 methyl groups.

The sites of C-methylation in both the type I and III series are also reminiscent of the biomimetic C-methylation of the hexahydroporphyrins discovered by Eschenmoser[52] and the regiospecificity is in accord with the principles adumbrated[53] for the stabilizing effect of a vinylogous ketimine system. In principle the methylases of the B_{12} pathway, which synthesize both natural and unnatural pyrrocorphins and corphins can be harnessed to prepare several of the missing intermediates of the biosynthetic pathway, e.g. (**20**) → (**21**). It is of note that the instability towards oxygen rationalizes our inability to isolate any new intermediates under aerobic conditions (> 5 ppm O_2).

The possibility was also examined that a complete corrinoid structure based on a type I template could be prepared using uro'gen I, or reduced sirohydrochlorin I, as substrates with the cell-free system capable of converting Uro'gen III to cobyrinic acid. However no type I cobyrinic acid was produced suggesting that Nature, although capable of inserting at least 4 methyl groups into Uro'gen I by C-alkylation is unable to effect the key step of ring contraction using the type-I pattern of acetate and propionate side chains. This result, although negative, is in accord with a suggestion[53] concerning the requirement for two adjacent acetate side chains in pre-corrinoids, one of which (at C-2) can participate in lactone formation to the C-20 meso position (see Scheme 6, precorrin-**6b**) a postulate supported by experiments[35] with ^{18}O labeled precursors, while the second acetate function (at C-18) is used as an auxiliary to control the necessary activation of C-15 for C-methylation. The second methyl transferase which introduces the C-20 methyl into the substrate precorrin-**2** (→ precorrin-**3**) has been described recently.[36]

Evolutionary Aspects and Further Outlook

Although outside the scope of this lecture, the steps from cobyrinic acid to cobalamin (Scheme 10) involve amidation to cobyrinic acid, attachment of 1-amino-2-propanol to the ring D propionate to give cobinamide and stepwise attachment of the nucleotide from ribose and dimethylbenzimidazole. A remarkable non-enzymatic self assembly of the nucleotide segment to ring D propionate has been disclosed.[53]

Genetic mapping of the loci of the B_{12}-synthesizing enzymes has been reported for *Pseudomonas denitrificans*.[44] This complements a most interesting study on the genetics of *Salmonella typhimurium* which cannot make B_{12} when grown aerobically.[54] A mutant

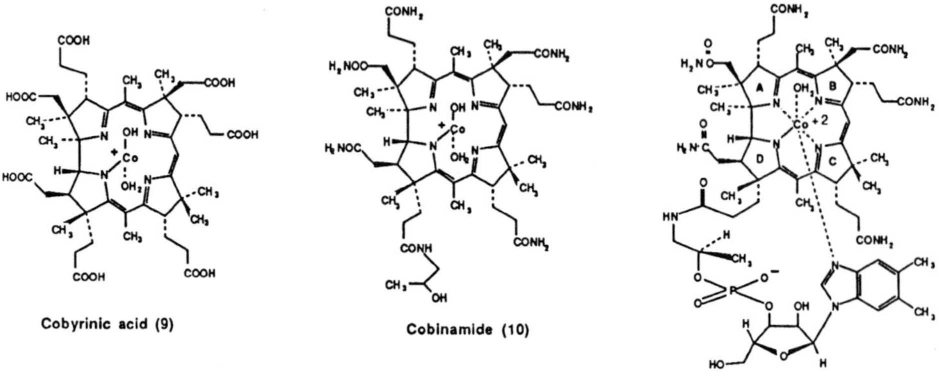

Cobyrinic acid (9)   Cobinamide (10)   cobalamin (11)

Scheme 10

requiring methionine, cobinamide or cyanocobalamin when grown anaerobically produces B_{12} *de novo* thus leading to the isolation of other mutants blocked in B_{12} synthesis including one which cannot make Factor II required for siroheme production. All of the cobalamin mutations lie close together on the chromosome and a cluster of several methyl transferases maps at 42 min. Thus rapid progress can be expected in the isolation of the remaining biosynthetic enzymes from *Salmonella*.

Until quite recently it had been assumed that the Shemin pathway (glycine-succinate) to ALA was ubiquitous in bacterial production of porphyrins and corrins. However it is now clear that in many archaebacteria (e.g. *Methanobacterium thermoautotrophicum*,[55] *Clostridium thermoaceticum*[56-57]) the C_5 pathway from glutamate is followed. Phylogenetically the C_5 route is conserved in higher plants, and it appears from recent work[58] that *hemA* of *E. coli* (and perhaps of *S. typhimurium*) encodes the enzyme for the glutamate $\rightarrow$ ALA conversion, i.e. the C_5 pathway is much more common than had been realized. In *C. thermoaceticum* it has been shown[57] that the B_{12} produced by this thermophilic archaebacterium is synthesized from ALA produced in turn from glutamate. Although *E. coli* does not seem to be able to synthesize B_{12}, the enzyme M-1 (M_r 54,000), a close relative of SUMT (M_r 36,000), is expressed as part of the genetic machinery (*CysG*) for making siroheme, as discussed above.

Eschenmoser has speculated that corrinoids resembling B_{12} could have arisen by prebiotic polymerization of amino nitriles and has developed an impressive array of chemical models[53] to support this hypothesis, including ring contraction of porphyrinoids to acetyl corrins, deacetylation and the C-methylation chemistry discussed earlier, which provide working hypotheses for the corresponding biochemical sequences. A primitive form of corrin stabilized by hydrogen,[53] rather than by methyl substitution may indeed have existed

>4 x 10^9 years ago, before the origin of life[59] or the genetic code[60] and could have formed the original "imprint" necessary for the evolution of enzymes which later mediated the insertion of methyl groups to provide a more robust form of B_{12}. Since B_{12} is found in primitive anaerobes and requires no oxidative process in its biogenesis (unlike the routes to heme and chlorophyll which are oxidative) an approximate dating of B_{12} synthesis would be 2.7 - 3.5 x 10^9 years, i.e. after DNA but before oxygen-requiring metabolism.[61]

If B_{12} indeed were the first natural substance requiring Uro'gen III as a precursor, the question arises "why type III?" Since the chemical synthesis of the Uro'gen mixture from PBG under acidic conditions leads to the statistical ratio of Uro'gens [I 12.5%; II 25%; III 50%; IV 12.5%] containing a preponderance of Uro'gen III, natural selection of the most abundant isomer could be the simple answer. It has been suggested[53] that the unique juxtaposition of two adjoining acetate and side chains in the type III isomer (which does not obtain in the symmetrical Uro'gen I) may be responsible for a self assembly mechanism requiring these functions to hold the molecular scaffolding in place via lactone and ketal formation as portrayed in Scheme 6. These ideas are being tested by ^{18}O labeling and by studying the possible biotransformation of types I, II, and IV uroporphyrinogens to "unnatural" corrinoids.

As discussed by Professor Roth[54] (Utah) and Dr. Crouzet[44] (Rhône-Poulenc) elsewhere in this volume the genetic mapping of B_{12} biosynthesis is now under way, and it should at last be possible to discover the remaining intermediates beyond-precorrin-**3**, together with the enzymes which mediate the methyl transfers, decarboxylation, ring contraction, deacetylation, and cobalt insertion. It is anticipated that the powerful combination of molecular biology and NMR spectroscopy which has been essential in solving the problems in B_{12} synthesis posed by the assembly and intermediacy of Uro'gen III and the subsequent C-methylations leading to precorrin-**3**, will again be vital to the solution of those enigmas still to be unraveled in the fascinating saga of B_{12} biosynthesis.

ACKNOWLEDGEMENTS

The work described in this lecture has been carried out by an enthusiastic group of young colleagues whose names are mentioned in the references. Financial support over the last 20 years has been generously provided by the National Institutes of Health, National Science Foundation, and the Robert A. Welch Foundation. It is a special pleasure to thank Professors Gerhard Müller (Stuttgart) and Peter Jordan (London) for their continued stimulating collaboration.

REFERENCES

1. A. I.Scott, *Acc. Chem. Res.* **11**, 29 (1978).
2. G. Burton, P. E. Fagerness, S. Hosazawa, P. M. Jordan, and A. I. Scott, *J. Chem. Soc. Chem. Comm.*, 204 (1979).

3. A. R. Battersby, C. J. R. Fookes, K. E. Gustafson-Potter, G. W. J. Matcham, E. McDonald, *J. Chem. Soc. Chem. Comm.*, 1115 (1979).
4. P. M. Jordan, G. Burton, H. Nordlow, M. M. Schneider, L. M. Pryde, A. I. Scott, *J. Chem. Soc. Chem. Comm.*, 204 (1979).
5. Review, F. J. Leeper, *Nat. Prod. Reports*, **2**, 19 (1985).
6. A. I. Scott, *J. Heterocyc. Chem.*, **14**, S-75 (1987).
7. A. I. Scott, C. A. Roessner, N. J. Stolowich, P. Karuso, H. J. Williams, S. K. Grant, M. D. Gonzalez, T. Hoshino, *Biochemistry*, **27**, 7984 (1988).
8. P. M. Anderson, R. J. Desnick, *J. Biol. Chem.*, **255**, 1993 (1980).
9. A. Berry, P. M. Jordan, J. S. Seehra, *FEBS Lett.* **129**, 220 (1981).
10. A. R. Battersby, C. J. R. Fookes, G. Hart, G. W. J. Matcham, P. S. Pandey, *J. Chem. Soc. Perkin Trans. I*, 3041 (1983).
11. P. M. Jordan, A. Berry, *Biochem. J.*, **195**, 177 (1981).
12. J. N. S. Evans, G. Burton, P. E. Fagerness, N. E. Mackenzie, A. I. Scott, *Biochemistry*, **25**, 905 (1986).
13. A. I. Scott, C. A. Roessner, K. D. Clemens, *FEBS Lett.*, **242**, 319 (1988).
14. A. I. Scott, N. J. Stolowich, H. J. Williams, M. D. Gonzalez, C. A. Roessner, S. K. Grant, C. Pichon, *J. Am. Chem. Soc.*, **110**, 5898 (1988).
15. P. M. Jordan, M. J. Warren, H. J. Williams, N. J. Stolowich, C. A. Roessner, S. K. Grant, A. I. Scott, *FEBS Lett.*, **235**, 189 (1988).
16. (a) M. J. Warren and P. M. Jordan, *FEBS Lett.*, **225**, 87 (1987).
(b) M. J. Warren and P. M. Jordan *Biochemistry*, **27**, 9020 (1989).
17. (a) G. J. Hart, A. D. Miller, F. J. Leeper, and A. R. Battersby, *J. Chem. Soc. Chem. Comm.*, 1762 (1987).
(b) A. D. Miller, G. J. Hart, L. C. Packman, and A. R. Battersby, *Biochem. J.*, **254**, 915 (1988).
(c) G. J. Hart, A. D. Miller, and A. R. Battersby, *Biochem. J.*, **252**, 909 (1988).
(d) U. Beifus, G. J. Hart, A. D. Miller, and A. R. Battersby, *Tetrahedron Letters*, **29**, 2591 (1988).
18. R. Radmer and L. Bogorad, *Biochemistry*, **11**, 904 (1972).
19. A. I. Scott, C. A. Roessner, P. Karuso, N. J. Stolowich, and B. Atshaves, manuscript in preparation; A. I. Scott, *17th IUPAC Symp. Nat. Prod.*, New Delhi, Feb. 1990, *Pure and Appl. Chem.*, 1990 (in press).
20. J. H. Mathewson, A. H. Corwin, *J. Am. Chem. Soc.*, **83**, 135 (1961).
21. S. Rosé, R. B. Frydman, C. de los Santos, A. Sburlati, A. Valasinas, and B. Frydman, *Biochemistry*, **27**, 4871 (1988).
22. A. R. Battersby,.J. R. Fookes, G. W. J. Matcham, E. McDonald, and R. Hollenstein, *J. Chem. Soc., Perkin Trans I*, 3031 (1983).
23. A. R. Battersby, K. Frobel, F. Hammerschmidt, and C. Jones, *J. Chem. Soc. Chem. Comm.*, 455-457 (1982).
24. H. C. Uzar, A. R. Battersby, T. A. Carpenter, F. J. Leeper, *J. Chem. Soc. Perkin Trans. I*, 1689-1696 (1987) and references cited therein.

25 C. Nussbaumer, M. Imfeld, G. Wörner, G. Müller, and D. Arigoni, *Proc. Nat. Acad. Sci. U.S.*, **78**, 9-10 (1981).
26. A. R. Battersby, M. J. Bushnell, C. Jones, N. G. Lewis, and A. Pfenninger, *Proc. Nat. Acad. Sci. U. S.*, **78**, 13-15 (1981).
27. L. Mombelli, C. Nussbaumer, H. Weber, G. Müller, and D. Arigoni, *Proc. Nat. Acad. Sci. U.S.*, **78**, 11-12 (1981).
28. G. Müller, K. D. Gneuss, H.-P. Kriemler, A. J. Irwin, and A. I. Scott, *Tetrahedron (Supp.)* **37**, 81-90 (1981).
29. L. Ernst, *Liebigs Ann. Chem.*, 376-386 (1981).
30. A. R. Battersby, C. Edington, C. J. R. Fookes, and J. M.Hook, *J. Chem. Soc., Perkin, I.*,, 2265-2277 (1982).
31. A. I. Scott, N. E. Mackenzie, P. J. Santander, P. Fagerness, G. Müller, E. Schneider, R. Sedlmeier, G. Wörner, *Bioorg. Chem.* **13**, 356-362 (1984).
32. A. I. Scott, H. J. Williams, N. J. Stolowich, P. Karuso, M. D. Gonzalez, G. Müller, K. Hlineny, E. Savvidis, E. Schneider, U. Traub-Eberhard, G. Wirth, *J. Am. Chem. Soc.*, **111**, 1897-1900 (1989).
33. C. Nussbaumer, D. Arigoni, Work described in C. Nussbaumer's Dissertation No. 7623,
E. T. H. (1984).
34. T. E. Podschun, G. Müller, *Angew. Chem. Int. Ed., Engl.*, **24**, 46-47 (1985).
35. K. Kurumaya, T. Okasaki, M. Kajiwara, *Chem. Pharm. Bull.*, **37**, 1151 (1989).
36. F. Blanche, S. Handa, D. Thibaut, C. L. Gibson, F. J. Leeper, A. R. Battersby, *J. Chem. Soc. Chem. Comm.*, 1117-1119 (1988).
37. A. R. Battersby, K. R. Deutscher, B. Martinoni, *J. Chem. Soc. Chem. Comm.* 698-700 (1983).
38. V. Rasettti, A. Pfaltz, C. Kratky, A. Eschenmoser, *Proc. Natl. Acad. Sci., U.S.A.*, **78**, 16-19 (1981).
39. B. Dresov, L. Ernst, L. Grotjahn, V. B. Koppenhagen, *Angew. Chem. Int. Ed. Engl.*, 20, 1048-1049 (1981).
40. A. I. Scott, N. E. Georgopapadakou, and A. J. Irwin, unpublished.
41. A. I. Scott, M. Kajiwara, and P. J. Santander, *Proc. Nat. Acad. Sci. U. S.*, **84**, 6616-6618 (1987).
42. A. R. Battersby, C. Edington, and C. J. R. Fookes, *J. Chem. Soc. Chem. Comm.*, 527-530 (1984).
43. G. Müller in Vitamin B_{12}, B. Zagalak and W. Friedrich, eds, de Gruyter, New York, 1979, 279-291, who described a methylase from *P. shermanii* capable of using Uro'gen I as substrate. The product was not characterized.
44. B. Cameron, K. Briggs, S. Pridmore, G. Brefort, J. Crouzet, *J. Bacteriol.* **171**, 547-557 (1989).
45. M. J. Warren, C. A. Roessner, P. J. Santander, and A. I. Scott, *Biochem. J.*, **265**, 725-729 (1990).

46. R. D. Brunt, F. J. Leeper, I. Orgurina, and A. R. Battersby, *J. Chem. Soc., Chem. Comm.*, 428 (1989).

47. M. J. Warren, N. J. Stolowich, P. J. Santander, C. A. Roessner, B. A. Sowa, and A. I. Scott, *FEBS Lett.*, **261**, 76-80 (1990).

48. A. I. Scott, M. J. Warren, C. A. Roessner, N. J. Stolowich, and P. J. Santander, *J. Chem. Soc., Chem. Comm.*, in press (1990).

49. G. Müller, J. Schmiedl, E. Schneider, R. Sedlmeier, G., Worner, A. I. Scott, H. J. Williams, P. J. Santander, N. J. Stolowich, P. E. Fagerness and N. E. Mackenzie, *J. Am. Chem. Soc.*, **108**, 7875-7877 (1986).

50. G. Müller, J. Schmiedl, L. Savidis, G. Wirth, A. I. Scott, P. J. Santander, H.J. Williams, N. J. Stolowich, and H.-P. Kriemler, *J. Am. Chem. Soc.*, **109**, 6902-6904 (1987).

51. A. I. Scott, H. J. Williams, N. J. Stolowich, P. Karuso, M. D. Gonzalez, F. Blanche, D. Thibaut, G. Müller, and G. Wörner, *J. Chem. Soc., Chem. Comm.*, 522 (1989).

52. C. Leumann, K. Hilpert, J. Schreiber, and A. Eschenmoser, *J. Chem. Soc., Chem. Comm.*, 1404-1407 (1983).

53. A. Eschenmoser, *Angew. Chem., Int. Ed. Engl.*, **27**, 5 (1988).

54. R. M. Jeter, B. M. Olivera, J. R. Roth, *J. Bacteriol.*, **159**, 206-213 (1984).

55. J. G. Zeikus, *Adv. Microb. Physiol.*, **24**, 215-299 (1983).

56. J. R. Stern and G. Bambers, *Biochemistry,* **5**, 1113-1118 (1966).

57. T. Oh-hama, N. J. Stolowich, A. I. Scott, *FEBS Lett.*, **228**, 89-93 (1988).

58. L. Jian-Ming, C. S. Russell, S. D. Coslow, *Gene*, **82**, 209-217 (1989).

59. K. Decker, K. Jungermann, R. K. Thauer, *Angew. Chem., Int. Ed. Engl.*, **9**, 153-162 (1970).

60. M. Eigen, B. F. Lindemann, M. Tietze, R. Winkler-Oswatitsch, A. Dress, and A. von Haeseler, Science, **244**, 673-679 (1989).

61. S. A. Benner, A. D. Ellington, and A. Taver *Proc. Nat. Acad. Sci.*, **86**, 7054 (1989).

POSTERS

Regulation of the *Vibrio Fischeri* by Luminescence Regulon
Thomas O. Baldwin, Jerry H. Devine, Robert C. Heckel and Gerald S. Shadel

Mechanistic Studies of an Enzyme, Phosphotriesterase
Steven Caldwell, George Omburo, Jennifer Newcomb and Frank Raushel

Structure and Dynamics of B_{12} Intermediates using Biophysical Spectroscopy
Mark R. Chance, Department of Chemistry, Georgetown University

Analysis of the Structure and Function of Bacterial *Luciferases* Using Mutants Generated by Random and Site-directed Mutagenesis
Lorenzo H. Chen, Lawrence J. Chlumsky, and Thomas O. Baldwin

Molecular Cloning of Methylmalonyl CoA Mutase
Fred Ledley, Howard Hughes Medical Institute, Baylor College of Medicine

A Simple Model Explaining the Specificity and pH Dependence of Cathepsin B/Catalyzed Substrate Hydrolysis
Robert Menard, Biotechnology Research Institute, Montreal

Processing of the Signal Sequence of Organophosphorus Hydrolase Encoded by the *opd* gene in *Pseudomonas diminuta*
Charles Miller, Department of Biochemistry & Biophysics, Texas A&M University

Enzyme Catalyzed Acylation and Deacylation of the Sugar Moieties of Nucleosides
Kenji Nozaki, Taiho Pharmaceutical Co. Ltd., Saitama, Japan

Functional Domains in Carbamoyl Phosphate Synthetase
Laura Post, Department of Chemistry, Texas A&M University

Mechanism of Urocanase Reaction
Janos Retey, Institut fur Org. Chemie/Biochemie, Universitat Karlsruhe

Analysis of Subunit Folding and Enzyme Assembly of *Vibrio Harveyi* by Bacterial *Luciferase*
Jun Sugihara, Jenny Waddle, Ke Wu and Thomas O. Baldwin

Selective ^{13}C Edited 1D-Noesy in Protein Using Heteronuclear Half Gaussian Pulse
Charles Tellier, Howard Williams, Claudio Ortiz, Neal Stolowich, and A. Ian Scott, Texas A&M University

Uroporphyrinogen-III Methylase Catalysis the Enzymatic Synthesis of Sirohydrochlorins II and IV by a Clockwise Mechanism
Martin J. Warren, Mario D. Gonzalez, Howard J. Williams, Neal J. Stolowich and A. Ian Scott, Department of Chemistry, Texas A&M University

Corynebacterium Acyltransferase: A New Enzyme for Chiral Synthesis

Gregg Whited, Eastman Kodak Company, Rochester, New York

Gram Scale Synthesis of Isotopomers of L-Tryptophan (^{2}H, ^{13}C, ^{15}N) Using Genetically Modified *E. coli*

Ellen M. M. Van den Berg, Department of Chemistry, Massachusetts Institute of Technology

INDEX

Printed in the United States
118278LV00002B/129/A

9 780306 438158